Life With Hubble
An insider's view of the world's most famous telescope

AAS Editor in Chief

Ethan Vishniac, John Hopkins University, Maryland, US

About the program:

AAS-IOP Astronomy ebooks is the official book program of the American Astronomical Society (AAS), and aims to share in depth the most fascinating areas of astronomy, astrophysics, solar physics and planetary science. The program includes publications in the following topics:

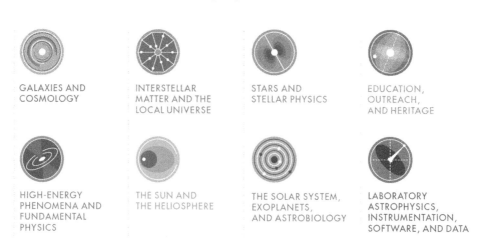

Books in the program range in level from short introductory texts on fast-moving areas, graduate and upper-level undergraduate textbooks, research monographs and practical handbooks.

For a complete list of published and forthcoming titles, please visit iopscience.org/books/aas.

About the American Astronomical Society

The American Astronomical Society (aas.org), established 1899, is the major organization of professional astronomers in North America. The membership (~7,000) also includes physicists, mathematicians, geologists, engineers and others whose research interests lie within the broad spectrum of subjects now comprising the contemporary astronomical sciences. The mission of the Society is to enhance and share humanity's scientific understanding of the universe.

Editorial Advisory Board

Steve Kawaler
Iowa State University, USA

Ethan Vishniac
John Hopkins University, USA

Dieter Hartmann
Clemson University, USA

Piet Martens
Georgia State University, USA

Dawn Gelino
NASA Exoplanet Science Institute, Caltech, USA

Joan Najita
National Optical Astronomy Observatory, USA

Bradley M. Peterson
The Ohio State University/Space Telescope Science Institute, USA

Scott Kenyon
Smithsonian Astrophysical Observatory, USA

Daniel Savin
Columbia University, USA

Stacy Palen
Weber State University, USA

Jason Barnes
University of Idaho, USA

James Cordes
Cornell University, USA

Life With Hubble

An insider's view of the world's most famous telescope

David S Leckrone

IOP Publishing, Bristol, UK

© IOP Publishing Ltd 2020

All rights reserved. No part of this publication may be reproduced, stored in a retrieval system or transmitted in any form or by any means, electronic, mechanical, photocopying, recording or otherwise, without the prior permission of the publisher, or as expressly permitted by law or under terms agreed with the appropriate rights organization. Multiple copying is permitted in accordance with the terms of licences issued by the Copyright Licensing Agency, the Copyright Clearance Centre and other reproduction rights organizations.

Permission to make use of IOP Publishing content other than as set out above may be sought at permissions@ioppublishing.org.

David S Leckrone has asserted his right to be identified as the author of this work in accordance with sections 77 and 78 of the Copyright, Designs and Patents Act 1988.

ISBN   978-0-7503-2038-2 (ebook)
ISBN   978-0-7503-2036-8 (print)
ISBN   978-0-7503-2039-9 (myPrint)
ISBN   978-0-7503-2037-5 (mobi)

DOI   10.1088/2514-3433/ab8ad0

Version: 20200701

AAS–IOP Astronomy
ISSN 2514-3433 (online)
ISSN 2515-141X (print)

British Library Cataloguing-in-Publication Data: A catalogue record for this book is available from the British Library.

Published by IOP Publishing, wholly owned by The Institute of Physics, London

IOP Publishing, Temple Circus, Temple Way, Bristol, BS1 6HG, UK

US Office: IOP Publishing, Inc., 190 North Independence Mall West, Suite 601, Philadelphia, PA 19106, USA

*Prais'd be the fathomless universe, for life and joy,
and for objects and knowledge curious.*

—Walt Whitman, 1865

Contents

Preface		xi
Acknowledgements		xiii
Author Biography		xiv
For Additional Information		xv
1	**A Small Error: The Origin of Spherical Aberration**	1-1
	References	1-19
2	**The Road to Redemption: Learning to Service in Space**	2-1
3	**The Sharper Image: Correcting Spherical Aberration**	3-1
4	**Leading the Way: Preparing to Service Hubble**	4-1
	References	4-24
5	**To the Rescue: The First Servicing Mission**	5-1
	References	5-30
6	**Transfiguration: Life Without Spherical Aberration**	6-1
7	**Once in a Thousand Years: Preparing for Comet S/L-9**	7-1
8	**When Worlds Collide: The Comet Campaign**	8-1
	References	8-22
9	**From Darkness to Light: Black Holes and Quasars**	9-1
	References	9-12
10	**How High Is the Sky? Measuring the Universe**	10-1
	References	10-17
11	**You Can Almost See Forever: The Hubble Deep Fields**	11-1
	References	11-17

12	A Bad Sunburn: The Second Servicing Mission	12-1
13	Resurrection: The NICMOS Saga	13-1
	Reference	13-18
14	Racing Y2K: Servicing Mission 3a	14-1
15	Dangerous Liaison: Servicing Mission 3b	15-1
	Reference	15-29
16	Stormy Weather: Hubble Under Siege	16-1
	References	16-32
17	Replenishing the Toolbox: The Science Instruments of SM4	17-1
18	A New Beginning: The Last Servicing Mission	18-1
19	Worlds Without End: Exploring Exoplanets	19-1
	References	19-20
20	The Fathomless Universe: Dark Matter and Dark Energy	20-1
	References	20-24
21	Epilogue	21-1

Preface

One definition of a memoir is, "a record of events written by a person having intimate knowledge of them and based on personal observation."[1] In that sense this book could be called a "memoir." But it's not my memoir so much as it is a Hubble Family memoir. I give voice to it; serve as a narrator if you will. It is a history of Hubble as I observed it and experienced it, beginning in 1990 with the discovery of Hubble's original optical flaw—spherical aberration—and continuing through the fifth servicing mission in 2009 and beyond. There are also a number of flashbacks to fill in some of the details from earlier years.

This narrative is based on my own files and recollections. In addition, I conducted interviews with 22 people—fellow scientists, engineers, astronauts, and managers—all of whom made major contributions to the Hubble legacy. When I was unable to do interviews, I simply observed key players in action, read their words or heard them speak about their encounters with Hubble. These are the stories of all of these people too. I view some of them as heroes who rose to the occasion when the situation looked dire. Some of them are pioneers who pushed back the boundaries of what is possible to do in space. Some are brilliant scientists who extracted the last ounce of performance out of the telescope and its instruments to change and clarify dramatically our understanding of how the universe works. Some of these people are well known. Only family, friends, and colleagues know many of them. All of them have great stories to tell.

I have found it frustrating that many of the documentaries about Hubble produced for television and numerous stories in magazines and newspapers have focused primarily on the early debacle of Hubble's spherical aberration and how NASA, the scientific community, and industry rallied to repair the problem. That's a great and compelling story of human determination and resilience. I tell the story once again in the early chapters of this book. But Hubble will be 30 years old on 2020 April 24, 6 months from now, and it's still going strong. The saga of spherical aberration came to an end 26 years ago! Hubble has come so far, accomplished so much, since then. There are so many other wonderful stories about Hubble to be told. That's the major reason I wanted to write this book.

I worked on the Hubble Project at the Goddard Space Flight Center in Greenbelt, Maryland for 33 years, beginning in 1976. For the last 17 of those years I was the Senior Project Scientist. My job was rather diffuse in nature. But put simply, I was the scientific conscience of the Project. I was there to participate in the major decision-making process of the Project or Program Manager (the title changed midstream), to assure that everything we did, within the constraints of budget and schedule, was the best we could do for science. I maintained strong collegial relationships with our partners at the Space Telescope Science Institute, with the scientists in the community who led the development of Hubble's set of high-performance instruments, and with our scientific colleagues from around the world

---

[1] Dictionary.Com based on the Random House Unabridged Dictionary, © Random House, Inc. 2020.

who consistently exploited the power of Hubble to probe Mother Nature's cosmic mysteries. I helped plan and I participated in the high adventure of all five of the shuttle-based servicing missions to Hubble. All of the above has given me a broad perspective that informs what I've written here.

The world of Hubble is vast, and I could not keep track of all of it by myself. Four other immensely talented astronomers worked with me within our Project Science Office at Goddard: my Deputy Senior Project Scientist, Dr Mal Niedner, the Project Scientist for Hubble Operations, Dr Ken Carpenter, the original Project Scientist for Flight Systems and Servicing, Dr Ed Cheng, and Ed's successor, Dr Randy Kimble. Each made essential contributions to Hubble's success in their own right, a few examples of which are woven into the stories told in this book.

I believe there is a large population of people around the world who have been fans of the Hubble Space Telescope—"Hubble Huggers"—for decades. I do not know for certain how large this population is. But everywhere I go, when I mention that I worked on Hubble, people I meet from every walk of life recognize what I'm talking about and tell me how much they love Hubble. They are awestruck by the beauty of the universe Hubble has revealed to them, and they are inspired by the new scientific discoveries made almost daily by astronomers using Hubble. When the observatory ran into technical problems, people were obviously aware of what was going on and worried with us about it. When NASA (temporarily) canceled our final servicing mission, people from around the world came to Hubble's defense. School children sent money to NASA to help pay for the mission!

This book is dedicated to all the Hubble Huggers of the world. I have tried to write both the space travel parts of it and the scientific explanations as clearly and as simply as I could while still conveying the facts and ideas accurately. I hope people who have followed Hubble over the years and who want to know more, as well as those who are coming to the subject for the first time, will find the book enjoyable and interesting.

David S. Leckrone
Silver Spring, Maryland, USA
2019 October 24

Acknowledgements

This book contains the personal stories of many people who worked on Hubble, who repaired and upgraded Hubble in orbit, or who used the observatory to acquire remarkable and unique data for their research. Many of these generously gave hours of their time for recorded interviews. Some additional interviews took the form of conversations in person, by phone, or by email. Where interviews were not recorded, they were documented in my hand-written notes. The book would not have been possible without the support of these consummate professionals. I am deeply indebted to the following: Scott Altman, Preston Burch, John Campbell, Frank Cepollina, Ed Cheng, Christine Cottingham, Jim Crocker, Larry Dunham, Holland Ford, Wendy Freedman, Tom Griffin, Heidi Hammel, Melissa McGrath, Keith Noll, Wes Ousley, Adam Riess, Joe Rothenberg, John Trauger, Brian Vreeland, Hal Weaver, Art Whipple, and Bob Williams.

Other people provided source material, information and guidance of various kinds. Special thanks go to: Jim Barcus, Justin Cassidy, Tony Ceccacci, Dennis Ebbets, Raja Guhathakurta, John Grunsfeld, Jim Jeletic, Buell Jannuzi, Stan Krol, Tod Lauer, Mike Massimino, Mark McCaughrean, Bruce Milam, Mal Niedner, Dave Parker, Marc Postman, Neill Reid, and Al Vernacchio.

It's difficult to find words to express sufficient gratitude to John Campbell, Frank Cepollina, Ed Cheng, Randy Kimble, Joe Rothenberg and several anonymous reviewers enlisted by the publisher for reading and critically reviewing large portions of my manuscript. I believe it's a much-improved book because of their insights.

I've spent the last seven years drafting this book. At first my pace was slow. But in the past two years or so, the process accelerated and became much more intense. I am very grateful to my wife, Marlene Berlin, for her love, support and patience during this period.

Author Biography

David S Leckrone

David S Leckrone worked as an Astrophysicist at NASA's Goddard Space Flight Center for 40 years. For over three decades, beginning in 1976, he served in various scientific leadership roles on the Hubble Space Telescope Project. From 1992–2009 he held the position of Hubble's Senior Project Scientist. In this role he and his Project Science Team provided scientific leadership for all aspects of the Hubble Project, including project management, spacecraft and science operations, development of new scientific instruments, and in-orbit servicing. He had overall Project responsibility to assure that the scientific performance requirements for the Hubble observatory were achieved and that the Hubble Space Telescope observatory remained scientifically productive and successful over its long lifetime. Leckrone helped plan and lead five highly successful Space Shuttle servicing missions to Hubble.

Concurrently with his Hubble duties, Leckrone held the position of Head of the Astronomy Branch in the Laboratory for Astronomy and Solar Physics at Goddard from 1981–1991. From 2003–2006 he served as Chief Scientist for the newly formed NASA Engineering and Safety Center, organized in response to the loss of Shuttle Columbia.

Leckrone's research career focused on ultraviolet astronomy, spectroscopic analysis of stellar atmospheres, and the abundances of the chemical elements in both normal and chemically peculiar stars. For many years he was strongly engaged in public outreach. He was much in demand as a public speaker, writer, and commentator on science, Hubble, and the space program.

Leckrone earned a BS with honors in Physics from Purdue University in 1964 and a PhD in Astronomy from UCLA in 1969. He holds an MAS/Management degree from the Johns Hopkins University, awarded in 1987. In 1996 he received an honorary doctorate in Natural Science from the University of Lund in Sweden. In 2013 Purdue University awarded him an honorary Doctor of Science degree.

In 2012 the American Astronomical Society awarded Leckrone the George Van Biesbroeck Prize in recognition of "long-term and unselfish service to Astronomy." In 2012 Purdue University recognized Leckrone as a Distinguished Scientific Alumnus. Leckrone has received the NASA Distinguished Service Medal, the highest honor the Agency bestows on a Civil Servant. He was also the recipient of NASA's Outstanding Leadership Medal and Medal for Scientific Achievement. He was a recipient of the U.S. Presidential Rank Award of Merit in 2008.

Dr. Leckrone retired from NASA in 2009. He currently is an Emeritus Senior Scientist at the Goddard Space Flight Center. He and his wife, Marlene, reside in Silver Spring, Maryland. He has a son, daughter, and two grandchildren.

For Additional Information

Two sites on the World Wide Web are particularly useful as sources of additional information:

www.hubblesite.org
www.nasa.gov/hubble

The first site contains a complete archive of news releases about Hubble and its discoveries; images, including Hubble observations, infographics, and artwork; and videos. Many of the figures in this book are taken from hubblesite.org. In the "Credit" statement at the end of each of these figures you will see the designation "STScI year-release number." For example, STScI 2002-06 refers to NASA/STScI news release number 6 from 2002. All of these news releases may be found by clicking on the "News" icon on this website. To search through images or videos, simply click on the corresponding icon.

The nasa.gov website contains a wealth of information about the Hubble observatory. A large archive of Hubble astronomical images may also be found there. The site contains a link to over 100 YOUTUBE videos discussing numerous topics related to Hubble, its discoveries, its history, and the people who work on it.

Many (though not all) of the images credited to NASA in the figure captions may be found online at www.flickr.com/nasa2explore. Search on the particular shuttle mission number to find images related to that mission. Alternatively, Google "NASA Photos Mission Number." The relevant mission numbers are STS-41C, STS-31, STS-61, STS-82, STS-103, STS-109, STS-125.

AAS | IOP Astronomy

Life With Hubble

An insider's view of the world's most famous telescope

David S Leckrone

Chapter 1

A Small Error: The Origin of Spherical Aberration

It was a lovely, brisk winter evening in Long Beach, California in January of 2013. At an evening "Public Policy" session organized by the American Astronomical Society—a regular event at the big winter national meetings of the society—a hundred or so astronomers gathered to hear presentations by members of Congress from the Southern California region. One of these was Representative Dana Rohrabacher from California's 48th congressional district. At the time, Mr Rohrabacher was the Vice Chairman of the House Committee on Science, Space, and Technology. As a well-known conservative Republican, he chose to emphasize in his remarks the sizes of the national deficit and debt, likening them for his astronomical audience to the "trillions of stars in the sky." He asserted the need for dramatic cuts in Federal spending, including spending for NASA. Some of his remarks seemed clearly provocative to an audience of scientists whose professional careers often relied on government support of research and technology development. Some of his remarks were inflammatory. He singled out the Hubble Space Telescope (HST), and its spherical aberration problem, as an example of the ineptitude of the Federal Government. "No one had to bear responsibility, no one was punished!" Sitting in the audience, I quietly muttered, "That's just not right!"

After his talk, a large cluster of astronomers surrounded Congressman Rohrabacher, each wanting to have a word with him about their own concerns. I waited for a while until the crowd had thinned out, then made my way to him. "Congressman, you're wrong about Hubble," I said. "In what way?" he asked. "What I said was based on testimony of a number of people, under oath, to my committee." "Well, I lived through it," I replied. At that point several senior astronomers who had gathered around backed me up, pointing out my role for many years on the Hubble Project. Michael Rich from UCLA spoke up. "This is one of the most respected members of the society. You should listen to him."

I explained that nearly a full decade (early 1981 to mid-1990) had gone by between the time the flaw had been ground into the primary mirror at Perkin-Elmer Corporation and the time the problem was discovered after Hubble was in orbit. Many of the NASA people involved in the "screw-up" in 1981, including the senior management at Marshall Space Flight Center, were no longer around.

More to the point, in October of 1993 Perkin-Elmer and Hughes Aircraft, which had acquired Perkin-Elmer's optics division in Connecticut four months before Hubble was launched, reached an out-of-court settlement with the United States Department of Justice. The two companies paid $25 million to the government—$15 million in cash, $3.5 million in waived fees under the telescope contract, and $6.5 million in reduced costs for future telescope work. In return they were released from all liability for the Hubble problem. The NASA Inspector General had found evidence that Perkin-Elmer had hidden clues of the mirror's flaws from the government, including in one case, clipping the edges from around a test image to hide the wavy fringes that indicated a serious problem. Officials at Perkin-Elmer denied any wrong doing. The terms of the settlement had originally been described to me by Hubble procurement specialist Gifford Moak at Goddard in late 1993 or early 1994. They were also documented in major media (Broad 1993).

I noted to Congressman Rohrabacher that, at the end of that decade, a large team of highly dedicated and competent people—scientists, engineers and managers from NASA, industry and academia—had brought Hubble back from disaster. And I didn't want those people's accomplishments to be diminished in any way by being conflated with the "screw-up" that had happened so many years before. At that point, his attitude changed. "Oh, I love Hubble. I've always supported Hubble," he said.

The original "trouble with Hubble" is ancient history now, replaced by the glory of the observatory's scientific and cultural contributions for over 30 years of operation. But how could NASA and its highly experienced industrial partner have gotten something so important so terribly wrong? There are important lessons from this episode that need to be retained as part of our collective cultural memory —as was the case in the Challenger and Columbia shuttle tragedies, as was the case in the Titanic tragedy of 1912 for that matter—lessons about human arrogance, hubris and the psychology of denial and rationalization; lessons about the need for humility, curiosity and a continual awareness that we humans are fallible creatures.

In June of 1990, I was pulling double duty at the Goddard Space Flight Center in Greenbelt, Maryland. I was Head of the Astronomy Branch, an organization of about 15 PhD astrophysicists in the Laboratory for Astronomy and Solar Physics. I was also the Deputy to the Hubble Space Telescope's lead Project Scientist, Al Boggess.

Eighteen years had elapsed since NASA Headquarters selected the Marshall Space Flight Center in Huntsville, Alabama to be the Lead Center for Hubble's design and development. During that period Goddard was a minority partner, responsible for providing Hubble's scientific instruments and its flight operations systems. The Goddard Project also managed a host of NASA contracts with aerospace firms and with the Space Telescope Science Institute in Baltimore, better known to everyone as the STScI. It was the STScI's job to plan and execute Hubble's scientific observations.

According to NASA's plan, full responsibility for the Hubble Project was to be handed over from Marshall to Goddard about two months after launch. By that time it was surmised that the observatory would be fully checked out and ready to begin its exploration of the cosmos. However, given the long delay in Hubble's launch resulting from the loss of Space Shuttle Challenger in January of 1986, the transition from Marshall to Goddard management was accelerated. The Hubble Project team at Goddard was fully in charge of running the observatory at the time it was launched on Space Shuttle Discovery on 1990 April 24.

In the weeks after launch, Goddard teemed with activity day and night. Engineers, scientists, and managers concentrated on powering up all spacecraft systems, checking them out and bringing them to full operational status. Marshall engineers and managers were still on the scene, deeply involved in commissioning the spacecraft and telescope whose development they had led. Astronomers and engineers from the five Hubble scientific instrument development teams had also taken up residence at or near Goddard to check out and calibrate the performance of their instruments. A community of experts from all across the US and Europe had come together to bring Hubble on line. These were people who had been immersed in the observatory during its 14 year gestation. They were impatient to see what this incredible machine could actually do once the full power of its capabilities had been brought to bear on the solar system, the Milky Way Galaxy, and the deep universe beyond.

At a press briefing at Kennedy Space Center two days before Hubble's launch NASA's Associate Administrator for Space Science and Applications, Len Fisk, promised reporters that they would be allowed to witness "first light" from the telescope, the first images of the sky beamed down from Hubble. The astronomers in the room quietly groaned. First light observations from any optical telescope usually consisted of very simple images of a few stars, unresolved point sources of light in the sky. These were needed to verify that the incoming light beam was successfully making it all the way through the telescope and instrument optics to the electronic sensors on which the images were focused. First light images were pedestrian to say the least. Initially, the stellar images might be out of focus or misshapen. To astronomers, that didn't matter. It was an engineering observation. If successful, it would bring joy to the hearts of the Project's scientists and engineers; but it was certain to bewilder journalists expecting immediate, glorious, spectacular images from the new telescope. Nevertheless, the commitment from NASA had been made and reporters were invited to Goddard to witness first light on 1990 May 20.

First light proved to be a strange and unsettling event. The first images were obtained with Hubble's primary camera—the Wide Field and Planetary Camera (WFPC, pronounced "wif-pic") developed by NASA's Jet Propulsion Laboratory in Pasadena, California. The target field was selected to provide images of stars in the open star cluster NGC 3532 in the southern constellation Carina.

The Principal Investigator for the camera, Jim Westphal from Caltech, was on hand at Goddard to explain the observations to reporters. A Co-Investigator on

Jim's team, Jeff Hester from Arizona State, had set up computers and software to process and display the images. For comparison with the Hubble observations, Jeff brought along some observations of the same star field obtained with a 100 inch ground-based telescope at a good mountaintop site, Las Campanas Observatory in Chile.

As seen on the computer screen, the Hubble images looked only slightly better than their ground-based counterparts (Figure 1.1). One of the stars in the field looked like an elongated blob in the ground-based image, but was slightly resolved as two distinctly separate stars in the WFPC image. The scientists and engineers present were perplexed and uncomfortable given the presence of news media in the room. Someone asked, "Is that the way it's supposed to look?" "It's better [than the ground-based telescope images], right?" The first test images of stars from Hubble seemed to be far out of focus.

The circular entrance aperture of any optical telescope produces an image of a star that has a well-defined pattern—a central bright peak surrounded by alternating bright and dark circular rings. This pattern is the result of a process we learn about in high school physics called diffraction. The size of the telescope's diffraction pattern is dictated by the diameter of its entrance aperture (94 inches in Hubble's case) and by the wavelength or color of the light being observed. It limits how finely the telescope can resolve spatial detail in astronomical objects. Hubble is a relatively small telescope, compared to the 8–10 m behemoths at some mountaintop observatories. The latter are at a great disadvantage, however, in having to put

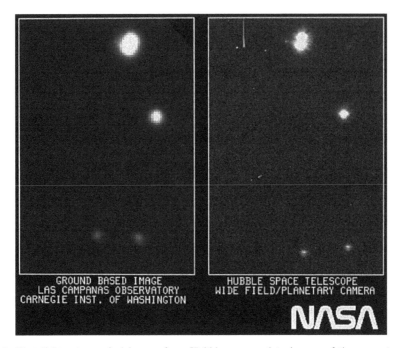

Figure 1.1. First light astronomical images from Hubble compared to images of the same stars from a mountaintop telescope in Chile. (Credit: NASA, ESA, STScI 1990-04.)

up with the turbulence of the Earth's atmosphere, which smears their images out. The problem is mitigated to some degree, particularly at infrared wavelengths, by active control of the telescope optics—moment-to-moment readjustment of the shape or position of the mirrors to compensate for the roiling motions of the air above them.

Their smeared images are also spread out over a background glow of light emitted by molecules in the Earth's atmosphere. So, a 10 m telescope on a mountaintop simply can't detect objects as faint as those at Hubble's limit, at least not in principle. But to work properly, it is absolutely essential that Hubble's point-source images be very sharply concentrated into a tight bundle of light rays projected onto the electronic sensors within its science instruments. Against the deep blackness of outer space, such tightly focused images can readily be detected even for extremely faint astronomical sources. Even the largest ground-based telescopes can't see as far across the universe and as far back in time as was anticipated for the modestly sized Hubble above the atmosphere.

Hubble's first images were much larger than what should have been produced by the telescope's diffraction alone. They had only about 15% of the starlight concentrated in their central peak, not the 70% specified for the telescope, and the remaining 85% of the light was going to waste, smeared out in a broad apron around the central core.

All of the Hubble Project people at that first light event assumed that these images were simply showing that a lot of work remained to be done to adjust the focus and optical alignment of the telescope. No one thought for a moment that what they were seeing in those first fuzzy images was evidence of a far deeper and more ominous problem. In the 1990 May 21 edition of the Washington Post, reporter Kathy Sawyer's article carried the headline, "Hubble Space Telescope Returns First Images; Quality Exceeds Initial Expectations." Well, not so much!

In the ensuing weeks, while all the other telescope and spacecraft systems were being checked out and brought up to speed, astronomers and engineers at the Goddard Space Telescope Operations Control Center (STOCC) grappled with the telescope's optical focusing and alignment controls. Hubble's secondary mirror (Figure 1.2) could be moved in and out to find the setting that produces the sharpest images—to focus the telescope. In addition the secondary mirror could be tilted in various directions to improve alignment of the telescope and to achieve round, symmetrical star images. The telescope also included devices called "wave-front sensors" that provided information about the specific nature of distortions and aberrations in the optical image. Despite the sophistication of these tools, and despite the extensive expertize and best efforts of the team, little improvement was achieved in the quality of the images the telescope was producing. The entire Hubble team was feeling beyond frustrated!

Late on the morning of 1990 June 19, Ron Polidan, the Hubble Project Scientist for Operations, walked into my office and said, "Boggess is at home tending to a plumbing problem. I really think we need to be at the status briefing this afternoon." "What's up?" I asked. Ron said, "I'm not sure, but I think it's grim."

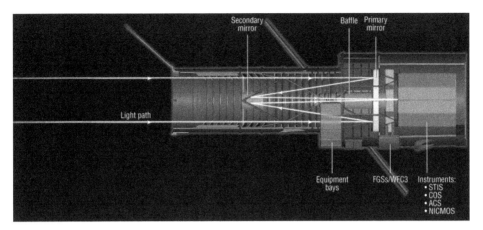

Figure 1.2. Artist's rendering of a cross section of the Hubble Space Telescope, showing the light path from the entrance aperture to the primary mirror, reflected to the secondary mirror, back through a hole in the center of the primary mirror into the apertures of the science instruments, ultimately coming to a focus on the electronic light sensors in each instrument. The instrument set shown is that following SM4 in 2009. (Credit: NASA, STScI Images.)

The conference room in Goddard's Building 3 was packed that afternoon as the lead engineers gave technical status reports about the various spacecraft systems for which they were responsible—thermal, electrical, pointing control, computers, science instruments, and telescope optics. The Optics Lead Engineer, Charlie Jones from Marshall, reported a continued lack of success in attempts to find an alignment and focus position of the telescope optics that would produce sharp stellar images. At that point, Chris Burrows, an astronomer and optics expert from the STScI, stood and said, with an edge of anger in his voice, "You have about a half wave of spherical aberration, and there's nothing you can do about it!"

Behind the scenes Chris Burrows, and independently Jon Holzman and Sandy Faber from the WFPC Team, had been analyzing the Hubble images. They had correctly deduced the nature of the problem with the telescope's optics, namely spherical aberration (Burrows et al. 1991). Up to that time, the Marshall engineers and the telescope's manufacturer, Perkin-Elmer Corporation in Danbury, Connecticut had ignored the astronomers' conclusions and denied that anything was wrong with the telescope. That's why Chris was angry. He had been frustrated for some time trying to convince the telescope makers that "the emperor had no clothes," that the telescope's primary optics were defective. They didn't believe it. They didn't want to believe it. But now Chris, Sandy, and Jon felt that their evidence was ironclad. The room resonated with a stunned silence.

The designers of Hubble had anticipated that, in going from $1g$ on the Earth's surface to the microgravity of low Earth orbit, the telescope's 2.4 m primary mirror might deform slightly so as to produce a variety of "higher order" aberrations or distortions of the focused image. For this reason, the telescope's design included 24 pads situated behind the primary mirror, that could be actuated in various combinations so as to apply gentle pressure to the back of the mirror to adjust its shape and remove the aberrations. The one aberration that could not be corrected in this way was the simplest of optical aberrations—spherical aberration.

Every amateur astronomer who grinds his or her own mirrors by hand tests for spherical aberration and removes it in the grinding process. But one of the world's most sophisticated telescope makers, Perkin-Elmer Corporation, who reputedly had produced a number of similar large mirrors for Earth-observing spy satellites for the U.S. Department of Defense, had failed to detect this most basic of aberrations in the primary mirror they had built for Hubble.

The realization not only that the Hubble telescope was optically flawed, but also that we were powerless at that moment to make it right, sank slowly, sadly, painfully that afternoon into the psyches of the scores of people gathered in that conference room. They had devoted substantial fractions of their professional lives to bringing the dream of the Hubble Space Telescope to reality; and now this—failure, defeat, and deep embarrassment.

Shortly after the meeting a small group of managers and engineers, primarily from Marshall, gathered in a nearby office. They needed to call the Hubble Project Manager, Fred Wojtalik, in Huntsville and inform him of the catastrophe. As I walked into that office, I heard the words, "Break out the hemlock, boys." That suicidal exclamation came from the lips of Jean Olivier, the Chief Engineer for the Hubble Project at Marshall. My reflexive response to Jean was, "Actually, I'd prefer that you break out the scotch. We're going to need you to help fix this problem." Wojtalik's response on the other end of the line was calm and measured. After probing with a couple of technical questions to be assured that we hadn't overlooked anything, Fred said, with sad resignation in his voice, "We'll need to let Headquarters know."

On 1990 June 21, a group of us from both Marshall and Goddard met with Associate Administrator Fisk at NASA Headquarters in Washington, DC. We gave him the terrible news and briefed him on some of the technical details. He was dismayed. His immediate comment was, "This is Space Science's version of Challenger."

The explosion of Space Shuttle Challenger on 1986 January 28 had, of course, taken the lives of seven astronauts. Nothing about the Hubble problem could compare to that level of disaster. But the loss of Challenger had also been an enormous blow to the entire NASA Human Spaceflight Program, setting its schedule back years and inflicting major financial impacts. In this latter regard, Len Fisk was right. The failure of the Hubble program would have devastating consequences to NASA's Space Science Program, not only financially, but also in terms of the esteem in which it was held by the government, the scientific community and the public. Fisk's painful task was now to explain what had happened to the NASA Administrator, to the White House and to Congress. It was left to the Hubble Program Office at NASA Headquarters and to the Hubble Project to break the news to the press, the public, and the scientific community.

The Hubble Science Working Group—20 or so scientists who had been selected by NASA in 1977 to oversee the observatory's development—together with the newly formed Hubble Users' Committee, met jointly on June 26 at Goddard. They needed to thrash out the situation and in particular to discuss the potential scientific impacts of the telescope's reduced capability. Sensing perhaps that something was afoot, at least one reporter sat quietly in the back of the room hoping to hear what was said. Given the sensitivity (and shock!) of what we needed to discuss openly, this had to be a closed meeting. The reporter was asked to leave and the door was locked.

Chris Burrows and Jon Holtzman discussed their independent findings that the telescope suffered from spherical aberration. Their two analyses agreed as to both the nature and magnitude of the problem. The Principal Investigator of each instrument was asked to describe the scientific investigations that were still possible with their respective instruments, given the actual on-orbit performance of the telescope.

Poor image quality from the telescope affected each of the five scientific instruments on board in different ways. The two cameras, WFPC and the European Faint Object Camera could not produce the clear, highly resolved, spectacular images of extended objects in the sky—planets, nebulae, star clusters, galaxies—that everyone had expected. Jim Westphal was understandably upset, saying that his WFPC science program was 90%—100% dead.

The two spectrographs, the Goddard High Resolution Spectrograph (GHRS) and the University of California's Faint Object Spectrograph (FOS), were not designed to produce pictures of the sky. Rather, they needed only to collect as much light as possible in a given period of time through their small entrance apertures. The spectrographs spread the incoming light out into its component colors or spectrum (analogous to water droplets in a cloud spreading out sunlight into a rainbow) and extracted from the spectrum the information needed to analyze cosmic sources.

Spectral data were used to deduce the physical properties of planets, comets, stars, interstellar gas and dust, galaxies, and the ephemeral gas between the galaxies—their temperature, density, chemical composition, velocity of motion, etc. Because the telescope's spherical aberration smeared the light in a stellar image out over a larger area than planned, the spectrographs had less light coming into their apertures and consequently less light to work with. They would lose efficiency and, depending on how much extra observing time was devoted to compensating for it, they would lose data quality.

The Goddard Spectrograph could achieve many of its original objectives, because it had been designed to work primarily on bright stars; loss of light wasn't a deal breaker. The FOS was more seriously diminished, because it couldn't acquire high quality data on objects at the level of faintness its name implied without implausible increases in observing time. The University of Wisconsin's High Speed Photometer, designed to measure the rapid flickering of light from certain exotic astronomical objects such as the accretion disks of gas spiraling into black holes, would be both less efficient and less precise.

In addition to its function in pointing the telescope and locking it onto its targets, one of the three fine guidance sensors was used by a team from the University of Texas to measure the distances and motions of stars with very high precision—a branch of astronomy known as astrometry. All three fine guidance sensors would be less efficient and require brighter targets for pointing control as well as for astrometry.

All of the instruments were still capable of producing some good, interesting science. Because the telescope's images of stars still contained a sharp, albeit small central peak, some of its originally expected resolution could still be realized, especially by using a mathematical correction technique called image de-convolution, or image sharpening. But Hubble's highest-level objectives to probe the universe far out across space and far back in time, and to resolve the finest details in planets, nebulae, and galaxies were now out of reach.

Just before the meeting began, John Trauger cornered Ed Weiler, the Hubble Program Scientist from NASA Headquarters, in the hallway outside the meeting room. John was the Principal Investigator for the backup or "clone" Wide Field and Planetary Camera 2 (WFPC2) that had been in development at JPL since 1985. A backup camera for the original WFPC was essential "so that Hubble would never go blind," as Weiler once said. "I think we know how to fix the problem," John told him. "We can change the shape of a set of small mirrors in our camera to exactly compensate for spherical aberration." Weiler asked Trauger to describe his idea to the group.

On June 27 NASA held a very painful press briefing to tell the media about the situation with Hubble. The heaviest burden of explanation fell on Ed Weiler's shoulders. He tried to be as hopeful as he could, stating that even a degraded Hubble could still do good science. Ed seized on the news John Trauger had given him the day before, that an optical remedy to the spherical aberration might be achievable internally within WFPC2.

Hubble was unique in being the only space science satellite to be designed explicitly for in-orbit servicing by space-suited astronauts in the payload bay of the Space Shuttle. Periodic servicing missions, nominally one every three or four years, had always been in the plans for Hubble. Possibly, in about three years, an emergency first servicing mission could be mounted to Hubble. Perhaps an optically corrected WFPC2 could take the place of the original WFPC, enabling 40%–50% of Hubble's originally planned science investigations—the portion that required high resolution imaging over a relatively wide field-of-view on the sky.

Events were unfolding rapidly at the time of the press briefing. The sketchy ideas being thrown out there were only based on a little bit of quick-look data and analysis from JPL. Their feasibility, and the practicality of rapidly initiating a servicing mission to the observatory, remained to be established. But they provided all of us with a reason to be hopeful and to portray an optimistic face to the public.

All of the scientists who had been working on Hubble for at least a decade were devastated by the turn of events, none more so than Westphal. Jim's instrument was to serve as the workhorse camera on Hubble, the instrument that would produce not only much of Hubble's primary science, but also its gloriously beautiful pictures. It was to give humans a really clear view of the universe for the first time in history.

Jim was especially revered among the scientists on his team and among his colleagues working on the Hubble Project. He was outspoken, a bit gruff, a bit homespun, but charming, witty, and awesomely bright. He was one of astronomy's leading instrument designers, but without the academic trappings of a PhD. He had originally worked as a geophysicist in the oil industry where he had proven his mettle as a technical genius.

Westphal could barely stand the pain of what had happened, but he stuck with Hubble. The same could not be said of some other astronomers. Jim's Deputy Principal Investigator Jim Gunn, a brilliant astrophysicist, was angry. Gunn had been the original driving force at Caltech, urging Westphal to propose a camera based on charge coupled device (CCD) solid state light detectors—the latest and greatest thing to come along at that time in detector technology. Westphal and Gunn pulled together a team of highly qualified astronomers and submitted a proposal for a Wide Field and Planetary Camera in response to NASA's 1977 Announcement of opportunity for instruments to be launched on the Space Telescope. Gunn had moved from Caltech to accept a faculty appointment at Princeton in 1981. From there he continued to work as Westphal's Deputy after their camera proposal had been selected for development. But he abandoned his Hubble role entirely after the announcement of spherical aberration. He reportedly explained his decision in a fiery email to his colleagues that strongly berated NASA. Other people in the science community who had not worked on Hubble chimed in with harsh criticism. Rumors and gossip were rampant about what had gone wrong and who was to blame.

The two Telescope Scientists who had been selected by NASA to serve on the Hubble Science Working Group as respected optical experts, Bill Fastie from Johns Hopkins and Dan Schroeder from Beloit College, came in for especially harsh criticism. Their role had been to represent the astronomical community in monitoring the design and construction of the telescope optics. As was later shown, the criticism was unfair. Bill and Dan had never been given direct oversight or access to the inner workings of Perkin-Elmer's Optics Division. Like everyone else, they had to rely on the flawed test data Perkin-Elmer presented to NASA.

Everyone working on Hubble, scientists, engineers, managers, technicians, secretaries—everyone—felt the humiliation deeply when a prominent comedian on late-night television raised his leg on a model of Hubble like a dog peeing on a fire hydrant, or when the cover of Newsweek magazine showed a picture of Hubble and referred to it as NASA's 1.5 billion dollar failure. Even Senator Barbara Mikulski from Maryland, a longtime champion of NASA and of Hubble, publicly characterized the new observatory as a "techno-turkey." It was so hard for any of us to admit what we did for a living to a neighbor, or to the inquisitive passenger sitting next to us on a plane. We worked on the Hubble Space Telescope but were loath to tell people about it.

Slowly, however, a core group of very stubborn people emerged from the hell that was Hubble at that time. The dream of a large, high quality optical telescope in orbit, above the atmosphere that so badly impedes telescopes on the ground, was worth fighting for. In a classified Rand Corporation report he wrote in 1946,[1]

---

[1] Reprinted in 1990 in The Astronomical Quarterly, Volume 7, pages 131–142.

Lyman Spitzer, then a professor of astrophysics at Yale (later at Princeton), who would in time be revered as the "father" of the Hubble Space Telescope, first described what kinds of incredible science such a space observatory could do. Later, other advocates in addition to Spitzer, including John Bahcall, George Field, Nancy Roman, and Bob O'Dell, had passionately articulated this dream to the scientific community, to Congress and to the general public. Even in the face of the disaster of 1990, Spitzer's ideas remained valid and critically important to the advancement of human knowledge. We could not give up on those aspirations in the face of adversity, no matter how severe.

At NASA Headquarters, Marshall, Goddard, and the STScI our thoughts first turned to questions of what had gone wrong. What was wrong with the telescope's optics? How did the problem arise? Could we precisely characterize the images Hubble was actually producing and use this knowledge to deduce the nature and cause of the flaw?

On 1990 July 2 Associate Administrator Fisk appointed the Hubble Space Telescope Optical Systems Board of Investigation chaired by Lew Allen, at that time Director of JPL. The Board was chartered to determine the cause of the telescope's flaw, how the flaw occurred and why it wasn't detected prior to Hubble's launch. Allen's team was to perform what might be called a forensic "crime scene investigation" at the plant in Connecticut where the telescope was designed and manufactured.

In parallel with the Allen Board's review, a second team of optics and telescope experts, the HST Independent Optical Review Panel, was commissioned by the Hubble Project to determine as accurately as possible the actual optical prescription of the as-built telescope in orbit. Duncan Moore, Director of the Institute of Optics at the University of Rochester, chaired this panel.

Moore's panel used several techniques, the most informative of which was "phase retrieval analysis." The telescope's secondary mirror was moved back and forth, inside and outside the best focus position. From the unfocused images of stars obtained with Hubble's cameras at each of these positions, the experts could derive the telescope's prescription and characterize whatever aberrations were present. Images of stars taken with the telescope could be simulated with computer models assuming various combinations of optical parameters and these could be compared to the actual images from the telescope to see which parameters gave the best fit (Figure 1.3).

The panel verified and extended the work on the telescope images done earlier by Chris Burrows, Sandy Faber, and Jon Holtzman. They were able to demonstrate that the secondary mirror of the telescope was not the source of the problem. It had been ground to the correct prescription. The culprit had to be the 94 inch primary mirror.

Coming at the problem from two different directions, the Allen Board and the Moore Panel reached the same conclusions about the optics (Allen et al. 1990). The primary mirror had been ground and polished too flat. Too much glass had been removed. At its worst the error was about 2.2 μm—about 1/50 the thickness of a human hair—near the mirror's outer edge. As tiny as this error seemed, it was

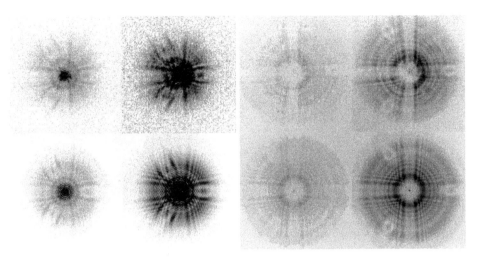

Figure 1.3. Example of a phase retrieval analysis of Hubble's optical flaw. The pair of frames on the upper left show an image of a star taken with the planetary camera on WFPC processed to two different contrast levels to bring out different levels of detail. These were taken with the telescope's secondary mirror at the best focus position. The pair on the upper right show images of the same star taken with the secondary mirror moved out of the best focus position, again processed to two different contrast levels. The pairs of images on the bottom left and right are computer model simulations of these observations that include spherical aberration. (Credit: NASA, J. Holtzman, The Allen Board.)

comparable to the wavelengths of light being focused by the telescope. Thus, it was huge in terms of the precision needed for Hubble to work properly. Light rays hitting the mirror at different radii out from the mirror's center came to a focus at different points. The focus positions were spread out over about 43 mm along the telescope's optical axis. There was no setting of the telescope's secondary mirror that allowed all the light rays to converge at a single focal point on-axis or on an off-axis focal surface. The telescope could not be focused and its images were perpetually smeared out. Both investigation teams derived very nearly the same numbers for the telescope's actual, as-built optical prescription. Hubble's primary mirror was exquisitely ground and polished, but to the wrong shape.

The tangible culprit that produced this error would have been laughable, had the consequences not been so dire—a missing chip of paint, or more precisely a missing chip of an anti-reflective coating—made all the worse by poor judgment on the part of technicians performing a critical test. It was not feasible to test directly the optical performance of the completed telescope. The optical fixture needed to perform such a precise test would have been as large as the Hubble telescope itself and prohibitively expensive. So, instead, the technicians who polished the big mirror at Perkin-Elmer had to rely on a smaller, cleverly designed optical device called a reflective null corrector, together with an optical instrument called a laser interferometer. They used these instruments to check to see that they were removing the right amount of glass and were achieving the correct shape for the mirror's surface. The primary mirror was supposed to be ground and polished into a concave "bowl" with a cross section the shape of a precisely defined hyperbola.

The reflective null corrector consisted of two simple spherical mirrors and a field lens of sophisticated design (Figure 1.4). It became the sole standard of truth at Perkin-Elmer and within the Hubble Project at Marshall, both for judging how much glass remained to be removed from the mirror's surface as the polishing proceeded, and also, in the end, for demonstrating that the final mirror was correctly shaped, that it met its very stringent design specifications.

We should have seen from the outset the logical flaw in this approach. The test device that provided the data to guide the grinding and polishing of the primary mirror was the very same device that was used to "prove" that the mirror had been ground and polished correctly (Figure 1.5). There was no independent check, or at least there was no independent set of measurements that Perkin-Elmer and NASA believed to be credible, to verify that the mirror was good. Full faith was put solely in the reliability and accuracy of the reflective null corrector. The defendant was allowed to be the jury at its own trial.

As fate would have it, this logical trap caught us. The reflective null corrector was not correctly assembled. Its field lens was mistakenly located below where it was supposed to be by 1.3 mm. How could that happen?

A metal rod cut to a precise length was used to set the distance between the field lens and the center of curvature of the lower spherical mirror in the null corrector (Figure 1.6). The location of the rod was determined with a laser beam shining on the rod's polished tip. To assure that the laser hit exactly the right point on the tip, a cylindrical metal cap was placed on the end of the rod. In the cap was a small circular hole through which the laser beam had to pass. The surface of the cap surrounding the hole was painted with an anti-reflective coating. In the darkened lab the laser beam would be moved laterally over the surface of the cap. The anti-reflective coating would prevent it from reflecting back upward and no signal would be detected until the laser hit the hole in the cap. The light would then be reflected brightly back upward off the tip of the rod and a strong signal would be seen in the laser interferometer above.

Things didn't go exactly as planned. Unknown to the technicians who were assembling the reflective null corrector, a small area of the anti-reflective coating just next to the hole in the cap was missing; it had somehow chipped off (Figure 1.7). So, the metal cap's shiny surface was exposed and could strongly reflect the laser beam back into the interferometer—mimicking the polished tip of the rod located 1.3 mm below.

That is exactly what happened! The technicians inadvertently placed the laser beam on the shiny area on the surface of the cap next to the aperture. They mistakenly thought they had located the rod's polished tip. So, they ended up placing the cap rather than the tip of the spacing rod at the center of curvature of the bottom mirror (Figure 1.8). They then needed to move the field lens to bring it in contact with the bottom of the rod. But something was wrong. They ran out of adjustment range. They couldn't quite get the lens to move far enough downward.

At this point the technicians might have (and should have) called "time-out" to inquire, "What the hell's going on here?" "Why can't we get this lens to move down far enough?" But for some unfathomable reason, they didn't do that. Instead they

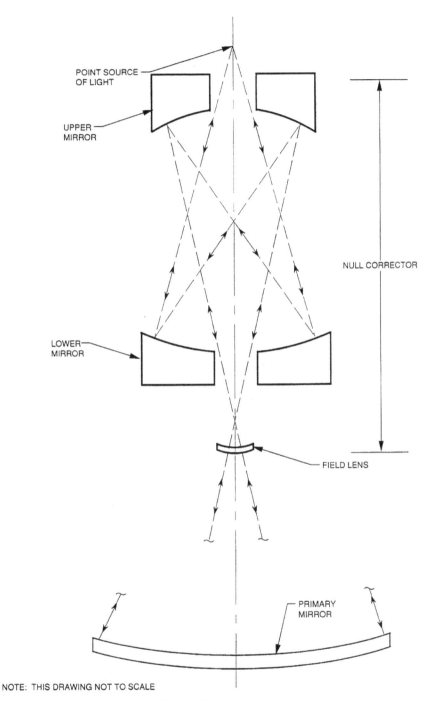

Figure 1.4. Paths of light rays through the reflective null corrector. Light from a laser at the top was reflected off of two spherical mirrors; it then passed through a special field lens, onto Hubble's primary mirror. From there the light was reflected back up along the same path through the null corrector into the laser interferometer, which created a pattern of fringes called an interferogram. If the mirror was properly ground and polished the fringes should be perfectly straight and parallel everywhere within the interferogram. (Credit: NASA, The Allen Board.)

Figure 1.5. Hubble's 2.4 m primary mirror during the final polishing process at Perkin-Elmer Corporation, probably in 1982. (Credit: NASA.)

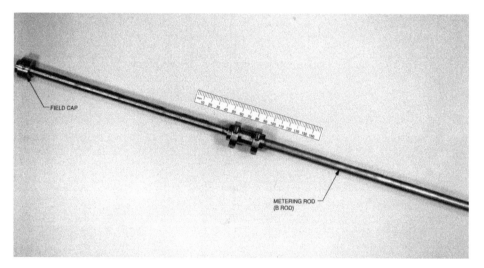

Figure 1.6. The metering or spacing rod designed to accurately position the field lens of the reflective null corrector with respect to the lower mirror. Note the field cap at the upper end of the rod. (Credit: NASA, The Allen Board.)

inserted a number of spacers so as to extend the range of motion of the field lens and achieve the needed extra 1.3 mm of spacing. In that way they forced the performance of the null corrector to conform to their expectations, although it was actually far off the mark.

Almost a decade later Lew Allen and his Board uncovered what had happened during their investigation on the scene at Perkin-Elmer. They demonstrated that the

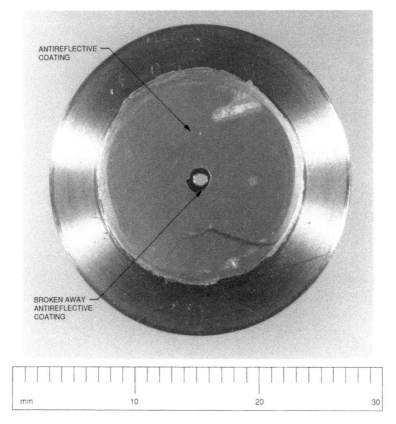

Figure 1.7. The field cap designed to precisely position the laser beam from the laser interferometer onto the shiny tip of the spacing rod within the reflective null corrector. The position of the rod was adjusted up or down until the laser interferometer produced the desired pattern of fringes. The field lens was then supposed to be brought into contact with the bottom of the rod. In principle that would assure that the field lens was at the correct position to allow accurate measurements of the figure of the primary mirror during polishing. Note the missing anti-reflective coating around the central aperture. The laser beam reflected off of the uncoated surface of the cap instead of passing through the hole at its center as intended, causing the field lens to be positioned 1.3 mm too low. (Credit: NASA, The Allen Board.)

1.3 mm error in the spacing between the lens and the lower mirror in the reflective null corrector, and the corresponding polishing error in the shape of the Hubble telescope's primary mirror, precisely explained the amount of spherical aberration observed in the camera images Hubble was sending back from space.

Ironically, two other optical test devices clearly showed the mirror's defect in data they provided at the time, independent of the reflective null corrector. But both Perkin-Elmer and NASA chose to ignore this, having become convinced that those other two instruments were not sufficiently accurate to provide credible data at the high level of precision required. They thought only the reflective null corrector could do that. The Allen Board demonstrated that this was not true. Both of the other devices were accurate enough to reveal a major, gross error of the magnitude that had been built into the Hubble mirror.

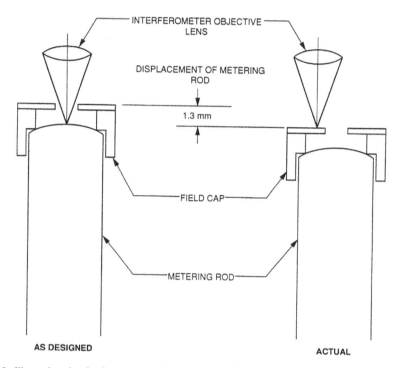

Figure 1.8. Illustration showing how the metering rod was positioned too far downward by 1.3 mm. The laser beam from the interferometer was inadvertently focused on a shiny spot on top of the field cap instead of passing through the central aperture to reach the tip of the spacing rod. (Credit: NASA, The Allen Board.)

The investigations revealed myriad failures of management oversight, technical processes, and quality control on the part of both Perkin-Elmer and the NASA Hubble Project. There were a lot of excuses. Perkin-Elmer was badly behind schedule and over budget in manufacturing both the telescope and the Hubble fine guidance sensors for which they were also responsible. So there was much pressure to work quickly and perhaps to cut corners. The division of Perkin-Elmer that had the job of grinding the mirrors and conducting the optical testing operated in a closed-door environment. According the report of the Allen Board, neither people from NASA nor from other divisions of Perkin-Elmer had easy access to the Optical Operations Division for purposes of oversight. Although the Department of Defense denied any responsibility for this state of affairs, the review team speculated that a culture of secrecy prevailed within that group because it reputedly had also produced optical systems for classified satellite payloads in the same plant. Finally, the small cadre of personnel the Hubble Project at Marshall assigned to monitor the work at Perkin-Elmer really did not have the experience or expertize needed in the area of large optical telescope systems.

The Hubble Science Working Group, of which I was a member, did not push hard enough on the Project to gain access to the facilities and testing programs at Perkin-Elmer. It was self-evident that this was necessary given the fragile nature of the testing procedures adopted for the telescope. We all took it for granted that Perkin-

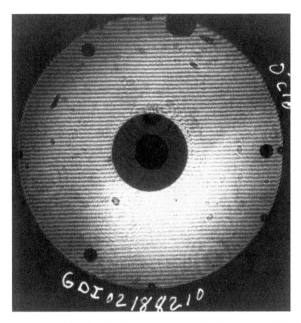

Figure 1.9. Interferogram produced by the reflective null corrector in February of 1982, demonstrating that the Hubble primary mirror had been ground and polished to fully meet its stringent design specifications. It was essentially "perfect." Unfortunately, the test device, the null corrector itself, had been incorrectly assembled, so that it was providing erroneous test data. Consequently, the shape of the "perfect" mirror was perfectly wrong. (Credit: NASA, The Allen Board.)

Elmer must have done this kind of work successfully before for programs involving national security and that they could be trusted to get it right.

At one meeting of the Science Working Group at Marshall, Telescope Scientist Bill Fastie—a large man, towering tall above most of us—proudly and with his usual bubbling enthusiasm held up a picture of an interferogram—the pattern of bright fringes or bands obtained for the completed and now certified primary mirror using the reflective null corrector and laser interferometer (Figure 1.9). To the knowledgeable observer, the interferogram looked gorgeous—all the fringes lay straight and parallel to each other as they were supposed to for a primary mirror that met or exceeded its design specifications. It seemed to be perfect. Unfortunately, it was perfectly wrong.

Hubble's inability to focus light into sharp images because of spherical aberration was exacerbated by the thermal flexing of the two ESA solar arrays that had been launched on the spacecraft. At sunrise and sunset, as the spacecraft moved from night to day or day to night, the extender rods that held the flexible blankets of solar cells in place would suddenly expand or contract, putting a mechanical impulse into the spacecraft and causing to vibrate. So the telescope's images were smeared both by defective optics and by unexpected jitter in its pointing. The vibrations would soon die away, but they made locking onto and observing astronomical targets all the more challenging. Life was complicated.

In the years immediately after its launch, Hubble produced some interesting science. Most notable was former Hubble Project Scientist Bob O'Dell's serendipitous discovery of a population of newborn stars in the Orion Nebula surrounded by ring-shaped disks of dust and gas. These were the environs where new planets might be forming. If technically advanced aliens could have seen our own solar system from a great distance shortly after its formation 4.6 billion years ago, it undoubtedly would have looked like this. O'Dell named these objects "proplyds," short for protoplanetary disks. He discovered them in images taken with the first WFPC, a capable camera notwithstanding the affliction of spherical aberration.

NASA worked hard to accentuate the positive in telling the public about the science the flawed telescope was able to do in orbit. But that science rarely approached the significance of what we all expected as Hubble was being designed and built. It quickly became a national embarrassment and the butt of jokes on late-night talk shows. Without the promise of a Shuttle-based repair mission, the scandal would likely have persisted for a while and then faded as Hubble's less-than-memorable science also faded from public interest. Still, Hubble would probably have persisted in popular culture for many decades as another in a series of legendary failures.

In the 1991 movie farce, "Naked Gun 2½: The Smell of Fear," Leslie Nielsen's bumbling Lieutenant Drebin walked into a bar called "The Blue Note." As its name implied, this was a place where the sad and depressed went to drink away their sorrows. The music was blue and downbeat, the lighting was dark, the ambience morose. Hanging on the walls of this mournful establishment was a series of photographs: the Titanic, the Hindenburg, the Ford Edsel, and ... the Hubble Space Telescope. Such was the popular view of our mission one year after Hubble's flaw was announced to the public.

Without servicing, Hubble likely would not have continued to operate for a long time—possibly only for five or six years. In particular, we learned later that the six gyroscopes, needed to accurately guide, point, and stabilize the telescope, were prone to failure. In Hubble's early years three of the six gyroscopes had to be operating at any given time for any astronomical observations to be done. Otherwise the spacecraft would cease operations, go into a self-protecting safe mode and wait for intervention from engineers on the ground. A non-serviceable Hubble would likely have failed sometime in the mid-1990s for want of working gyros. Today it would probably still be orbiting the Earth as a large, expensive, and embarrassing piece of space junk. Thankfully, redemption was possible.

References

Allen, L., et al. 1990, The Hubble Space Telescope Optical Systems Failure Report (Washington, DC: NASA)

Broad, W. J. 1993, Telescope Makers to Pay $25 Million for Flaw, New York Times, October 5

Burrows, C. J., et al. 1991, ApJ, 369, L21

Chapter 2

The Road to Redemption: Learning to Service in Space

In 1972 May John Naugle, the Head of the Office of Space Science at NASA Headquarters, sent a provocative memo to NASA's number two man, Deputy Administrator George Low. In it Naugle stated his conclusion that, "it is technically feasible to develop this 3 m optical telescope, and … it can be placed in operation in the decade of the 1980s as an essentially permanent observatory in space through the marriage of automated spacecraft technology and the unique capabilities of the Shuttle transportation and maintenance systems."[1] This was a watershed moment in the history of the Hubble Space Telescope that foresaw a future partnership between the human spaceflight and the robotic space science arms of NASA. But at that point in 1972 Naugle's conclusion was just a statement of plausibility—just words. The shuttle did not exist. No human-tended spacecraft had ever launched a satellite into orbit. No one had ever serviced, repaired, or upgraded a satellite in orbit. There was no experience within NASA of implementing a program merging human space flight with robotic space science. Converting this management idea into future reality required the creation of complex technological, engineering and management disciplines, as well as a major cultural shift within the Agency.

Hubble was by no means the first astronomical observatory in space. It was preceded by numerous smaller space telescopes and spacecraft bearing scientific instruments designed to collect electromagnetic radiation from cosmic sources across the spectrum from gamma-rays, X-rays, ultraviolet, infrared, microwave, and even radio wavelengths. What made the concept of Hubble unique was its ability to take clear, sharp pictures of planets, stars, nebulae, and galaxies both near and very far away. No prior space instrument had been designed to clearly image the

---

[1] As quoted in The Space Telescope by Robert Smith (1989; Cambridge: Cambridge Univ. Press), p 85.

universe in ordinary visible light (as well as ultraviolet and near-infrared wavelengths), completely unimpeded by the shimmer and glow of the Earth's atmosphere.

Hubble's most direct predecessors were the Orbiting Astronomical Observatories (OAOs) launched from 1966 to 1972, and the International Ultraviolet Explorer, a small observatory designed to collect spectra of a variety of astronomical sources at ultraviolet wavelengths, launched in 1978. It is with the OAO series that the story begins of the development of NASA's capability to service orbiting spacecraft, ultimately leading to the rescue of Hubble in 1993.

Three OAO missions were planned. Each drew scientists and their scientific instruments from across the US, Europe, and the United Kingdom. The Goddard Space Flight Center managed the missions, with spacecraft designed by Grumman Aerospace Corporation and scientific payloads developed primarily in academic and government laboratories.

The first of these observatories, OAO-A, launched in 1966 April was a total failure. Its guidance system Star Trackers were turned on prematurely, before the residual atmospheric molecules carried into orbit within the spacecraft had time to dissipate into the vacuum. As a result a high-voltage power supply produced a powerful arc of electric current, which passed through the thin gas to the structure of the spacecraft, creating a short-circuit. One failure led to another and the end result was that the spacecraft batteries began to overcharge and to increase rapidly in temperature. Fearing the hot batteries would explode, spacecraft engineers purposely put OAO-A into an uncontrolled tumble to reduce the electrical power coming from the solar arrays and charging the batteries. This was intended to buy time to allow the batteries to cool down and to allow the engineers to troubleshoot the problem. However, they were never able to re-stabilize the tumbling spacecraft. The OAO-A scientific payload was never turned on and the mission was terminated three days after launch. It was a very bad day for NASA's fledgling space astronomy program.

In the face of this embarrassing failure Goddard's Center Director, John Clark, appointed Joe Purcell, who had chaired the OAO-A Failure Review Board, to take over the OAO Project and lead it to success. Joe had the appearance and demeanor of a military officer—tall, crew cut, square jaws that he clinched when he talked in a soft southern accent. In fact he had originally enrolled at the Naval Academy, but left after a football head injury rendered him prone to seizures. He studied physics at the University of Richmond.

One member of Joe's OAO Project team was a 30 something mechanical engineer from California and the University of Santa Clara, Frank Cepollina. Frank, or "Cepi" as we all called him, would go on over the next several decades essentially to invent the new aerospace engineering discipline of in-orbit servicing of spacecraft by Shuttle-based astronauts and by robots. He would become the impresario, the producer, the technical management leader, and the key driving force in five highly successful Shuttle servicing missions to Hubble from 1993 to 2009. During that period he also developed a reputation as a maverick, a brilliant inventor, a "used car salesman," an astute politician, a relentless driver of the people who worked for him, and a builder of can-do teams. Cepi regularly became a thorn in the side of higher level NASA management, both at Goddard and at NASA Headquarters. NASA couldn't do without his passion, technical savy, and leadership, but had a terrible time putting up with him. Noel Hinners, the Center Director at Goddard from 1982 to 1987, was quoted as saying, "Every NASA Center should have a Cepi ... but only one."

In the mid-1960s, however, Cepi was just getting started at Goddard working for Joe Purcell on trying to recover from the OAO-A debacle. The OAO Project had a backup OAO-A spacecraft in shops at Goddard, a replica that had been used originally to test and verify the design of the flight unit. Purcell and Cepi quickly organized the effort to convert this "qualification unit" into a fully capable flight spacecraft. Launched in 1968 December, OAO-A2 carried seven small ultraviolet photometric telescopes designed to measure the brightness of planets, stars, and nebulae at ultraviolet wavelengths (Figure 2.1). This Wisconsin Experiment Package (WEP) was developed at the University of Wisconsin under the leadership of Principal Investigator Art Code (who, later in his career, would play a seminal role in establishing the Space Telescope Science Institute). Additionally, OAO-A2 carried four cameras in a package called Celescope designed to snap low-resolution pictures of our near environs in the Milky Way Galaxy to measure the ultraviolet brightness of a large number of stars simultaneously. Principal Investigator Fred Whipple and his team created the Celescope at the Harvard Smithsonian Center for Astrophysics. OAO-A2 was a success, producing a number of important scientific findings over its lifetime, which concluded in 1973.

NASA originally hired me in 1969, fresh out of graduate school at UCLA, to do scientific research with the instruments to be carried on the next observatory in the series, OAO-B. With the low-resolution ultraviolet spectrum scanners in the

Figure 2.1. Artist's visualization of Orbiting Astronomical Observatory-A2 in orbit in 1968. (Credit: NASA.)

Goddard Experiment Package (GEP) we hoped to make the most accurate measurements ever of the energy output of a wide variety of stars. The GEP's Principal Investigator at Goddard was Albert Boggess—my de facto first boss and mentor at NASA.

Sadly, for Al and the entire OAO team, the rocket launching OAO-B on 1970 November 3 failed to eject the heavy shroud at its top that enclosed the observatory payload. The booster didn't have the thrust needed to carry all of that weight into orbit and it fell back to Earth, probably somewhere over the Indian Ocean. On that day, I saw many grown men and women cry.

After the OAO-B launch failure I was re-assigned to work with the science operations astronomer team on the Wisconsin Experiment Package. What a wonderful education that was! I learned hands-on how to properly plan and implement the operation of a space observatory, as well as how to do significant research with a remote orbiting instrument. And I developed many friendships with other astronomers in the Wisconsin group; some of them would also go on to work on Hubble.

The most successful of the OAO missions was OAO-C, nicknamed Copernicus. Its telescope, sensitive to far-ultraviolet light, was the largest and most challenging space telescope built up to that time. This seems fitting, as the Principal Investigator was none other than Lyman Spitzer, the father of the Hubble Space Telescope. He and his group from Princeton designed a very high-resolution spectrograph to measure the composition, temperature, density, and distribution of the gas between the stars in the Milky Way, the interstellar medium. The young astronomer hired by Princeton to run the observatory science operations at its control center in Building 3/14 at Goddard was Ed Weiler, a PhD graduate of Northwestern University.

In 1971, following failures in two out of the three flights in the OAO series up to that time, George Low, NASA's Deputy Administrator, challenged Goddard management to come up with more reliable "spacecraft buses" that would also take advantage of the potential two-way capability of the Agency's new "space airplane" (the Space Shuttle) then being designed. Perhaps looking ahead to a future Large Space Telescope, Lyman Spitzer also openly advocated servicing spacecraft that had suffered failures in orbit. Purcell and Cepi paid visits to Low, Naugle, and NASA's Chief of Astronomy, Nancy Roman, at Headquarters, as well as to Spitzer at Princeton, to discuss how this might be realized. Out of those discussions, and some engineering studies at Goddard, emerged the concept of the Multi-Mission Modular Spacecraft, or MMS.

The MMS would be a modular, universal bus that could provide the spacecraft functions for a wide variety of missions. It would be a standardized back end to which could be attached telescopes, instruments, and sensors of all kinds. It would include separate, detachable modules each of which would provide a critical utility, such as a communications system and antennas, command and data handling computers, electrical power, temperature control, attitude control, guidance and navigation, propulsion and the mechanical structure to which all of these modules, as well as the primary mission payload could be attached. The number and selection of modules could be tailored to fit the unique needs of a specific mission. Because the modules were produced in quantity, the costs could be kept relatively low. Because the same module designs were used again and again, the MMS would be very reliable.

An MMS-based mission could be launched with an expendable rocket, but it could also fly in the payload bay of the future Space Shuttle. Because of the spacecraft's modularity and the ease of accessibility of its component systems, either people or robots could repair it in orbit. The MMS concept laid a practical foundation on which would later be built the modular, serviceable spacecraft at the back end of Hubble, the Support Systems Module (SSM).

In the early 1970s, when the MMS concept was gestating at Goddard, NASA and its scientific working groups had not agreed even to an optimum size for the Large Space Telescope or LST (as Hubble was then called). A mature conceptual design did not exist. Nor had a management structure within NASA for a potential future LST mission been defined. Both Goddard and the Marshall Space Flight Center were commissioned by NASA Headquarters to do "Phase A" (conceptual design) studies. In essence they were competing for the lead responsibility for designing, building and operating the LST and its scientific instruments. But the slate was essentially blank on which each Center could propose creative concepts.

Joe Purcell, Frank Cepollina, OAO Project Scientist Jim Kupperian, John Mangus, Goddard's lead optical engineer, and several others formed an *ad hoc* engineering study team to work on a concept for the Large Space Telescope. Given the recent work done by Purcell and Cepi on designs for the MMS, they were in a strong position to advocate to NASA management that the LST should have a modular spacecraft that could readily be serviced by astronauts.

The Goddard LST Study Team strongly emphasized the need to create a space observatory that was both very capable in terms of optical performance and relatively low in cost. To demonstrate how this might be done, John Mangus pushed the team to build an actual full-scale model LST—a telescope structure and a modular spacecraft. Such a brash scheme was the perfect mechanism to galvanize support for a Goddard concept and to provide an infrastructure on which to test engineering ideas.

In 1971 the United States Congress terminated funding for the Supersonic Transport (SST) airplane being developed by the Boeing Corporation in Seattle. To fabricate two prototypes of the SST, Boeing had accumulated a large stockpile of titanium. When the SST program was canceled, the Federal Aviation Administration (FAA), which owned the metal, had no further need for it, and advertised it as "government excess property." As such, other government agencies had first priority in claiming it. That gave Purcell and Cepi the bright idea to build the LST structure out of titanium.

There are technical drawbacks to using any metal like titanium for a structure that must remain perfectly stable without expanding or contracting as its temperature varies from warm to cold or vice versa. Even small changes in the dimensions of the structure would throw a perfect set of telescope optics out of focus and alignment, rendering a sharp image blurry. And during the course of one orbit, in going from day to night and back again, the external skin of a spacecraft is subjected to swings in temperature of several hundred degrees. In sunlight, one side of the spacecraft is illuminated and hot, while the other side is shaded and cold. Affixed to the outer skin of any spacecraft is a thin blanket of multi-layer insulation (MLI), designed to protect the interior from the wide swings of temperature in orbit. Inside the spacecraft, where the telescope structure and optics reside, those wide variations in temperature are further controlled with insulation and electric heaters.

Maintaining the optical stability in the telescope rests ultimately on the composition and design of the "metering structure" that separates and holds the primary and secondary telescope mirrors in place. A "perfect" thermally stable structure could be created using newly developed, high tech materials fabricated from carbon fibers—"graphite epoxy." However, in the early 1970s this was a relatively new technology and certainly would be expensive to incorporate into the LST. Purcell, Cepi and Mangus thought they had a simpler and cheaper idea in using titanium.

Titanium is less susceptible to thermal expansion and contraction than some other lightweight metals such as aluminum. However, it is not nearly as stable as the carbon composite materials. To even out the thermal gradients within the structure that would cause it to change its size and shape, they envisioned using heat pipes embedded in the thermal shield on the outside of the spacecraft. Heat pipes are cylindrical tubes filled with a fluid, such as ammonia, that rapidly move heat from one location to another to smooth out temperature differences. They are commonly used for thermal control in spacecraft of all kinds. In this way, the Goddard engineering concept for a 3 m diameter telescope structure that was both thermally stable and inexpensive could be realized with a metering truss built from the readily available supply of titanium encircled with standard heat pipes.

The Goddard engineers quickly seized the opportunity to acquire the large stockpile of the titanium on offer. They negotiated a deal with the FAA. The Department of Defense had first claim on the excess titanium. But after the DOD needs were met, Boeing or its sub-contractors would ship the remaining metal to Goddard, only requiring Goddard to pay any extra shipping costs that were not already included in the original SST contract. Goddard had to agree not to re-sell the material. The Goddard team assured the FAA, "Oh, don't worry. We're going to use it all at NASA to build satellites."

What ensued was an almost comical scenario. As Cepi described it, "Before we even realized what we were signing up for, thousands and thousands and thousands of pounds of titanium began coming into Goddard from all parts of the country. So, all of sudden trucks started arriving, and this went on for month after month. You've got to picture the Center. In those days it was weeds and grass, open woods everywhere. The titanium would come in and the receiving people would say, 'we've got another truck, where do you want it?' And we would say, oh just put in Field B or Field C. It never rusted of course. But the stacks of titanium became the favored haunt of field mice building their nests."

The team put the in-house machine shops at Goddard and an outside contractor to work building a full-scale mockup of a 3 m diameter Large Space Telescope. They welded together titanium rings and rods to create an open cylindrical metering truss structure, some 35–40 feet long (Figure 2.2). It looked like something one might

Figure 2.2. Full-scale metering truss for a mockup 3 m Large Space Telescope built from titanium at Goddard Space Flight Center in 1971. (Credit: NASA, GSFC, F. Cepollina.)

Figure 2.3. Titanium 3 m primary mirror support ring fabricated at Goddard in 1971, part of a full-scale mockup of the structure of a Large Space Telescope. (Credit: NASA, GSFC, F. Cepollina.)

build with Tinker Toys. Within that structure the telescope secondary mirror would be mounted and held in place to maintain the focus and alignment of the telescope. They fabricated a 3 m titanium ring stout enough to support a large primary mirror (Figure 2.3). An aluminum MMS mockup was built to which the telescope could be attached (Figure 2.4).

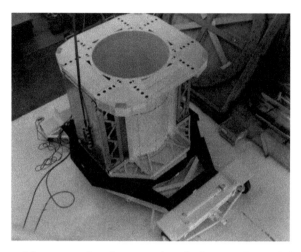

Figure 2.4. Mockup modular spacecraft to be mounted at the bottom of a stack including the titanium metering truss and primary mirror support ring to simulate a full-scale 3 m Large Space Telescope, fabricated at Goddard in 1971. (Credit: NASA, GSFC, F. Cepollina.)

Stacked together, the long metering truss structure, the primary mirror support ring and the modular spacecraft would provide a realistic structural representation of the LST. In fact, the three units were never actually stacked together. John Mangus took the metering truss into a large thermal-vacuum test chamber and began using it for optical stability tests. The mockup support ring and MMS were connected, together with a plywood dummy front end, to represent a generic payload so that tests could begin on techniques for spacecraft servicing.

Meanwhile, bad news came the team's way. For a variety of reasons John Naugle had decided that Marshall should manage the LST development. Naugle's scheme entailed leaving responsibility for managing the development of the observatory's scientific instruments and ultimately for operating the LST in orbit at Goddard. These were, after all, areas in which Goddard had the greatest expertize. Naugle explained his decision in the same 1972 May memo to George Low in which he described the idea of launching and servicing the LST with the Shuttle.

After the loss of the LST development competition to Marshall, Purcell and Cepi pressed on using the mockup modular spacecraft they had built for the mockup LST to continue to test concepts for in-orbit maintenance and repair. To the spacecraft they attached other kinds of simulated large payloads, including Earth-observing instruments. The silver lining for them in Naugle's decision was that they still retained responsibility at Goddard for developing Multi-Mission Modular Spacecraft and for servicing modular Shuttle-based payloads, including the LST.

In 1972 or 1973 representatives of the Canadian Department of Industry, Trade, and Commerce (from which would later emerge the Canadian Space Agency) approached John Yardley, the Associate Administrator for Manned Space Flight and the head of the Space Shuttle Program at NASA Headquarters, expressing interest in a possible Canadian involvement in the Shuttle program. In particular,

Figure 2.5. Demonstration of servicing a modular spacecraft in the payload bay of a full-scale plywood mockup of the Space Shuttle at the Rockwell plant in Downey, California in 1974. (Credit: NASA, GSFC, F. Cepollina.)

they believed Canadian industry had a future in robotics and in the application of robotic techniques to Shuttle missions. Yardley and Deputy Administrator Low suggested that they go to Goddard to observe the tests already being carried out on servicing large telescope structures and spacecraft.

In due course, the Canadian company SPAR Aerospace was commissioned by the Industry, Trade, and Commerce Department to build an automated robotic servicer, a remotely controlled machine capable of manipulating boxes and drawers built into a modular spacecraft. The spacecraft would rest on a rotating, lazy Susan style platform sitting in the Shuttle payload bay, while the robotic servicer would sit adjacent to it. The machine's mechanism for grabbing hold of a module on the spacecraft was similar to that of a mechanical pinsetter in a bowling alley.

The Canadians sent this device to Goddard for testing with the mockup LST modular spacecraft in 1974. Subsequently, the SPAR machine, the lazy Susan platform, the aluminum modular spacecraft, the titanium support ring, and a dummy front-end payload were shipped to the Rockwell Corporation plant in Downey, California, where the Shuttle itself was being designed and built. All of these components came together in the payload bay of a full-scale plywood mockup of the Shuttle orbiter in the Rockwell plant (Figure 2.5). Using this configuration, Cepi's Goddard team and their Canadian colleagues conducted high fidelity simulations of the servicing process, as it might actually unfold in a real Shuttle mission.

In 1974 Rocco Petrone, former Director of the Marshall Space Flight Center, became the chief Associate Administrator of NASA—the number-three position under Administrator James Fletcher and Deputy Administrator George Low.

Petrone was responsible for the major program offices at Headquarters, including Manned Space Flight; Space Science; Aeronautics and Space Technology; and Tracking and Data Acquisition.

In 1975 Petrone accompanied John Yardley to one of the routine monthly reviews of Shuttle development progress at Rockwell. During the meeting the two Headquarters managers were invited to observe the ongoing servicing tests being conducted with the Goddard and Canadian hardware in the payload bay of the mockup Shuttle orbiter.

Petrone was deeply impressed with what he saw, so much so that he directed his Headquarters program Associate Administrators to fly to Downey to see how it would be possible to maintain, repair and upgrade modular spacecraft taken onboard the Shuttle in orbit. That chance encounter between high-level NASA Headquarters officials and the NASA/Canadian engineering team that had been working at Goddard for several years to develop the concepts and techniques of Shuttle-based servicing sealed the deal for the Large Space Telescope. Top NASA management solidified the decision that the LST would be designed to be launched on the Shuttle and would be serviceable in orbit by teams of Shuttle astronauts.

And so, nearly from the beginning, Hubble (né LST) and the Space Shuttle were brothers. They were conceived as NASA programs at about the same time—the late 1960s to early 1970s. Both were advocated to Congress and to the Executive Branch at about the same time, each to the benefit of the other. Hubble provided a prime example of a major science payload that could benefit from a Shuttle launch and Shuttle-based servicing by astronauts in orbit. The prospect of becoming serviceable on a periodic basis meant that Hubble could be viewed as a "permanent" observatory in space, analogous to the major mountaintop observatories on the ground, as John Naugle had noted in his 1972 memo. Not only would it have a long lifetime, with all the possibilities that long-term and iterative research could provide, but also its scientific instruments and spacecraft systems could be repaired or replaced with the latest technological upgrades, making it a progressively more powerful scientific tool as time went on. A regularly refreshed Hubble could perpetually represent the state of the art. The Shuttle would allow the Hubble observatory always to operate at the frontiers of science, even to re-define the frontiers of science.

Hubble was designed to fit exactly (and barely) in the Shuttle's payload bay, while still being large enough—having enough light-collecting area on its main mirror—to accomplish its most important scientific tasks, as exemplified by a key research project to accurately determine the size scale, rate of expansion and age of the universe. Shuttle designers incorporated systems and design standards for power, communications, equipment storage, cleanliness etc that were compatible with Hubble when it was in the payload bay.

Hubble became a prime example of the value of the Shuttle in providing regular access to low Earth orbit for a large crew and large, massive payloads. The Shuttle enabled ways of working in space that were not otherwise feasible, and the Hubble Program was both the proof of that concept and its immediate beneficiary. The two programs represented a nexus between human spaceflight and robotic exploration of the universe. The Shuttle Program brought Hubble "perpetual youth" and Hubble brought major scientific significance to the Shuttle Program.

Nancy Roman came to NASA Headquarters in 1959 to organize the Agency's program in space astronomy. She worked there as Chief of Astronomy for two decades. Nancy was a pioneer in her era, in part by being one of the very few professional women to be a successful leader in the male-dominated world of America's space program. She was also the "Mother" of the Large Space Telescope.

To advise her on policy issues, Nancy organized a standing committee of prestigious senior astronomers from around the academic community. Among other topics, they focused intently on the possibility that NASA might build a Large Space Telescope to do high-resolution optical and ultraviolet astronomy in orbit above the absorption and turbulence of the Earth's atmosphere. Nancy needed some help running the committee meetings and sent a request to Goddard management for a staff astronomer to become its "Executive Secretary," basically a note taker and writer of meeting minutes. As a "fresh-out" young astronomer, I was well suited for the job. I was asked to do it and readily agreed. After the loss of OAO-B, I had the time.

That turned out to be a good decision. I was immediately immersed into a wide range of scientific and technical topics, as represented by the diverse membership of the committee. I also was given an insider's view of the "astro-politics" that were a routine part of everyday life inside the Washington beltway, as a particular community tried to advocate for the funding of its own programs in competition with many other constituencies. Finally, it allowed me to develop a network of senior leaders within the astronomical community and within NASA with whom I became acquainted and who came to know me at least by name and reputation.

When the Space Shuttle Program began in earnest in 1972, Nancy's advisory committee took up the question, "What kinds of astronomy might be done on brief sorties to orbit in the payload bay or in the near vicinity of the Shuttle?" I was asked to put together and lead a new team specifically to consider the opportunities for astronomy presented by the new Shuttle Program. We designated this concept "Spacelab Astronomy," as it would most likely use hardware and infrastructure provided by the European Space Agency's Shuttle Spacelab program.

My committee came up with a substantial list of science programs and concepts for new telescopes and scientific instruments that could make excellent use of this brand new spaceflight capability. Some of these were later converted into longer-term free-flying missions, the Spitzer Space Telescope being one example. Several others became part of a diverse package of ultraviolet telescopes that ultimately flew in the payload bay on two Shuttle missions, Astro-I (STS-35) in 1990 and Astro-II (STS-67) in 1995. In leading this scientific study team I was first exposed to what was to me an entirely new world—human space flight.

In 1975 Nancy Roman urged me to apply for a position at NASA Headquarters as her assistant (and likely successor when she retired). I was grateful for her regard, but politely declined. I still considered myself first and foremost a research scientist. At Headquarters I would not be allowed to do personal research (as opposed to providing funding support for the research programs of others). My career would have veered down a path toward being a full time manager/bureaucrat. I far preferred the more academic and research oriented environment at a NASA field center. Besides, I loved Goddard's rural setting and informal life style. Wearing a necktie and starched collar every day, while sitting at a desk surrounded by marble and concrete was just not for me.

In 1978 NASA Headquarters hired Ed Weiler away from his Copernicus operations job at Goddard to work under Nancy Roman in the position I had declined. Ed's professional goals in life were focused more on management than on research. So, the position at Headquarters was perfect for him.

Early in 1976 the Goddard scientist, Stan Sobieski, who had overseen the early feasibility studies of astronomical instruments that might ultimately be flown on the Space Telescope (by that time "Large" had been dropped from the name as an act of political prudence), withdrew from that role. He wanted to be able to propose an instrument of his own design, without any conflict of interest, when the time came for the community to submit contract proposals for the actual flight ST instruments. The Project needed a replacement. The lead Project Scientist at Marshall at that time was a well-known astronomer, Bob O'Dell. Bob suggested my name, based I think on the reputation I had developed in my role supporting Nancy Roman and working on studies of Shuttle payloads.

One day in July my boss's, boss's boss, Director of Space Sciences at Goddard George Pieper, walked into my small office in a building across the campus from his far grander digs, which we all referred to as "Pieper's Palace." It was startling to see this high level, senior Goddard manager walk into the office of a humble GS-13 staff astronomer. He had come to talk to me face-to-face about the Space Telescope position left open by Sobieski's departure, Scientific Instruments Project Scientist. He strongly urged me to take the job. It was a matter of maintaining Goddard's high visibility and high level of support within the Marshall-led ST Project.

I demurred. "I am not an instrument designer or builder," I said. "I'm simply an observational astronomer who uses many instruments designed by other people as tools for my research. I don't know that much about their technical aspects." Pieper said it didn't matter. I was the guy O'Dell wanted. It would make Goddard look bad if we refused. And so, I said, "yes," but with one critical stipulation. Mindful of the ongoing and sometimes bitter jealousy and inter-center rivalry between Goddard and Marshall I said, "If I take on this job, my loyalty will always be to the dream and the promise of the Space Telescope first and foremost, not to NASA or field centers or particular astro-political factions in the community." Pieper agreed.

In the mid-1970s the Multi-Mission Modular Spacecraft initiative became a full-fledged flight project at Goddard. Cepi was named the MMS Project Manager. He, his Goddard team, and their hardware contractor, Fairchild Space and Electronics Division, set about creating an inventory of subsystem modules sufficient to assemble five modular spacecraft. These were earmarked for the Solar Maximum Mission (SMM) to be launched in 1980, Landsat 4 (1982), Landsat 5 (1984), and the Upper Atmosphere Research Satellite (UARS) to be launched in 1991. The final set of modules, originally intended as spares for these four missions, eventually found their way onto the bus for the Extreme Ultraviolet Explorer (EUVE) satellite in

1992. Other MMS buses were commissioned by the Department of Defense and by other NASA centers. In all 16 were built.

The MMS-based satellites were designed to be launched on the Shuttle. For a variety of reasons, including delays in Shuttle development, the first ones, including the Solar Maximum Mission, Landsat 4 and Landsat 5, were launched on expendable Delta rockets. However, in the case of SMM the Shuttle ended up playing a pivotal role that helped prepare the way for the first Hubble servicing mission to come later.

Although the SMM was long delayed, it was launched just in time to observe the Sun at the peak of its 11 year cycle of activity in 1980. It carried seven scientific instruments to observe solar activity across a wide spectrum of wavelengths from visible light through the ultraviolet and X-rays to the more energetic gamma-rays. The majority of these instruments had to be precisely pointed at the sun. Unfortunately, about a year after it was launched, SMM's pointing control module blew a fuse, then another fuse and another in quick succession. In all, three reaction wheels that were needed to point the observatory stopped running as the fuses blew. SMM limped along doing what science it could in a spin mode, with three of its seven instruments observing along its longitudinal axis as the spacecraft rotated at 1/3 rpm.

Harkening back to the Shuttle servicing demonstration given to Rocco Petrone at Rockwell in 1975, NASA made the decision to attempt to recover and repair the SMM during a Shuttle flight. Cepi was assigned to lead this Solar Maximum Repair Mission (SMRM) as Project Manager. What had previously been a conceptual demonstration rapidly turned into technical and managerial reality. It was on the SMRM that Cepi and his team began to develop the tools and techniques that ultimately would allow them to service, repair, and upgrade Hubble.

The SMRM was launched on Shuttle Challenger in 1984 April. In the payload bay was a Flight Support System (FSS), a lazy Susan like platform to which the SMM's spacecraft could be latched, rotated, and tilted to enable easy access by space-walking astronauts. Originally funded by the Human Space Flight Program for the SMRM, the FSS would later become the platform to which Hubble would also be latched during each of its servicing missions.

During the first spacewalk of the SMRM one of the EVA astronauts, George "Pinky" Nelson, donned a manned maneuvering unit (MMU) and used its thrusters to move to the SMM spacecraft. Pinky's spacesuit had a docking adapter attached to its mid-section. The plan was for him to dock himself to the spacecraft and then to use the MMU thrusters to transport it back to the shuttle's payload bay. But the docking mechanism didn't function properly. Apparently the documentation available on the ground that had been used to design the docking adapter wasn't accurate. There was some king of undocumented structure on the as-built SMM spacecraft that got in the way, so that the docking adapter couldn't make the connection.

Nelson then decided to improvise. He thought he could simply grab hold of one of the spacecraft's solar arrays with one hand, and operate the MMU controls with the other hand to drag the SMM back into the payload bay. But when he touched the solar array, he set the satellite tumbling. This had the unfortunate outcome that the spacecraft began losing electrical power, as its solar arrays could not maintain a lock on the sun to keep its batteries charged. Just in the nick of time, controllers at Goddard succeeded in using the magnetic torquers—electromagnets mounted on the spacecraft that interacted with the Earth's magnetic field—to stop the satellite from spinning.

The SMM was the first satellite built with a grappling fixture, a specially shaped rod that could be grabbed and held firmly by a mechanism called an "end effector," at the end of the Canadian remote manipulator arm, the RMS. For the first time in spaceflight history the RMS was used to grapple a satellite, setting the aft end of the SMM spacecraft precisely into the latches on the FSS. In that arrangement two space-suited astronauts had easy access to the areas of the satellite where they needed to do work (Figure 2.6).

In one spacewalk it took astronaut James "Ox" van Hoften only about 40 min, using a special, space-qualified, battery-powered tool designed by Cepi's team, to remove and replace the faulty Attitude Control System module on the SMM bus.

Figure 2.6. Astronauts James van Hoften and George Nelson making repairs to the Solar Maximum Mission observatory during STS-41C on space shuttle Challenger in 1984. (Credit: NASA.)

This was the first power tool to be used during an EVA. Pinky Nelson had considerably more difficulty changing out a failed electronics circuit board.

One of the SMM's scientific instruments, the ultraviolet spectrometer/polarimeter, had suffered an electronics failure. It needed a new interface control board. Nelson, in a cumbersome inflated spacesuit with big bulky gloves, used the power tool to unscrew the lid to the box housing the instrument. He then had to disconnect and remove the faulty card, replace it with and reconnect the new version, and refasten the lid. The problem was that the card had miniature electrical connectors, which he had "a devil of a time" unscrewing. Eventually he succeeded but the task took three grueling hours to finish. In so doing, Pinky Nelson blazed a trail for future EVA astronauts who, 25 years later in the final Hubble servicing mission, would successfully carry out the more complex replacement of failed internal electronics boards in two of Hubble's advanced scientific instruments, using a far more sophisticated set of power and manual tools that Cepi's team designed.

When I first joined the Project in 1976, NASA had planned a launch date for the Space Telescope in late 1983. However, when that year rolled around the Project was in serious budget and schedule disarray. The Project began as a "new start," officially funded by Congress in late 1977 (Fiscal Year 1978). At that time NASA Headquarters and Marshall had projected the cost to develop the observatory (not counting the cost of its Shuttle launch) in the $400–500 million range. But the story of the Space Telescope Project over its first four years was one of an "irresistible force"—cost growth as the magnitude of the job came into sharper focus and serious technical problems arose—versus an "immoveable object"—the stubborn determination of managers at Marshall and at NASA Headquarters not to admit that the budget and schedule for the Space Telescope Project were inadequate, that we "couldn't get there from here."

Various possibilities for reducing the capabilities of the observatory to cut costs were seriously considered. Examples included deleting the spacecraft's aperture door, which protected the telescope from exposure to direct sunlight, deleting the pressure pads that could adjust the shape of the primary mirror in orbit, and even delaying the completion of several of the science instruments, the two spectrographs and the High Speed Photometer, assuming they could be flown up to the telescope on a later servicing mission. All of these ideas were ultimately rejected as being penny-wise and pound-foolish.

In the early 1970s the community of astronomers advising NASA about the scientific rationale for a Large Space Telescope had acquiesced to a decision to reduce the aperture size of the telescope from 3 m to 2.4 m. That would make it considerably simpler to build, as well as make it compatible with the Shuttle. Any smaller, and the telescope would no longer be able to achieve some of the primary scientific objectives that were its basic rationale in the first place. The scientific community went all in for a 2.4 m telescope of superb optical quality to be launched on and serviced by the Shuttle. Anything less than that was deemed not worth the cost and effort.

It is always easy (and glib) to blame the cost growth and schedule slips of large and challenging space missions on "poor management" or conscious "buy-in," purposely underestimating the true cost of a mission in order to get it approved and started. There may sometimes be a kernel of truth in such allegations, but they don't capture the reality of why cost and schedule problems come along so frequently. Space hardware of any kind has to work reliably, often for many years, in an extraordinarily harsh environment. Robotic satellites like the Space Telescope have to operate autonomously under computer control and ground command. Even when scientific instruments and observatories are designed to be as "simple" as possible, they are surrounded by the great complexity of control electronics and other spacecraft systems, all of which have to work without failure for years. Even in a Shuttle-serviced spacecraft like the Space Telescope, frequent failures of onboard systems would be intolerable. Though we regularly have our cars serviced and repaired, we do expect them to operate well for long periods of time with only the occasional need to replace a fuel pump or an alternator.

To achieve a high level of reliability and scientific excellence in the Space Telescope or any other important space mission takes a great deal of time and effort in design, manufacturing, and testing. Not even the most brilliant and prescient manager, engineer, or scientist can anticipate all the technical and programmatic problems that will inevitably arise when the project moves from initial concept to the thousands of details of its implementation. Resolving unanticipated problems requires unanticipated expenditures that may go beyond the funds that have been set aside as reserves. The best teams succeed in controlling the problems to some degree, but almost never remain unscathed by cost and schedule issues. That's the price we pay to push back the frontiers of knowledge and to achieve scientific discoveries that impact entire generations.

The Space Telescope mission was really one of a kind, notwithstanding some similarities it may have borne to an earlier generation of secret Earth-observing reconnaissance satellites. To save money, it was developed as a "protoflight" system. There was no "prototype" Space Telescope on which to work out the technical kinks. So, when unanticipated problems arose—and a number of them did—they were first seen on flight hardware components that had already been designed and sometimes built. Work that was well underway sometimes had to be re-done—two steps forward and one step back. This was especially hard on the scientific instruments that were constructed as "black box" modules to fit precise spacecraft optical, electrical, mechanical, thermal, and computer interface specifications that were agreed to in advance. When a spacecraft interface had to be changed to solve a technical problem of some sort, the effect rippled through the suite of instruments, which in turn had to be re-designed or modified in some way. Clearly this was an inefficient and expensive way to work, but was inevitable given the way the management of the Project was structured.

By late 1982 morale among the members of the Space Telescope Science Working Group (Figure 2.7), and especially among the Principal Investigators, the leaders of the US and European science instrument teams—Jim Westphal, Jack Brandt, Bob Bless, Rich Harms, Duccio Macchetto, and Bill Jefferys—was exceedingly low. There was great pessimism that the Project would not be completed. An especially irritating factor was the "emperor has no clothes" syndrome, wherein the Marshall

Figure 2.7. Hubble Science Working Group and other colleagues at the STScI in the mid-1980s: left to right—John Clarke, John Bahcall, Al Boggess, Jim Westphal, Sally Heap, John Trauger, David Lambert, Bill Jefferys, Phil Crane, Ed Weiler, Bob Bless, Dan Schroeder, John Caldwell, Ed Groth, Jim Odom, Malcolm Longair, Neta Bahcall, Garth Illingworth, Richard Harms, Bob O'Dell, David Leckrone, Bill Fastie, Duccio Macchetto. (Credit: STScI.)

Project and Center management continued to struggle with irresolvable budget and schedule dilemmas, seemingly unable to admit that they were failing. They apparently were worried that, if they brought their problems up to higher authority at NASA Headquarters, the Space Telescope Project would be canceled.

This situation fomented what I call "the great revolution of 1983." Behind the scenes, the scientists took their worries directly to NASA Headquarters. Program Scientist Ed Weiler quietly invited them to submit whatever their concerns were, in writing, to him. Meanwhile, upper management at Headquarters had already deduced that the Marshall Project Office was in a nearly hopeless state. An independent investigation by a team from Headquarters had drawn conclusions about the Space Telescope's budget, schedule and technical situation very similar to those enunciated by the scientists.

Out of the "revolution of 1983" emerged a re-start for the Space Telescope Program. Its priority as a NASA mission was elevated. New money was infused into the program, much needed schedule relief was granted, and a new, stronger management hierarchy was initiated at Headquarters, Marshall, and Goddard. The nascent observatory was re-named the Hubble Space Telescope, in honor of American astronomer Edwin Hubble. It was Hubble who set modern observational cosmology on its present course with the discovery that galaxies like our own Milky

Way are individual entities separated by vast distances across the universe, and that the universe itself is expanding, growing in size over time.

The rejuvenation of the mission also stimulated a fresh emphasis on the major engineering and scientific efforts that would be needed, if the future Shuttle servicing of Hubble after its initial deployment in orbit was to be more than a pipe-dream. This topic had largely been neglected as the Marshall Project wrestled with the more immediate concerns of getting the observatory built and launched in the first place.

In 1980 the Project Manager had deferred work on the spacecraft's servicing capabilities and used the associated funds to resolve the budget crisis of the moment. The number of spacecraft "black box" modules that could be removed and replaced in orbit was greatly reduced from 124 to less than 20, to save money. As a result of the "revolution of 1983" the number of serviceable components was increased back to about 49.

It was Hubble Program Scientist, Ed Weiler, who took the initiative to prod NASA Headquarters and the Marshall Project to begin to get serious about servicing. In particular Ed pointed out the long lead-time required for the process of selecting and developing the second generation of Hubble's scientific instruments. He was especially concerned about WFPC, the observatory's primary, workhorse camera. It was anticipated to be the instrument of scientific choice for 40%–50% of the available observing time in orbit. It also would be producing the first really clear images of the universe that human eyes had ever gazed upon. Its importance in revealing what Hubble could see to the general, tax-paying public could not be underestimated. Failure of WFPC soon after it was launched would be tantamount to Hubble "going blind," and that was simply unacceptable.

In 1983 Weiler became a passionate advocate for the near-term development of a backup "clone" camera, WFPC2. He reasoned that a clone, a nearly exact duplicate of the original WFPC already being developed at JPL, could be built for considerably lower cost than a brand new instrument that had to be started from scratch. The latter would necessarily have to go through the yearlong NASA process of competitive selection initiated with an Announcement of Opportunity from NASA Headquarters. In the normal process science and industrial teams from all over the country are invited to propose new scientific investigations and the flight instruments needed to carry out those investigations. A selected instrument team would have to go through an extended contractual negotiation with NASA before beginning a preliminary design of the instrument. After passing a series of reviews at various stages of the process, the instrument would enter into full-fledged detailed design and development phases, culminating in an extensive testing program to assure its suitability for space flight. This entire process for a Hubble instrument could easily require five–six years or longer—simply too long (not to mention too costly) for an instrument intended to be put on the shelf for use as a backup, if needed, perhaps as early as two–three years after Hubble's launch (which at that time was scheduled for 1986).

By 1985 Weiler, supported by the Hubble Science Working Group, had won the backing of the Project, NASA management, and even Congress to proceed with a "clone," WFPC2. The Jet Propulsion Lab was asked to build a second version of the

Wide Field and Planetary Camera. John Trauger, an astronomer on the staff at JPL was assigned the role of lead instrument scientist for the backup camera.

Trauger, supported by a small team of scientists, reviewed the original technology going into the flight WFPC and found a few areas where it was already deficient, even though WFPC had not yet been launched. This is a typical situation for flight instruments whose designs usually have to be "frozen" years in advance of their orbital missions. So WFPC2 would not precisely duplicate WFPC. It would have, for example, greatly improved CCD light sensors, and some newly designed filters to isolate particular, scientifically important color bands of light.

The WFPC2 was nearing the end of its development at JPL when the Project was afflicted with the spherical aberration problem in 1990.

For several years after the successful completion of the Solar Max Repair Mission, Cepi continued as MMS Project Manager. He and his SMRM team provided support to the Space Shuttle Program Office at Johnson as it undertook previously unplanned rescues of three major communications satellites. Originally intended to be boosted to geosynchronous orbits at about 22,000 miles altitude, the satellites had been left stranded in low Earth orbit as a result of the failure of kick stage rocket motors.

In 1984 November the astronauts of Shuttle mission STS-51A successfully wrestled both the Westar 6 and Palapa B2 satellites into the payload bay of Discovery. Neither spacecraft possessed a grappling fixture for the Shuttle's remote arm, as neither had been designed for Shuttle servicing. The astronauts improvised a manual grappling procedure after the originally planned technique for grabbing hold of the two satellites was unsuccessful. Both satellites were fastened securely into the payload bay and brought back to the ground.

Similarly, in 1985 August the astronauts of STS-51I, also on Discovery, captured the Syncom IV/Leasat 3 satellite that was adrift in the wrong orbit. They had to reposition a switch that had originally been installed backward, thus preventing the kick motor from igniting. Once again the satellite had not been built with a grappling fixture for the RMS. Astronaut "Ox" van Hoften, was assigned the task of grabbing it by hand as it passed by the orbiter. It almost tore his arm off. Adequate attention had not been paid to the vast difference in inertia between the enormously massive orbiter, to which Ox was attached, and the spacecraft bus of Leasat 3. On a second more cautious attempt, van Hoften was successful. He attached a tool that resembled a butterfly net with a slot at its center to the balky switch and used it to pull the switch to the correct position for motor ignition. After Discovery had backed several miles away, the kick motor was successfully ignited and Syncom IV/Leasat 3 was off to its proper geosynchronous orbit.

These stories of the ad hoc rescues of very valuable communications satellites in the early days of the Shuttle program simply illustrate that the Shuttle managers, the astronauts, and the teams of engineers supporting them on the ground were really new at this game. They were just coming up to speed and in a sense making it up as they went along. Cepi with his SMRM experience was an invaluable ally.

From the early 1970s through the 1980s, interrupted for a time by the catastrophic loss of Challenger in 1986 January, an extended NASA team including

Cepi's MMS and SMRM Projects at Goddard, the Shuttle program and astronauts at JSC, and the Shuttle launch personnel at KSC had gradually acquired the knowledge, skills, and experience needed to execute a Shuttle-based servicing mission. These geographically separate and disparate organizations had learned to work together toward achieving highly complex ends.

They had learned that in a servicing mission methodical, intricate step-by-step flight and EVA procedures, carefully planned and practiced almost endlessly, were essential. An entire toolbox of custom-designed, specialized tools was needed to allow ease of execution by people in unwieldy spacesuits for every servicing task in orbit. Well-designed carriers and fixtures, electrically and mechanically hooked up to the Shuttle in the payload bay were necessary both for carrying new equipment up to orbit and for bringing retrieved equipment back to the ground. The primary payload equipment—scientific instruments, gyroscopes, computers, electronics boxes, solar arrays, batteries, etc—had to be robustly designed and built so that they could be launched in the Shuttle without being damaged. Each piece of satellite hardware had to work properly for many years in the harsh and unforgiving environment of space. Most importantly, all of these components collectively had to yield a scientifically excellent and productive long-term mission. The result was a NASA team that was "ready to go" when the urgent business arose of planning a previously unscheduled servicing mission to Hubble to correct its faulty eyesight.

Leading the broad Hubble servicing enterprise and making it successful became Cepi's responsibility. He and his team had effectively created a new discipline of aerospace engineering—satellite servicing. With the approach of the first Hubble servicing mission in December of 1993, they were about to be tested in a high-risk, high-stakes, highly visible expedition into poorly charted waters. Success was by no means assured. But failure was unimaginable.

Life With Hubble

An insider's view of the world's most famous telescope

David S Leckrone

Chapter 3

The Sharper Image: Correcting Spherical Aberration

It was late in May of 1990, only a few days after the Hubble "first light" press event at Goddard. John Trauger, the lead scientist for WFPC2, was sitting in his office at JPL. He heard a knock, looked up, and standing there were Aden and Marjorie Meinel. The Meinel's were much revered in the astronomical community, a close husband and wife team of astronomers who had a long-standing, almost legendary reputation as optical experts and designers of astronomical telescopes and instruments. Aden led the team that founded the Kitt Peak National Observatory near Tucson, Arizona and served as its first Director. In the 1980s the Meinel's developed an interest in the design of optics for space telescopes and instruments. As a result, JPL hired them as Distinguished Visiting Scientists in 1987. It was natural—and as it turned out, highly fortuitous—that they would pay a visit to Trauger's office, curious to see the "first light" images that had just come down from JPL's new WFPC camera on Hubble.

Trauger brought some of the images up on his computer monitor. Aden looked closely at them and said, almost immediately, that the images showed the telltale signature of spherical aberration! This insight came weeks before the Project team at Goddard officially reached the same conclusion. With casual brilliance, this senior and highly experienced expert in astronomical optics had deduced the flaw in Hubble's optics, simply by eyeball inspection of a few images. Of course at that moment it still remained to be shown that the aberration arose in the telescope optics and not in the optics of the WFPC camera. But soon thereafter the optical problem was also seen in images from Hubble's other camera, the European Space Agency's Faint Object Camera, and in data from the other instruments as well, clinching the conclusion that the telescope was delivering images of poor quality to all the scientific instruments.

A few weeks later, in mid-June, Trauger was attending a meeting of the Optical Society of Southern California. In line for a buffet dinner one evening, he found himself standing next to Aden Meinel. Their conversation turned to the Hubble images Aden had seen earlier in John's office. Meinel asked, "In the optical layout of your WFPC2, isn't there a small mirror in each of the 'repeaters' sitting close to the telescope pupil? If so, then fixing the spherical aberration should be pretty simple. Just put a new figure on each of those little mirrors that exactly offsets a well-characterized error in the telescope's optics."

This line of reasoning from an optics expert can be translated as follows. The repeaters or relays in WFPC2 are four miniature reflecting-telescope-like devices that bring the light coming into the camera from the Hubble telescope to a focus on four electronic image sensors—charge coupled devices or CCDs, similar to those in commercial digital cameras. The tiny, nickel-sized secondary mirror of each of these small telescopes is the last surface the light hits before it focuses onto a CCD (Figure 3.1). Each sits very close to an optical "pupil" of the Hubble telescope. At a pupil the telescope optics create a focused image of Hubble's 2.4 m primary mirror. If you placed a sheet of paper at that location, and could somehow see it being illuminated by light passing through the telescope and through the WFPC2 optics, you would see a nicely focused picture of the primary mirror, as well as the heads of the large bolts that hold the big mirror in place. It is at a pupil that the surface figure of the mirror is precisely reproduced. Grinding an accurate negative or equal-and-opposite representation of the error in the Hubble mirror onto each of the small

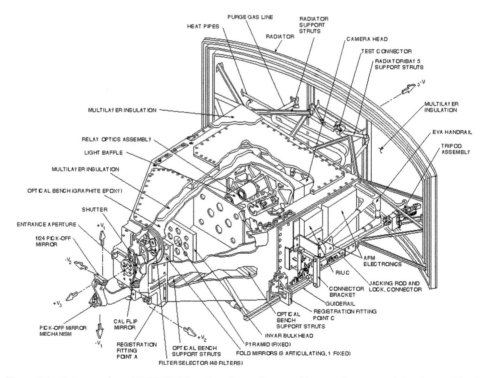

Figure 3.1. Cut away view of WFPC2. Note the relay optics assembly near the center of the picture. That is where the correction for spherical aberration was implemented. (Credit: NASA, JPL-Caltech, J. Trauger.)

mirrors within WFPC2 should, in principle, exactly offset the mistaken shape of Hubble's mirror. As far as WFPC2 was concerned, there would be no spherical aberration. The clear images of the sky Hubble was originally expected to produce would come sharply into focus.

Aden Meinel was not a member in any official way of the Hubble team. He was simply an interested professional bystander. And yet, his brilliant insight, conveyed in such a casual way to John Trauger that evening in June of 1990, while moving through a dinner line at an optical society meeting, was perhaps the pivotal event that set us on a course ultimately to rescue the Hubble Mission and NASA's reputation.

Trauger spent the next several days in a basement lab at JPL, working with optical designer Norm Page and a computer model Norm had developed to simulate with high fidelity Hubble's telescope optics. They tested three different scenarios, assuming the spherical aberration was caused by an error in the surface shape of the telescope's primary mirror, the secondary mirror, or both. In each case they computed what kind of shape would have to be ground into each of the four small WFPC2 relay secondary mirrors (Figure 3.2) to offset the spherical aberration in the beam of light coming from the telescope. John and Norm found a workable solution for each of those three possible situations. The simplest fix was for the case in which only the 2.4 m primary mirror was at fault.

On June 26, during the Science Working Group meeting at Goddard, Trauger cornered Ed Weiler in a hallway and told him about the possibility of correcting the aberration problem within the optics of the WFPC2. That hint of good news in the midst of such a seriously bad and embarrassing situation gave Weiler an opening to

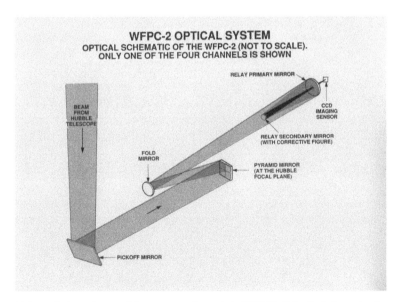

Figure 3.2. Light path through one of four camera channels in WFPC2. Spherical aberration is corrected by the figure of the relay secondary mirror, as suggested by Aden Meinel. Articulating fold mirror allows adjustment of optical alignment from the ground. (Credit: NASA, JPL-Caltech, J. Trauger.)

tell the press at the next day's news conference that, although the Hubble telescope wasn't working properly, there might be a way to fix the problem during a servicing mission in a few years, at least within one of the instruments, Hubble's workhorse camera.

In the days immediately after the confirmation that the Hubble Space Telescope was deeply flawed by spherical aberration, NASA Headquarters and the Hubble Project seemed stunned, rocked back on their heels, as it were. I don't recall anyone in a high-level position exhibiting defiance or stubborn determination to press on toward recovery and ultimate success. The early direction from the top emphasized performing investigations of what went wrong and why. Aden Meinel's ideas for recovering full imaging performance with a relatively simple modification of WFPC2's four small relay imagers fell into NASA's lap unexpectedly, like manna from Heaven. It wasn't something the Agency's management or the Hubble Project proactively sought. Nevertheless, it was a blessed glimmer of hope and NASA seized it.

From this stroke of luck evolved a long-term strategy for recovery that minimized additional costs. The first servicing mission would be flown, perhaps slightly ahead of the normal schedule, three to four years hence. The optical modifications would be made to WFPC2 and it would be inserted into the spacecraft, replacing the original WFPC, during that servicing mission. Two other advanced instruments were also in development at the time at Ball Aerospace Systems Division in Boulder, Colorado: The Space Telescope Imaging Spectrograph (STIS) and the Near-Infrared Camera and Multi-Object Spectrometer (NICMOS). Their internal optics could also be re-designed to fully compensate for spherical aberration. They could fly to Hubble on a second servicing mission at some unspecified time farther in the future.

Meanwhile, Hubble's five other original instruments would simply have to do the best they could, hindered as they were by the faulty telescope. So, Hubble would begin to provide sharp, clear images of the sky three to four years later than originally planned, accommodating about 50% of the selected scientific research programs. But the other 50% of the science required different tools—the Faint Object Camera, the Faint Object Spectrograph, the Goddard High Resolution Spectrograph and the High Speed Photometer. In this scheme, those areas of astrophysics would simply be put off for a long time.

After a technical review, NASA authorized JPL to proceed with implementing the optical fix in WFPC2 (Figure 3.3). However, they couldn't start in earnest until they had in hand the conclusions of the two NASA-appointed investigation boards who were determining the actual, as-flown prescription of Hubble's primary mirror. The corrective figure could not be defined nor ground into the WFPC2 repeater mirrors until the error in the shape of the primary mirror had been numerically specified.

JPL was instructed to aim for a servicing mission to be flown in 1993. John Trauger was authorized to enhance the capabilities of his science team by adding new members with specific areas of scientific expertize, to help assure that the revised camera would meet the performance requirements now being levied on it. The

Figure 3.3. The completed WFPC2 in a clean room at JPL in the spring of 1993. (Credit: NASA, JPL-Caltech, J. Trauger.)

Goddard Hubble Project Office began working with JPL managers and engineers to implement this much more aggressive and date-certain schedule. Frank Cepollina, newly appointed as head of the Hubble Flight Systems and Servicing Project, was the government manager now in charge of this effort.

In 1980 NASA and the Space Telescope Project conducted a competition to select an organization to build and operate a Space Telescope Science Institute (STScI). The astronomical community had expressed a strong desire that the science program of what was to be the most important astronomical facility of our time should be under the direction and control of an "independent" organization outside of NASA that had strong ties to the scientific community. This viewpoint was formalized in a study organized by the Space Science Board of The National Research Council.[1] Donald Hornig, former Science Advisor to President Lyndon Johnson, chaired the study. Hornig's committee met at Woods Hole, Massachusetts, 1976 July 19–30. Their extensive list of conclusions and recommendations included the creation of a new institution, the Space Telescope Science Institute, "to provide long-term guidance and support for the scientific effort...." Further they recommended that the STScI "be operated by a broad-based consortium of universities and non-profit institutions ... under contract to NASA." The committee concluded that, "We have not found any compelling data-handling, managerial, or cost reasons for locating the Institute at an existing NASA center." At Goddard we took that as a rebuff to our own stated ambitions to host the STScI.

---

[1] "Institutional Arrangements for the Space Telescope," 1976 (Washington, DC: National Academy of Sciences).

Vocal members of the community argued that this critical science operations function should not be subject to the constraints of a government bureaucracy. Actually, a government "bureaucracy" at Goddard successfully and productively served the broad science community for many years in operating the small, pathfinding International Ultraviolet Explorer astronomical satellite from 1977 to 1996. But a strong antipathy against scientists employed by NASA, going back to the beginning of the Agency in the late 1950s, was widespread in the community. This was based on the perception that NASA showed favoritism in funding the research of its own employees. That perception was not entirely fair. Nevertheless, when it came time to define the institutional arrangements for managing the Space Telescope's research program, the astropolitics of the day prevailed. Not only did NASA Headquarters and the Marshall Space Telescope Project (strongly influenced by Project Scientist Bob O'Dell) determine that control of the "prize"—the science observations to be obtained with the Space Telescope—should be placed in the hands of a consortium of universities or similar organizations outside of NASA, but also that internal NASA entities such as Goddard would not even be allowed to compete in the selection process. Certainly, this led to bitterness on the part of some of the Civil Service astronomers at Goddard. They referred to the published report of the Hornig Committee as "the purple peril," in reference to the color of its cover. However, most of us swallowed hard, accepted the decision and moved on in our jobs, which included establishing collegial working relationships with whomever was selected to build and operate the Science Institute.

I was a member of the team of Goddard engineers and scientists who were tasked with providing a technical review of the Institute proposals submitted to NASA for the competition. Renowned senior astrophysicist Jesse Greenstein from Caltech led the team reviewing the scientific merit of the various proposals. From my perspective one proposal stood out far above the others in terms of thoughtful detail and realism. As it happened, that proposal, from the Association of Universities for Research in Astronomy (AURA), was ultimately selected to form the Science Institute. Their chosen location was the Baltimore campus of the Johns Hopkins University. As Science Working Group member Bill Fastie, a Hopkins faculty member noted, Johns Hopkins was just close enough to Goddard to allow good day-to-day working interactions, but far enough away to encourage the independence the community so desired—a "Goldilocks" solution.

In 1981 AURA appointed Riccardo Giacconi to serve as the first permanent Director of the Space Telescope Science Institute. Riccardo was a widely celebrated pioneer of X-ray astronomy from space, and a future Nobel Prize winner in physics for that work. I had never encountered him personally before. But he had a reputation among some of my fellow scientists at Goddard who considered him a difficult person to work with if you were not an insider on his team. As rumors flew at Goddard that he was AURA's likely choice, some of us who would, after all, be responsible for managing the AURA contract and for working every day with the Institute as we operated the Space Telescope in orbit, expressed reservations about this choice for Director. However, NASA Headquarters acquiesced to AURA and concurred in Giacconi's selection.

Indeed, Riccardo frequently did prove challenging for us to work with. Sometimes, in my opinion, he was arrogant and judgmental, sometimes gracious and charming. He always expressed himself in a colorful, strong Italian accent. I came to believe that his temperament made him the perfect man for the job of defending Hubble science in the age of spherical aberration.

At the time of the public announcement that the Hubble telescope was flawed in June of 1990, Riccardo was out of the country. When he returned and was briefed on the situation and on NASA's strategy for responding to it, he reportedly boiled over. It was as though, figuratively, he wanted to grab the Agency by the "scruff of its neck," get in its face and shout angrily, "This is unacceptable!! We cannot allow this to stand!" I have no doubt that Riccardo was bitter in blaming NASA ineptitude and that of the NASA-appointed Space Telescope Science Working Group for causing the problem. I heard him vehemently express the opinion that the Science Working Group had not done its oversight job and its members should not be rewarded with the guaranteed observing time on Hubble that they had been awarded when they were competitively selected in 1977 (a view that NASA quickly dismissed). But he succeeded in persuading the Agency to abandon its conservative approach to recovery and to adopt a more comprehensive plan that would restore the totality of Hubble's scientific capabilities sooner rather than later.

Jim Crocker was sitting in his office at the Space Telescope Science Institute one day soon after the public announcement of Hubble's optical debacle, feeling particularly glum. That "late night host with the gap between his teeth," had gotten to him. Public ridicule of Hubble by David Letterman, Jay Leno, and others was unbearable. Jim, like so many others, had devoted a decade of his life to this mission. Now it had become difficult even to admit any affiliation with Hubble to friends, neighbors, or perfect strangers sitting next to him on an airplane.

At that time Crocker was the Head of the STScI's Projects Office, responsible for leading the development of Hubble's software and computer-based ground system at the Institute. His office was also responsible for verification of Hubble's operational science capabilities after it was launched. Giacconi hired Crocker to bring engineering expertize and experience with human space flight to the STScI. Crocker had previously worked as a junior engineer on Apollo 17 and had done electronic design work at Marshall Space Flight Center for three astronaut missions to the Skylab space station. Born in South Carolina, Jim spoke with a soft southern accent that embellished his personal charm and genteel manner.

As Crocker sat moping that day, in walked STScI astronomer Holland Ford. "We have to fix this!" Holland declared. "There are some really smart people on this program. We can't just go down in flames. Just doing the WFPC2 correction is not enough; we've got to fix the other instruments as well. We may not get another servicing mission. This may be our only shot." Obviously, these thoughts had been churning in Ford's head for a while and he just had to spill them out to a like-minded professional.

In addition to his duties at the Institute, Holland Ford served as Deputy Principal Investigator on the Hubble Faint Object Spectrograph team. Principal Investigator Richard Harms at the University of California, San Diego was the leader of the team. The FOS was seriously hampered by the telescope's spherical aberration. As its name implies, it was designed to collect and analyze as much light as possible from very faint objects—primarily faint galaxies and quasars at great distances out across the universe. Although the FOS could not take two-dimensional pictures, it could be used to construct two-dimensional spectroscopic maps of extended objects, one point at a time. So, it required a lot of light to be concentrated on its small entrance aperture and it required outstanding angular resolution from the telescope. In other words it required the promised sharp images that Hubble was now unable to provide. Thus, Holland had a strong scientific motivation in seeking a fix for the optical problem, beyond that to be provided to WFPC2. Simply achieving 50% of the planned science for the foreseeable future wasn't acceptable. All of the science was important, not just the half that produced pretty pictures. Hubble was supposed to tackle some of the biggest scientific mysteries of our time. It would be badly handicapped without a complete set of tools. It would be like a carpenter trying to build a house with a hammer and saw, but without a tape measure or a level in his toolbox. Holland's career and research was on the line. He wouldn't take "no" for an answer.

Ford's passionate outburst to Jim Crocker left them both energized. They agreed to go talk to Riccardo, to figure out how to work with NASA to get a servicing mission designed to restore the entirety of Hubble's scientific capabilities. They figured with NASA's blessing and Giacconi's prestige it should be possible to rally the international community of astronomers and the American political establishment to take a bold step, to commit to fully restoring the most important scientific facility of the late twentieth century.

Of course, Riccardo did not need convincing. He embraced Holland and Jim's initiative immediately. He called Charlie Pellerin, the head of Space Science missions at NASA Headquarters, and persuaded him to endorse an independent study of Hubble repair options to be led by the Institute.

Giacconi asked Ford and Institute astronomer Bob Brown to co-chair a Strategy Panel with a mandate to identify and evaluate possible remedies to the spherical aberration problem that would be applicable to the observatory as a whole. They were to pull together the very best astronomers, optical design experts and engineers from both the US and Europe who were capable of addressing the problem. Particularly at Brown's insistence a ground rule was adopted that the study was to begin with a "blank sheet of paper." Nothing was to be left off the table, no matter how far-fetched it might be. The study was not to jump immediately to a "solution" that might lead down the wrong path. Brown and Ford asked, "What is the physics of the problem?" "What is the physics of correcting the problem?" And then, "let a thousand flowers bloom." Even the optical solution already proposed for WFPC2 was not to be taken for granted. Perhaps a more global solution could correct WFPC2 as well. This STScI HST Strategy Panel met four times between August and October of 1990—three times in Baltimore and once in Germany.

Notably absent from the Strategy Panel was any member from the NASA Hubble Program. Retired NASA astronaut Bruce McCandless was a formal member, however. NASA had already engaged its own study teams, primarily to assess what caused the optical problem and to quantify the problem accurately so that it might be fully cured, at least in WFPC2 and later-generation instruments. More to the point, Riccardo was concerned that, if the strategy panel study was not divorced as much as possible from official NASA, it would lead inevitably to the solution the Agency had already embraced.

Okay, so you gather together an elite team of 15 experts in the same room to brainstorm about what could be done to restore Hubble's eyesight. What possibilities might they think of, taking advantage of the fact that Hubble was designed to be serviced in-orbit by Shuttle astronauts? Consider the options from the outside in. Start with potential solutions that could be implemented by modifying the incoming light at or near the entrance aperture of the telescope before the wavefront hit the primary mirror—a desirable approach in that it would involve minimal intrusion inside Hubble.

One might place a corrective lens over the full entrance aperture of the telescope, above the secondary mirror. But such a refractive lens, made up of thin segments of glass, would be hard for astronauts to handle safely and would introduce its own optical problems, such as chromatic aberration and loss of ultraviolet light. A diagonal mirror with a shape designed to correct spherical aberration might be placed at a 45° angle outside the front of the spacecraft to relay the incoming light into the telescope tube. But such a structure would be heavy and likely would affect Hubble's ability to slew with reasonable speed across the sky. A circular mask might be placed in front of the telescope to block any light that would hit the outer portion of the primary mirror. It is there near the outer edge that the shape of the mirror is most in error. Preventing light from reflecting off this part of the mirror would reduce, but not completely eliminate, the smear in the telescopes image. However, such an aperture mask would block up to half the light coming into the telescope and reduce the ultimate sharpness of the image that was theoretically possible with a 2.4 m telescope. It would be as though a smaller telescope had been launched, and the scientific program of the observatory would be compromised.

An even more challenging class of possible fixes involved direct modifications to the telescope's optics—the primary and secondary mirrors. Perhaps the secondary mirror could be replaced. Or the primary mirror could be over-coated with a metallic film the thickness of which would be controlled to exactly replace the thickness of glass inadvertently ground off the primary mirror when it was being manufactured. Or corrective lenses could be placed inside the central baffle of the primary mirror—a conical tube mounted over the central "doughnut hole" of the mirror to keep out unwanted light. The fundamental flaw with all of these ideas is that the telescope's two mirrors were deeply buried inside its mechanical structure and would be very difficult to reach. One member of the panel—telescope designer Roger Angel—suggested (with tongue in cheek) that NASA might train a cadre of dwarf astronauts who could crawl down the telescope tube to implement such modifications.

That then brings us to the correction of the telescope's aberrated beam of light separately for each individual Scientific Instrument. This approach had already been demonstrated to work at JPL for the new WFPC2. Internal optical corrections were already being designed at Ball for the two future instruments, STIS and NICMOS. But such solutions, internal to the structure of individual instruments, could not be implemented for the current suite of instruments already on board Hubble without retrieving them and returning them to the ground. The instruments were not designed to be modified by astronauts in orbit. Their downtime back on Earth and their resulting lack of scientific productivity would be measured in years and most likely some or all of them would never fly again. This violated the stricture that the solution to spherical aberration must work for the entire observatory, if at all possible.

In all, the STScI Strategy Panel came up with 26 scenarios for restoring Hubble's faulty vision, 19 of which were deemed feasible for implementation by astronauts using the then-existing tools and procedures for in-orbit maintenance and refurbishment, or minor extensions thereof. Ultimately, however, the discussion converged on the only solution that could practically be implemented for the remaining axial scientific instruments during the first servicing mission to Hubble—the mission NASA had already begun planning in which WFPC would be replaced with the optically corrected WFPC2.

Each of the four axial instruments mounted in the aft shroud of the spacecraft near the telescope's focal surface was designed to intercept a separate portion of the beam of light relayed by the telescope back through the central hole of the primary mirror. It would be necessary to intercept those portions of the beam before the light entered each instrument's aperture and make the correction there for spherical aberration.

The back end of the Hubble spacecraft is the "business end," insofar as astronaut servicing is concerned (Figure 3.4). The Aft Shroud at the rear is an aluminum cylinder 14 feet in diameter and 11.5 feet tall. It contains three sets of tall double doors, providing access to the four Axial Scientific Instruments, gyro packages, star trackers, etc, and four radial-bay doors through which the astronauts can reach the WFPC and the three Fine Guidance Sensors. Around the circumference of the spacecraft, between the aft shroud and the forward light shield that houses the telescope, is a ring of ten enclosed bays—the equipment section, containing computers, reaction wheels, an electrical power switching box, and hardware needed for pointing control, data management, mechanism control and communications.

All of these "black boxes" and the bays that enclose them were designed to be easily accessible and "astronaut-friendly." Bulkier boxes, such as the scientific instruments and fine guidance sensors, slide in and out on guide rails, are latched and firmly bolted down. Smaller boxes are seated on metal pins and bolted in place. Electrical and communications cables have connectors that were relatively easy to attach or detach with bulky spacesuit gloves. Astronaut handholds and foot restraints encompass the aft shroud and equipment section. The two crewmembers working outside on Hubble during an EVA are tethered to the shuttle structure, to the Canadian remote arm, or to the Hubble spacecraft.

Inside the aft shroud is the focal plane assembly that holds the five scientific instruments and three fine guidance sensors firmly in place and properly aligned to the telescope's beam of light. The four axial instruments are rectangular boxes three

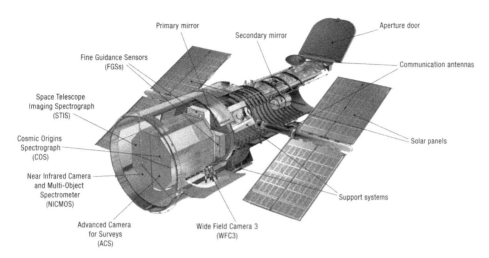

Figure 3.4. Cut-away view of the Hubble spacecraft. The forward section, containing the telescope optics is called the optical telescope assembly or OTA. The rear section consists of the equipment section and the aft shroud. Collectively it is called the Support Systems Module or SSM. The SSM contains the scientific instruments, fine guidance sensors, gyroscopes, and many other subsystems. (Credit: NASA, GSFC, STScI Images.)

feet wide and seven feet tall, about the size of an old-fashioned telephone booth (does anyone in the 21st century remember telephone booths?). The WFPC and its successors, WFPC2 and WFC3, and the fine guidance sensors are roughly the shape and size of baby grand pianos. The focal plane assembly (FPA) is made of rigid, light-weight graphite epoxy beams, held together with metallic fasteners—looking much like a supersized erector set toy. All the instruments and other modules in the aft shroud are mounted onto the FPA.

Correcting the telescope's spherical aberration individually for each axial instrument (assuming the WFPC2 had already been corrected internally) would require gaining access to the volume between the top surfaces of the axial instruments and the pickoff mirror arm of the WFPC2 protruding into the space above them. Corrective optics must somehow be placed above the entrance apertures of each axial instrument to receive the beam of light from the telescope, correct it, and relay it into each instrument separately. The telescope was not designed with such a possibility in mind. It's too crowded in there for an astronaut's bulky gloves.

The field of view presented by a correctly manufactured Hubble telescope to the scientific instruments and fine guidance sensors is supposed to contain images brought to a sharp focus over a gently curved surface—a circle about 18 inches in diameter, the size of a large pizza (Figure 3.5). Projected on the sky that corresponds to about 28 min of arc—a bit less than half a degree, roughly the size of the full moon. The inner 12 inches of this field (18 min of arc on the sky) centered on the optical axis of the telescope is called the science field and provides the best quality images. The additional ring-shaped three-inch-wide portion of the field of view outside the science field is called the tracking field. It provides good images of relatively bright guide stars to the three fine guidance sensors.

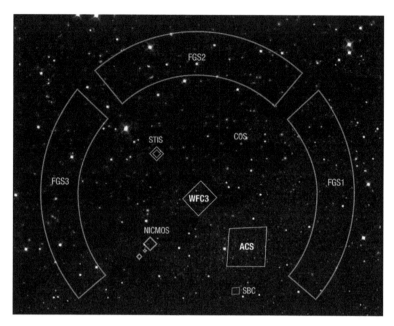

Figure 3.5. The layout of Hubble's telescope focal surface, projected on the sky, as it appeared after the last servicing mission, SM4, in 2009. (Credit: NASA, STScI Images.)

The "sweet spot" in terms of the very sharpest images is at the center of the science field on the optical axis of the telescope. It is reserved for the observatory's workhorse camera, WFPC and its successors. A pickoff mirror is mounted on a thin arm extending inward from the WFPC to the center of the telescope field of view. It picks off what is supposed to be a very sharp, pristine image from the central 3 arcmin of the field and reflects it at a 90° angle back into the housing of the camera. The four axial scientific instruments have to "make do" with images that are slightly degraded, mostly by astigmatism which they correct internally, residing a few minutes of arc away from the optical axis. Nevertheless, these images are still of excellent quality. The entrance apertures of the axial instruments are located so as to avoid being blocked by the WFPC pickoff mirror above.

So, all five large scientific instruments have to share a crowded tract of real estate within a science field area of only about three quarters of a square foot. At any given time each instrument sees a different small area of the sky. Add to this the fine guidance sensors, each picking off its own portion of the tracking field, and you have eight large, bulky instruments sharing about 1.8 square feet of physical area within which the telescope image is nominally focused. Any new optical elements designed to correct for spherical aberration must somehow be mounted in place within an imaginary cylinder no larger than 12 inches in diameter (the science field) and roughly 18 inches high above the entrance apertures of the axial instruments. How to do this within the available volume was a major challenge.

Two brilliant men solved the problem—Jim Crocker, whom we've met earlier in this chapter, and optical physicist, Murk Bottema (Figure 3.6). Both were members

Figure 3.6. Jim Crocker (left) and Murk Bottema (right) who together conceived and designed COSTAR, the instrument inserted into Hubble in 1993 that corrected spherical aberration for three axial instruments—Faint Object Camera, Faint Object Spectrograph and Goddard High Resolution Spectrograph. (Credit for Crocker photo: NASA, J. Kowsky; Credit for Bottema photo: Ball Aerospace.)

of Brown and Ford's Strategy Panel. At the time, Bottema had been working for over two decades at Ball Aerospace Systems Division in Boulder, Colorado as their lead optical designer. Bottema and Crocker had some traits in common. Both were soft-spoken, gray-haired, mid-career gentlemen dapper in their dress and genteel in their manners—charming and polite to a fault. Both had penetrating intellects, capable of seeing through to the heart of an engineering problem without being overly distracted by the details.

Murk was in some ways an "old-school" optical designer. He had a strong intuitive grasp of optics and could sketch out a conceptual solution to an optics problem on the back of an envelope. On the other hand he was a thoroughly modern optical physicist, fully skilled at utilizing the latest optics design computer codes, but not at their mercy. He was Dutch by birth, and spoke with that accent. I loved the fact that one of his favorite pastimes was shooting pool (or billiards). After all, the same physical laws apply to the path of a pool ball bouncing off the rail of a table, as apply to the reflection of light bouncing off a mirror. At the time of the Strategy Panel meetings in 1990, he had already designed the optics of three Hubble scientific instruments—the Goddard High Resolution Spectrograph already on board the orbiting telescope and the STIS and NICMOS instruments nominally being developed for flight on the first servicing mission. The latter two would of course be delayed to make room for a mission to address the spherical aberration problem.

Armed with a precise specification of the erroneous optical prescription of the as-flown telescope officially sanctioned by NASA's Independent Optical Review Panel, Bottema created an elegant optical design that would exactly correct the grinding error in the primary mirror. Above each entrance aperture of an Axial Scientific Instrument would be placed a pair of small mirrors, each about 20 mm or 0.8 inches in diameter—roughly the size of a US nickel coin (Figure 3.7). Depending on the

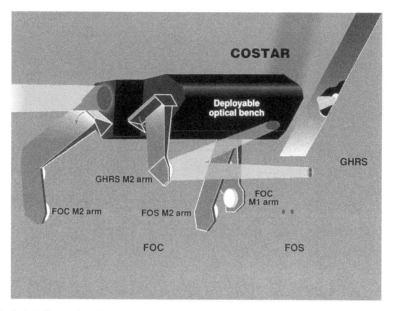

Figure 3.7. Artist's illustration of layout of COSTAR with its optical bench deployed. The light path to allow correction of the aberrated image for the Goddard High Resolution Spectrograph is shown. The M1 mirrors for the GHRS and FOS are mounted inside the COSTAR enclosure. (Credit: NASA, Ball Aerospace.)

Axial Instrument being corrected, the mirror pairs would be separated by distances along the direction of the telescope axis ranging from about 10 to 18 inches.

The first mirror of each pair, M1, had a simple spherical curvature; its job was to intercept the light beam from the telescope before it reached the instrument and to reflect it back out along the telescope axis to the second mirror, M2. The M2 mirror was located precisely at the telescope pupil where a sharp image of the full primary mirror itself came to a focus. In other words M1 would project a greatly miniaturized image of the primary mirror onto M2. The M2 mirror had a curvature that shortened the distance over which light rays must travel by an amount that exactly compensated for the extra path length introduced by the too-flat primary mirror. The M2 mirror had a steep curvature to offset the primary mirror's shallow curvature. The shape of M2 was approximately that of a potato chip. Formally it was called an anamorphic fourth-order asphere on a toroidal blank!

One complication stands out in this discussion of Murk Bottema's design. It takes an image of a large primary mirror, about 94 inches in diameter, and reduces it down to the size of a nickel. A 12 inch ruler lying along the radius of the primary would have a projected image length on the M2 mirror of only a tenth of an inch! Imagine how precisely M2 would need to be shaped in order for its corrective curvature at a given point to exactly line up with and to exactly correct the corresponding location on the curved surface of the full-sized primary mirror. Imagine how precisely M1 would have to center the pupil image onto M2 in order for each tiny spot on its curved surface to map point-by-point exactly onto the curved surface of the primary mirror. And then the M1–M2 pair of mirrors hovering over each Scientific

Instrument aperture must be precisely positioned so that the light reflecting off M2 is directed exactly down the center of the hole, into the instrument. The mirrors had to be positioned within 1/10,000th of an inch of the desired location. Dennis Ebbets, an astronomer at Ball Aerospace, noted that this would be equivalent to parking your car within 1/50th of an inch of a particular spot in your garage. Finding a way to position multiple pairs of Bottema's mirrors with such great precision in front of the axial scientific instruments was a daunting challenge.

As NASA's partner in the Hubble program, and given its interest in correcting the spherical aberration problem for its own instrument, the FOC, it was appropriate that ESA should be represented on the STScI Strategy Panel and should serve as the host for one of the Panel's meetings. That meeting took place at the European [Hubble] Coordinating Facility (ECF) in Garching, a suburb of Munich Germany, in September of 1990.

Prior to the meeting at the ECF Jim Crocker had gotten drawings of the layout of the Focal Plane Structure in the area in front of the Axial Instruments. He used the drawings to construct a 3D foam-board full-scale model of the structure in that area and glued the engineering drawings to it. This allowed him to visualize the volume where the optical correction needed to be placed. He found that, relatively speaking, "it was huge." There was actually plenty of space available, even taking into account that any structure had to be limited in size so as not to contact the WFPC pickoff mirror or to reflect stray light into WFPC's aperture. That gave him a cylindrical volume the diameter of a large pizza and about 18 inches long to work within.

While attending the meeting, Crocker stayed at a charming little place called Hotel zur Mühle (to the mill) situated on a stream running through the town of Ismaning, not far from the ECF. One evening he climbed into the shower in his room. It was a European style shower with the head mounted on a vertical bar on which it could slide up and down. The maid had left the showerhead in the completely down position. Jim grabbed it to pull it upward to accommodate his considerable height. As he did so, he noticed that the head could be tipped up or down and right or left. Eureka! Why not mount pairs of corrective mirrors on some kind of a column or optical bench that could be moved "up and down" above the axial instruments, deployed perhaps from within a new phone-booth-shaped box to be placed in one of the axial instrument bays? Like the showerhead, the mirrors could be folded downward out of the way to allow the column to be moved in and out of the new box. And like the showerhead they could be tipped up or down and rotated to the right or left to properly align them to the beam of light from the telescope. To accommodate the new box, one of the four current axial scientific instruments would have to be sacrificed—removed to make room for the new corrective optics enclosure. But that would still leave three axial instruments rescued from optical aberration and made available for their originally intended, cutting-edge astrophysics.

Before the meeting the next day, Jim described his idea to Murk Bottema. They assumed that the High Speed Photometer, a very specialized instrument unlikely to be in high demand by astronomers using Hubble, would be the instrument to be replaced. Crocker suggested three pairs of small corrective mirrors, one each for the

Goddard High Resolution Spectrograph, the Faint Object Spectrograph and the Faint Object Camera. (In the end five pairs of mirrors would be necessary, two each for FOS and FOC, and one pair for the GHRS.) Murk had already worked out a preliminary optical design for mirror pairs of this kind, as described previously. Jim measured where the layout would go on his foam model. "Oh boy, we've got lots of room!" he exclaimed.

When the meeting convened Jim presented the concept to the assembled Strategy Panel. He was met with stunned silence. Then astronaut Bruce McCandless spoke up. "They'll let me do that," he said, meaning the people in the Shuttle Program who plan EVA's and astronaut tasks. The idea of replacing one axial instrument with another was completely within the bounds of the normal servicing tasks already planned for Hubble.

The meeting adjourned to the local Hofbrau House, satisfied that the solution to the spherical aberration problem for three axial instruments was in hand. Crocker retained a coaster from the pub as a souvenir of that seminal moment. Jim, Murk, Holland Ford, and Wally Meyer from Ball Aerospace called Frank Cepollina at Goddard and described the concept to him. Cepi was enthusiastic. "Let's go sell this; let's go make it happen," he proclaimed.

Back in the US Crocker and his colleagues, supported by Cepi and the Goddard Project, presented the idea to Riccardo Giacconi at STScI and to Charlie Pellerin and Ed Weiler at NASA Headquarters. They called the concept "COSTAR," a clever acronym for "Corrective Optics Space Telescope Axial Replacement." All agreed it was a great idea. However, some remaining obstacles had to be overcome.

Weiler relied on an advisory committee made up of senior astronomers from the community, chaired by Bruce Margon from the University of Washington. When the committee reviewed the COSTAR concept, Margon in particular expressed strong reservations. Among other things, he worried that somehow this deployable column with its moveable mirrors could get stuck in the wrong position and block light from entering one or more of the axial instruments, ruining any opportunities they might have to keep doing some good science even if impaired by spherical aberration. As Bruce, like Holland Ford, was also a member of the Faint Object Spectrograph science team, he took such possible risks personally.

Back at Ball Aerospace, Wally Meyer machined an accurate, full-scale, working model of the proposed deployable optical bench of COSTAR. He and Crocker tried to anticipate everything that might go wrong, tried to counter every objection that had been raised by Margon. The motors would have redundant electrical wiring; the electronics controlling COSTAR would be redundant. The optical bench would be deployed while the astronauts were still present during their spacewalk. If it got stuck, the design was failsafe in that an astronaut could use a socket tool to manually deploy the bench or to pull it back into its enclosure by hand (Figure 3.8). When all of this was presented to Margon's committee, he relented. "Let's go do it."

The Hubble Project Office at Goddard proceeded to organize the COSTAR effort. NASA Headquarters, which was responsible for the selection of all the scientific instruments on Hubble, approved Holland Ford as the lead scientist and Jim Crocker as the lead engineer for COSTAR. Ball Aerospace was asked to design

Life With Hubble

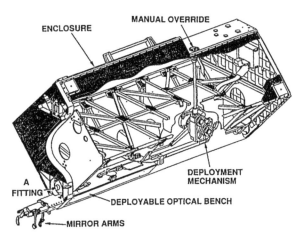

Figure 3.8. Cut-away view of the interior of COSTAR. Note the Manual Override fitting on the exterior of the enclosure. If the deployment mechanism failed, the EVA astronauts would use a tool either to manually deploy the bench or to pull it back into the COSTAR enclosure. (Credit: NASA, Ball Aerospace.)

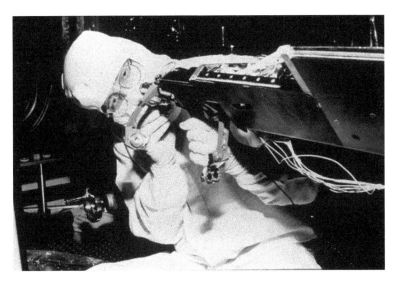

Figure 3.9. Technician at Ball Aerospace working on the COSTAR optical bench. Note the elliptical opening in the bench above and to the right of the center of the picture. This allows light to reflect from the M1 mirror of GHRS through the wall of the bench out to its M2 mirror mounted on a deployed arm. (Credit: NASA, Ball Aerospace.)

and build COSTAR (Figure 3.9). Ball had developed the GHRS already on board the observatory, and it was in the process of designing and building STIS and NICMOS. So, this relatively simple new axial module was a straightforward extension of work the Ball team already had underway. As always, Cepi's Flight Systems and Servicing Project assumed responsibility for overall technical and contractual management. The approximately $60 million of new funding required was drawn out of the funding originally earmarked for STIS and NICMOS development, as those two new instruments would now be delayed to a later servicing mission. In the end Senator Mikulski of Maryland worked to restore this money to the HST account in congressional appropriations.

As the design of COSTAR proceeded in late 1990 and early 1991, another technical challenge arose and another hero came to the fore. I've already described how incredibly precisely the M1 and M2 mirror pairs had to be aligned to the telescope light beam, to each other, and to the entrance apertures of the three axial scientific instruments. The first two of these alignments could be controlled and adjusted by operators on the ground with motors driving each mirror in very fine steps in tip and tilt. But to align COSTAR's mirror pairs to the entrance apertures of the instruments, the designers had to know with high accuracy the exact physical location of each entrance aperture, relative to COSTAR's location. If the mirror pairs didn't send the corrected light beam straight down an instrument's aperture, the optical correction would be for naught. And this precise positioning of each mirror pair had to be built into COSTAR from the start; it couldn't be adjusted from the ground once COSTAR was in orbit. The problem was Jim Crocker and the Ball designers didn't have that information. They didn't know where to position the small arms carrying the mirror pairs so that they would accurately send the light beams "down the hole" of each axial instrument.

A couple of times before—when Aden Meinel bumped into John Trauger at a dinner buffet and suggested to him how to correct the telescope's spherical aberration within WFPC2, and when Jim Crocker climbed into that shower in a hotel room in Germany—Hubble people had stumbled serendipitously, almost casually, onto solutions to the aberration problem. Now Hubble's guardian angel was at work again.

One day Jim was having lunch in a cafeteria at Goddard. He happened to bump into Olivia Lupie, a sharp young astronomer with seemingly limitless enthusiasm and energy. Olivia was a member of the Hubble High Speed Photometer team, ironically the team that would lose their instrument when COSTAR was installed into the spacecraft. Over lunch she asked Jim how things were going with COSTAR. He lamented that they were stuck on the issue of the locations of the five instrument apertures, two in FOS, two in FOC, and one in GHRS, which had to be accessed by the five M1–M2 mirror pairs. Then Olivia responded, "Oh, I can tell you that!" It seems that, unbeknownst to the COSTAR team, she had been working on accurately mapping the locations of all the instrument apertures, primarily to enable her own HSP instrument to accurately acquire its target stars. The HSP had no autonomous target acquisition mode of its own and had to rely on "blind pointing" offsets from the other instruments to the HSP. Jim Crocker considers Olivia Lupie an unsung hero of COSTAR and of the Hubble observatory. With the aperture maps she provided, she made a properly working COSTAR possible, and that in turn brought about the restoration of Hubble's sharp vision for three critical Hubble instruments.

Sadly Murk Bottema, who along with Jim Crocker had brilliantly created one of the two devices that would correct Hubble's eyesight, wouldn't be around to see his work come to fruition. Murk died of cancer on 1992 July 3, a year and a half before the first servicing mission got off the ground. Hubble Project Manager Joe

Rothenberg hurried to Boulder when he heard of Murk's condition and presented the NASA Public Service medal to him while he was on his deathbed. His colleagues at Ball Aerospace attached a small plaque to the exterior of COSTAR before it was shipped to Goddard and later launched into orbit. It bore the inscription, "The COSTAR Instrument is dedicated to Dr Murk Bottema in recognition of his contributions to COSTAR and the entire HST Program."

Chapter 4

Leading the Way: Preparing to Service Hubble

I served for 11 years, from 1976 to 1987, as the Hubble Scientific Instruments Project Scientist. After the Challenger disaster in 1986, NASA decided to accelerate the transfer of management responsibility from Marshall to Goddard. My mentor Al Boggess became the Senior Project Scientist and I became his Deputy.

From 1981 to 1991 I also had a management role in the Laboratory for Astronomy and Solar Physics at Goddard as Head of the Astronomy Branch. I had taken that job mainly to earn a higher salary. Being a mid-level manager in the Federal Government really wasn't much fun. I was responsible for the care and feeding of about 16 PhD scientists. Each had a strong personality and strong opinions about the federal bureaucracy within which they had to work. I was perpetually caught in the middle between the entanglements of rules and policies imposed by the administration over my head and the gripes of these highly educated, high achievement-level individuals residing in my Branch. But I've always been a good, patient listener, a skill that made me reasonably adept at the job.

One of the five original scientific instruments to be launched on Hubble, the Goddard High Resolution Spectrograph (GHRS), was developed under the science and engineering management of my Lab. The Principal Investigator was our Laboratory Chief, Jack Brandt and his Co-PI was my old fellow grad-student from UCLA, Sally Heap. Sally was also a member of my Branch. Early on, the GHRS Science Team adopted me as a Co-Investigator. I was pretty well known at that time for my research as a stellar spectroscopist. I was also a Hubble insider who could help guide the team through what was sometimes a morass of requirements laid on by NASA and by the Hubble Project. So, my joining the GHRS Science Team was beneficial both for the team and for me. It would give me the opportunity to do research I had longed to do with Hubble, and it would help them keep ahead of the game, as they had to deal with Project management.

After Hubble was launched and spherical aberration was discovered in its telescope optics, the GHRS Science Team was still able to proceed with much of its originally planned scientific research. Of all Hubble's instruments the optical flaw affected the GHRS the least. Each Instrument Science Team had been guaranteed hundreds of hours of observing time on the telescope in return for devoting many years of their lives to providing these powerful new instruments to the astronomical community. The other Instrument Teams had to "go back to the drawing board" to determine what science they could still do in the presence of spherical aberration. The GHRS Team had only to make minor adjustments to large portions of its planned research. With Ed Weiler's advocacy NASA Headquarters pitched in by allocating more observing hours to the instrument teams to allow them to compensate as much as possible for losses in observing efficiency, an inevitable consequence of the telescope's poorly focused images. As it turned out, I was one of the first astronomers to acquire high quality observations with Hubble—very high-resolution ultraviolet spectra of a bright star with a bizarre chemical composition called chi Lupi, rich in gold, platinum and mercury among other elements.

I will never forget the first time I looked at my new Hubble data displayed on a computer screen. It was Columbus catching a first glimpse of the New World. I was the first human in history to set eyes on what I was looking at that afternoon in the autumn of 1990 (Figure 4.1; Leckrone 1991). Admittedly, only another astrophysicist could understand and appreciate the squiggles plotted on that graph and why I was so excited by them—a spectroscopic "face only a mother could love," as it were. There is an old saying among spectroscopists that, "if a picture is worth a thousand words, a spectrum is worth a thousand pictures."

To me it was a moment of self-actualization. I went tearing down the hall showing anyone who would look a hard copy of my computer plot. Basically the whole periodic table of chemical elements, with abundances in this star set by forces of nature—gravity in a tug of war with radiation pressure—was spread out in front of me. At last perhaps I could begin to understand why this peculiar star, and others like it, were so weirdly different from, say, our own Sun.

But now I faced a professional quandary. Armed with these spectacular new Hubble observations, I had an opportunity to really sink my teeth into some important research. But this would require a lot of time and most of my time for more than a decade had been devoted to NASA company business—running the Astronomy Branch and serving on the Hubble Project. Something had to give.

I decided to leave my Branch Head job and take a one year sabbatical in place, doing research in Jack Brandt's laboratory. I was still enthusiastically devoted to Hubble and wanted to continue as Deputy Project Scientist. Al Boggess and I negotiated a deal. We agreed that he would cover most of the reviews and science team meetings that required time-consuming travel. I would stay back at Goddard to mind the store while he was away. I would do Hubble-related travel when Al was not able to, or when we both needed to attend an especially critical meeting. These arrangements freed up a lot of time, allowing me to concentrate on the analysis of my GHRS observations.

It was an exciting year. Several outstanding scientists came to Goddard to work with me, most especially my friend atomic physicist Sveneric Johannson and his students from the University of Lund in Sweden. We made some truly exciting and

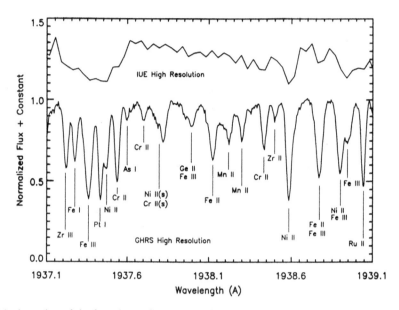

Figure 4.1. A portion of the first observation I made with the Goddard High Resolution Spectrograph on Hubble. This is a small segment of the ultraviolet spectrum of the highly chemically peculiar star, chi Lupi. The "squiggly" plot at the top is the best observational data we had prior to Hubble. It was obtained with the International Ultraviolet Explorer satellite. Its information content is negligible. Below the IUE spectrum is the same segment observed at much higher resolution with the Hubble/GHRS. The Hubble observations revealed a level of spectroscopic detail never before observed at ultraviolet wavelengths for any star other than the Sun. The elements and ions indicated by the labels are absorbing background light emitted by the star, making the star look darker within each line. A Roman numeral indicates the electrical charge or ionization state of each element: I for neutral atoms, II for atoms missing one electron, III for atoms missing two electrons. (Credit: Reproduced from Leckrone 1991. © 1991. The American Astronomical Society. All rights reserved.)

unexpected discoveries in the ultraviolet spectrum of chi Lupi. We began to create an inventory of all the chemical elements that we could identify, and perhaps more intriguingly, elements that we expected to see that were missing. I organized a multi-national research team of about 30 people into something I dubbed, "The chi Lupi Pathfinder Project." We wanted to achieve the most complete and accurate analysis of the chemical makeup of a peculiar star ever done. We figured we should characterize at least one of these weird stars as completely as possible. Most prior research done from the ground had settled for measuring trends of a few individual elements among many stars of this type. That approach had left a lot of questions unanswered.

Over the course of that year we experienced many exciting moments, had a lot of success and published numerous papers. But for better or worse, the call of a higher duty intervened. In September of 1991 Al Boggess retired from NASA. I was urged to apply to be his replacement as Hubble Senior Project Scientist. After a brief competition, I was selected for the position. The first servicing mission to restore the observatory to its proper function as the world's finest astronomical instrument loomed two years in the future. So, in January of 1992 I left my research in the hands of my collaborators and took up residence across campus in the office next door to the Hubble Project Manager. From that position I would provide scientific leadership to the Hubble Project for the next 17 years, including helping to plan and execute five successful Shuttle servicing missions to Hubble.

Over a decade earlier, in December of 1981, I was asked to give a talk at a technical interface meeting (or TIM) of engineers from Marshall, Goddard and the industrial contractors who were creating the system of software and computers, both on the ground and on the spacecraft, that would be used to operate the Space Telescope when it was in orbit. Each organization was responsible for different pieces of the system. They had to be sure that the completed operations system worked properly as a whole when all these individual pieces were finally integrated together. So, they met periodically to talk about their "interfaces." In the midst of their deeply technical and detailed discussions they wanted to take a break and hear about why they were doing all this work. What would the Space Telescope be able to do to change our understanding of the universe when it was finally in orbit and being operated with the system they were creating? They invited me to give a briefing on potential Hubble science.

As the Space Telescope Scientific Instruments Project Scientist at the time, I had been deeply involved in the competition NASA ran to select the initial science programs that the observatory would carry out and, especially, the new scientific instruments that would be used for that purpose. I had developed the scientific performance requirements that NASA expected for each instrument. And I had been working closely with the science teams for all five instruments. So, I was well qualified to talk about the scientific future of the observatory.

In my talk I discussed some of our overarching goals such as determining the history and evolution of the universe and whether the laws of nature are universal in space and time. Then I gave about a dozen specific examples of the kinds of observations we would be making and why, starting in the solar system and working my way out across the universe and back toward the beginning of time. We planned to: monitor the weather on the other planets in our solar system with a clarity similar to fly-by spacecraft when they're only a few days from closest approach; study how new stars are formed and observe old stars in the midst of their death throes; determine how individual stars in other galaxies differ from or are similar to those in our own Milky Way; and study how the chemical composition of stars and galaxies has evolved over cosmic time. We planned to greatly improve the measured distances to other galaxies, the rate of expansion of the universe and its age. We would look for the giant black holes that astronomers suspected, but could not prove from the ground, were lurking in the centers of distant galaxies. If we found them, we could determine if the mysterious, highly energetic quasars were in fact the outbursts of those black holes as they swallowed up stars and gas around them. I went so far as to speculate that, "we may even be lucky enough to detect early galaxies in the process of their original formation, since the Space Telescope will be able to observe objects at distances comparable to those traveled by light over the age of the universe," which at that time was overestimated to be 15–18 billion years.

Sitting in the audience that day was a mid-career aerospace engineer and manager named Joe Rothenberg. Joe was a diminutive and dapper man, as bright and energetic as a figurative supernova. He had worked his way up as an engineer and team leader at Grumman Aircraft Corporation over a period of 18 years. At Grumman, as a contractor to Goddard, he helped develop and manage the Orbiting Astronomical Observatory (OAO) series of satellites, the Solar Maximum Mission and Frank Cepollina's Multi-Mission Modular Spacecraft (MMS)—all of which ultimately would provide a strong background for work on the Space Telescope.

In 1980 Joe was offered an opportunity to come to work at Goddard as a government employee in the Electrical Systems Division. But that fell through when President Reagan froze all federal hiring within a few days of his inauguration in January of 1981. Nonetheless, Joe left Grumman and went to work for a small start-up company in Maryland called Computer Technology Associates (CTA). He led a major expansion of CTA's business, including winning a contract to provide integration, testing, and independent validation of the Space Telescope's flight and ground operations system.

It was in that role that Joe was attending the TIM that day and heard me speak about the awesome scientific potential of the Space Telescope. I apparently blew him away. Many years later he told me that my talk inspired him and was the primary motivating factor in his ultimately deciding to leave CTA, where he had been very happy, and in 1983 come to work at Goddard as the Mission Operations Manager (MOM) on the newly renamed Hubble Space Telescope. He said, "that [the talk] did really get my heart and mind pumping, and obviously changed my career path forever." A seasoned and successful engineer and manager like Joe Rothenberg might have many reasons for accepting a new job—career growth, increased salary, visibility and status, etc. It is utterly remarkable that he was motivated mainly by the science. His heart and mind were clearly in the right place, considering what the future would hold for him and for Hubble.

Soon after Rothenberg came to Goddard as Hubble's new MOM, he spotted several serious problems. The most important of these had to do with the way spacecraft contractor Lockheed, under contract to Marshall, had designed the system to load software and commands into the DF-224 spacecraft computer during ground testing and later in orbit. After the routines and commands had been loaded, it was necessary to verify that they were working properly. If a bug was found, the software, which had been uploaded to the computer, had to be downloaded again so that the problem could be diagnosed and fixed. But there was a "catch 22." The system was designed to upload and download algorithms and data entirely via software commanding. But if the software were faulty to begin with, the ground controllers might not be able to command a follow-up download to look for the problem. Joe reckoned that there needed to be a hard-wired capability to download the contents of the computer's memory without relying on software to do it.

In that era there was an intense inter-Center rivalry between Marshall and Goddard, with almost palpable suspicion and distrust; each Center believed itself superior to the other. Joe first brought up his issue with downloading the spacecraft computer at one of the major, high-level Hubble Project reviews held each quarter in the Center Director's conference room at Marshall. He presented his thoughts as part of the Hubble operations briefing and was ignored, and later derided by the Marshall managers. At the next quarterly review, the computer developers reported that they had tried to load software onto the DF-224; it wouldn't turn on and they didn't know why. It came Joe's turn to present his operations briefing once again and he noted that the problem was most likely the one he had described three months earlier. Again, he was ignored and derided. Three more months passed. At the next quarterly, the computer guys reported once again that they couldn't get the computer to boot up after they had loaded the software into it. And once again they didn't know why. At that point Bill Lucas, the Director of Marshall looked at Joe, sitting at the side of the room and then turned to the Lockheed engineer doing the presentation. He said, pointing at Joe, "If you solve his problem, it seems a good bet that you will solve your problem." That was taken as high-level direction that a hard-wired, fail-safe download capability had to be built into the computer. If a similar problem had happened with Hubble in orbit, rather than in ground testing, it would have become the early cause célèbre on the Hubble mission, requiring months to fix somehow, if at all, and completely obscuring Hubble's most fundamental flaw—spherical aberration. With patience Joe Rothenberg achieved credibility with his counterparts at Marshall and Lockheed.

After the Challenger disaster in January of 1986, Hubble's launch was delayed indefinitely. Joe took on the managerial job of Chief of the Mission Operations Division in the Goddard Mission Operations and Data Systems Directorate, while still managing the remaining work needed to tie up loose ends on the Hubble flight and ground operations systems. He loved what he was doing in his new Division, and so was reluctant to agree when the call came for him to move up to Deputy Director of the entire Directorate. Over the years he had established a strong reputation as a team builder, motivator, negotiator, and solver of tough problems. So, his rise into higher levels of management responsibility was inevitable.

In the summer of 1990, news of Hubble's optical flaw hit NASA, and the whole country for that matter, like a bolt of lightning. As the weeks went by NASA began to lay plans for an early, call-up servicing mission to address that and several other technical problems with the observatory. One day that August, Joe received a call from Pete Burr, Goddard's Deputy Director. Pete told Joe that the Agency wanted him to take over the main leadership role of the servicing mission as the new Hubble Project Manager. With typical modesty, Joe replied that he loved Hubble and would be happy to help out; but it wasn't really necessary to make him the Project Manager. However, there was no debate to be had. Joe was the man Goddard and NASA needed to lead this critical, high-visibility mission. So, he agreed, on one condition. He wanted full responsibility and authority for Hubble to be transferred immediately from Marshall to Goddard. Later in a phone conversation the Project Manager at Marshall, Fred Wojtalik, made a simple request of Joe. Could the transfer be effective at the end of September, the end of the federal fiscal year, so that it would not have the appearance of the job being taken away from Marshall? That was a face-saving step for all the people at the center in Alabama who had worked diligently for so many years getting the Hubble mission started. Joe readily agreed.

Before Hubble was deployed and spherical aberration was diagnosed in 1990, NASA had penciled in a first servicing mission for some time around 1994 or 1995. After the optical flaw was discovered, the need for a mission became more urgent. It was not an emergency, but it needed to be accomplished with all deliberate speed. Several factors allowed the Hubble program to aim for a mission in three years rather than four or five or more. Flight Systems and Servicing Manager Frank Cepollina (Cepi) and his servicing team had worked previously with JSC and KSC, carrying out several rescues of ailing satellites in low Earth orbit, most notably the Solar Maximum Repair Mission in 1984. So, the Hubble Project at Goddard already had a lot of pertinent experience.

In 1990 July engineer Tom Griffin was hired by NASA to work on Cepi's team. Tom had just completed a tour of duty working as a contractor for MSFC and JSC on a study of a "launch-on-need" shuttle mission—that is, a mission that could be launched within a year of being called up. The study was a paper exercise where a generic payload was to be "flown." But it entailed careful and detailed engineering analyses, a thoroughly mapped out mission timeline, step-by-step spacewalk procedures, etc. The relevant documentation was prepared and put on the shelf ready for use. These written plans provided high-fidelity templates for later use in the event of an urgent call for a previously unscheduled shuttle mission—Hubble's first servicing mission, SM1, for example.

The first servicing mission to Hubble was different from the generic launch-on-need mission because it didn't have to be flown within one year. There was a considerable amount of Hubble hardware to be designed, built, and tested. That would take much longer than a year. But there was also relief within the Hubble Project that so much planning and experience already existed on the human spaceflight side and that Hubble Project engineers and managers were experienced with working in that world. The shuttle would be ready when we needed it. We just needed to populate the shuttle payload bay with space-qualified hardware that would solve Hubble's technical problems and redeem our mission.

We had a leg up on some of the payload hardware. One of the two key elements of the optical repair, the Wide Field and Planetary Camera 2, was already far along in its design and construction when SM1 was called up in 1990. But it needed to be outfitted with a new set of properly shaped and aligned small corrective mirrors. The second key element, COSTAR, had to be designed and built from scratch. Since it was not a full-up scientific instrument, COSTAR appeared at first to be simple. That was misleading. The required precision of its complex, potato-chip-shaped mirror surfaces had proven very challenging. In fact the Project could only find one manufacturer who was willing and able to take on the job for both COSTAR and WFPC2—a small company called Tinsley Laboratories in Richmond, California. The new, improved ESA solar arrays were similar to the original set, but with some extra thermal insulation that, hopefully, would make them less likely to jitter. The replacement gyroscopes were spares that were already in hand. A Co-processor needed to augment the spacecraft's computer and various other electronics boxes were well within the state of the art. Creating all the new hardware for SM1, therefore, should not take as long as it usually did for robotic space science missions.

Figure 4.2. Hubble payload carriers and protective enclosures mounted in the payload bay of a shuttle orbiter. Note the Flight Support System with its circular lazy Susan platform near the bottom. This will be rotated 90° in orbit to clamp onto the base of Hubble during the rendezvous and docking process. (Credit: NASA, KSC.)

Every piece of Hubble flight hardware required some kind of protective enclosure that would provide a benign environment for optics, electronics, light sensors, and other delicate components so that they would withstand the rigors of launch and the intense heat and deep chill of space during the servicing mission. The protective enclosures had to be attached to large platforms or carriers mounted to the shuttle itself within the payload bay (Figure 4.2). The arrangement of carriers and enclosures had to allow easy access to the equipment for space-suited astronauts while they were tethered and floating in microgravity. As had been learned from the Solar Maximum Repair Mission, an astronaut wearing bulky, inflated gloves had little manual dexterity. He or she needed specialized tools to overcome that limitation and to have the capability to get every job done within a limited amount of time. Creating these tools and training the astronauts to use them, a responsibility of Cepi's team, was a time-consuming, iterative process involving a good deal of trial-and-error and feedback from the EVA crew.

In principle, it looked feasible to have all this equipment designed, built, tested, de-bugged, and delivered to the Cape for integration into Shuttle Endeavour's payload bay in time for a late 1993 launch. Pulling this three-ring circus together, managing a tight schedule and the taxpayer-provided finances for the mission, providing technical oversight, direction and support to myriad contractors, and working in close coordination with the Shuttle managers at JSC and KSC to bring SM1 to a fully successful outcome was the responsibility of the Hubble Project at Goddard, led by Joe Rothenberg and Cepi. It was my job specifically to provide scientific leadership across this entire universe of people to be sure that, in the end, the rescue of Hubble would accomplish scientifically what the world was expecting.

From the autumn of 1990 up to Christmas time of 1992, the Project focused on the work of individual contractors. Each major piece of payload hardware was the full-time preoccupation of a dedicated lead Goddard engineer. In addition, WFPC2 and COSTAR each had a dedicated project Instrument Scientist keeping track of their development. About once per month the senior Project leadership would travel to the home bases of the major hardware developers to review their progress. The idea was to catch and solve problems, both technical and financial, before they blossomed into major headaches. So, it became a regular routine for Joe Rothenberg, Al Boggess, and Cepi to lead a small entourage on periodic trips to Ball Aerospace in Boulder Colorado, to JPL in Pasadena California and occasionally to the European Space Agency's field center in Noordwijk, the Netherlands.

Ed Weiler, the Hubble Program Scientist at NASA Headquarters, frequently told the story about how his advisor in graduate school at Northwestern had said that he would do better in life as a manager than as an academic, research-oriented scientist. High-level management of NASA space astronomy programs was the career path he chose. It was unusual for him to show interest in the "nitty-gritty" scientific, technical, or management details of any of the SM1-related activities for which the HST Project was responsible. The major exception was WFPC2.

Weiler could correctly be called the "Father" of WFPC2, going back to the middle 1980s, when he strongly advocated its addition to the Hubble Program to assure that "Hubble would never go blind." The possibility that WFPC2 might ameliorate the spherical aberration debacle in the near term was an unexpected bonus that allowed Ed to deflect some of the heat being directed at NASA by the media in the early 1990s.

From June of 1990 to the first servicing mission in December of 1993 Ed was preoccupied with getting as much favorable publicity for Hubble as possible, notwithstanding the observatory's disappointing performance. He judged the research being done with Hubble, and the value of each of its scientific instruments in large part on the newsworthiness of the results they produced. He wanted Hubble to be on national television news and in the newspapers regularly. Of greatest value was coverage of Hubble science results in the New York Times and the Washington Post—preferably on the front page, preferably above the fold, preferably illustrated with gorgeous imagery. The latter required WFPC in the beginning and the dramatic improvement of images to be obtained with WFPC2 later. This was

Ed's obsession. He obviously had a major personal stake in the outcome of WFPC2's development.

Starting in late 1990, Ed and the Headquarters Hubble Program Manager, Doug Broome, regularly accompanied Rothenberg, Al Boggess, Cepi, and the HST Project team to the WFPC2 monthly progress reviews at JPL. During the summer of 1991, Al announced that he would be retiring from NASA at the end of that September. He asked me as his Deputy to take his place on those monthly sojourns to the contractors' plants and to JPL. My first trip filling in for Boggess for a status review at JPL was in August of 1991. It was at that meeting that the wheels very nearly came off the WFPC2 bandwagon.

As I walked into the conference room at JPL around 8:30 am on the first day of the review, the nominal starting time, the meeting was already well underway. I heard Cepi shouting. He was very angry. (Years later I teased him that I could see the smoke coming out of his nostrils and his eyes glowing red.) Since this was my first monthly review substituting for Boggess I had only a superficial knowledge of what had transpired in previous months. At that moment I mainly felt embarrassment that Cepi seemed to be so abusive toward our JPL colleagues. That was naïve on my part as I was soon to discover.

At every monthly review up until that time, the WFPC2 managers had given the Project a specific dollar amount and schedule to complete the instrument's development. But now they were dramatically changing their story, with the projected budget to complete development roughly doubling and no guarantee of when WFPC2 would be delivered. When he heard this, Ed Weiler's reaction was visceral. He said that we had to complete COSTAR and couldn't afford the extra money JPL was requesting. At that point Ed, Cepi, and Joe held a side conversation and agreed that something drastic needed to be done. They instantly had lost all confidence in JPL's management and ability to deliver in time for SM1.

Jim Westphal, the original WFPC Principal Investigator, was a spectator in the room as this was being discussed. He asked Ed and Joe to take a walk with him during a break in the meeting. Jim had been musing about how to get WFPC2's cost and schedule back in line without significantly reducing the scientific productivity of the instrument. His idea was to remove four of the eight CCD channels, leaving three Wide Field Camera chips and one Planetary Camera chip. The mechanism that was rotated to select between the two cameras would be locked in place. An image produced by this arrangement would be L-shaped with the legs of the "L" formed by the three Wide Field Camera fields of view and with one smaller, higher resolution Planetary Camera field of view tucked into the intersection of the two legs.

Some tiny positioning mechanisms recently proposed by JPL engineer Jim Fanson could be attached to three critical mirrors inside the instrument to allow fine adjustment by ground controllers to the alignment of the optics in orbit. This would greatly reduce the risk that the instrument might be misaligned in such a way that it would not correct spherical aberration after it was launched. (Imagine the scandal *that* would incite!) The addition of this in-orbit alignment capability was viewed as crucial by the WFPC2 science team and might "sweeten" the deal a bit as they contemplated the loss of four CCD channels.

Westphal had done some back-of-the-envelope calculations that indicated the efficiency of WFPC2 in executing its scientific observing program would be reduced to about 85% of what it would have been otherwise—not really a severe loss at all. Jim had not planned to bring up this idea at the review because he had not as yet discussed it with WFPC2 Principal Investigator John Trauger. But at the same time, he felt a sense of urgency that people should be discussing the possibilities. He suggested that Weiler propose it when the meeting resumed.

Ed did bring it up in his usual, somewhat antagonistic style. He said if the revised, de-scoped design met budget and schedule constraints he would support it. Otherwise NASA would remove WFPC2 from the manifest for SM1, perhaps to be flown on a later mission, or perhaps not to be flown at all. He in effect directed JPL to implement Jim Westphal's approach.

The root cause of the cost and schedule growth probably was not the technical complexity of the instrument design. Actually, in some ways the original WFPC's design had been more complicated and problematic, and JPL had managed to pull it off in time for Hubble's deployment mission in 1990. The problems that lay before us at that August meeting resulted from the way the WFPC2 Project management was organized at JPL. It was a matrix management approach, with multiple organizations given responsibility for various pieces of the instrument design, fabrication, and testing. The WFPC2 Project Manager, Larry Simmons, had little insight into or control over what each individual Division was doing. Every August each Division would analyze its budget and schedule requirements for the following year and simply lay those in Simmons' lap. He had little recourse but to add up the numbers and present those to the Hubble Project. This August the "sticker shock" caused NASA to push back, resulting in a de-scoped WFPC2.

Joe Rothenberg was "pissed" at JPL because of this situation. He felt there was no Program control and that cost and schedule were open ended. He couldn't allow that to cause him to miss the SM1 launch schedule or overrun his budget, if there were any way to avoid it. The de-scope plan seemed to fill the bill. Joe carried this issue through Director of Space and Earth Science, Charles Elachi, up to JPL Director Ed Stone. They agreed both with Ed Weiler's direction to de-scope the WFPC2 design and with Joe's request that the WFPC2 Project structure be re-organized so that all the technical work, as well as control of budget and schedule was directly under the leadership of a single Project Manager, Larry Simmons.

Years later Ed Weiler liked to brag about how he had held firm on a budget and schedule problem by requiring the WFPC2 team to de-scope their instrument. Every WFPC2 image had a peculiar shape, something like the profile of a stealth fighter aircraft (Figure 4.3). Those who knew the background dubbed that shape, with the missing quadrant, "Weiler's Wedge," or "Weiler's Wing."

Being the new kid in the room on that August day in 1991, I did not try to interject myself into the discussion of whether or not to de-scope WFPC2. Since then I have sometimes wondered if the de-scope was really necessary. Could the budget and schedule problems have been cured simply by improving the way the instrument

Figure 4.3. Iconic Hubble image of the Eagle Nebula, "The Pillars of Creation," taken with the WFPC2 in 1995. Note the peculiar shape of the image with the missing Wide Field Camera channel leaving a black quadrant in the upper right. A single small Planetary Camera channel is tucked in near the middle of the image. This shape, referred to as "Weiler's Wing," resulted from the descoping of the instrument and the removal of four of the eight original CCD camera channels in 1991. (Credit: NASA, ESA, J. Hester, P. Scowen, STScI 1995-44.)

development had been managed, going from a matrix management scheme to one in which a single Project Manager was fully in control, as that seemed to be the root cause of the problem. We'll never know the answer to that question. But one thing is beyond debate. The WFPC2 turned out to be a remarkable scientific instrument, achieving scientific results well beyond all our expectations, producing glorious images of the cosmos that left ordinary people breathless and wanting more. It operated for an incredible 16 productive years in orbit. It was still working well when we removed it during SM4 in 2009, replacing it with a much more capable, high-tech Wide Field Camera 3.

As the end of 1992 approached, the key pieces of the SM1 payload began funneling into Goddard for their final testing and certification for space flight. They all sat, well protected from dust and chemical contaminants, in the world's largest "Class 10,000" clean room. In that environment, there were no more than 10,000 particles larger in size than 0.5 µm in every cubic foot of air. For comparison, in a typical outdoor urban environment, there may be 35 million or more such particles per cubic foot. Dust particles or organic vapors sticking onto optics like mirrors, filters and grisms, or within electronics and mechanisms, or other sensitive components of the payload could degrade their ability to work properly in orbit. The entire payload, including protective enclosures and carriers, had to remain pristinely clean.

The Hubble Project worked for years with Shuttle builder Rockwell and with KSC, building into their engineering culture the ethos of keeping the Shuttle, and especially the payload bay, as clean as possible. People who worked in the clean room had to don "bunny suits," ultra-clean overalls, boots, gloves, hats, and facemasks every day. They couldn't wear makeup, aftershave or cologne. An enormous wall of HEPA filters extracted contaminants from the air as it entered the clean room, flowed smoothly over the payload hardware and the people working on it, and was sucked out of the room at the rear. Particles and volatile condensable vapor were blown out of the room, not into it.

Almost from the moment Joe Rothenberg took over as Hubble Project Manager, he was besieged by Cepi wanting to build a full scale, high fidelity replica—both structural and electrical—of the Hubble spacecraft, to be placed in the clean room for use in testing all the payload hardware to be flown on SM1. Cepi had a good point. It was important to know before they were launched that each new component could easily be installed by space suited astronauts, that each would fit in its designated location, that each would power on and communicate properly with the spacecraft computer, and that each would function electrically as it was designed. Cepi's mantra was "test, test, and re-test." If a new instrument or new gyro package, for example, didn't fit into Hubble as expected, or didn't work properly after it was turned on in orbit, it would be too late to do anything about it; a precious opportunity and millions of dollars would be wasted. It would be imperative to find and resolve all problems of that sort on the ground before the mission. Moreover, the astronauts who were going to be working on Hubble in orbit during SM1 had not had the opportunity to see it or become familiar with it before it was launched in 1990. They needed a means to become familiar with the spacecraft.

Cepi wanted a High Fidelity Mechanical Simulator (HFMS), a full-scale replica of the Aft Shroud, the cylindrical spacecraft structure at the back end of Hubble, for testing the ease of insertion and proper fit for each new payload unit. He also wanted an accurate, fully operating replica of all the electronics boxes housed in the equipment bays within the Hubble spacecraft. That would be called the VEST, or Vehicle Electrical Simulator for Testing. All of the new payload units could be cabled into the VEST and operated from the Space Telescope Operations Control Center across the Goddard campus, as if they were actually in orbit. The problem was, these new test facilities carried a high price tag, about $25 million.

Joe Rothenberg read the mood of the scientific community, represented by the Hubble Science Working Group. They had basically lost all trust in NASA after the debacle of spherical aberration. He wanted to win their confidence back. He promised three things—that we would find the problems with Hubble's performance, including other technical problems in addition to spherical aberration, that we would fix those problems, and that we would do that without "eating our young," that is without depriving the next generation instrument teams (STIS and NICMOS) of the money they needed to build the instruments for the second servicing mission three to four years after SM1. He offered budget transparency to the Science Working Group, giving them the opportunity to review his budget to make sure he was keeping that promise. If he needed to borrow funds from the next generation instrument accounts to develop the payload for SM1, he would give the scientists the opportunity to review openly how he was juggling the money year by year. In effect he gave them budgetary veto power, or something close to it.

Joe took the case for the HFMS and the VEST to the Science Working Group seeking its input. The scientists largely panned the idea of spending so much money on test facilities. Cepi was continuously pressuring the instrument builders to keep their costs down. The Principal Investigators of the next generation instruments, STIS and NICMOS, bridled at the suggestion that so much money be spent on what they cynically called "Cepi's Sandbox," while they were being asked to constrain their instrument designs to keep them relatively simple and less expensive.

In the end, however, Rothenberg overruled the scientists and found funds in the Project budget for Cepi's test facilities (Figures 4.4 and 4.5). Events to come would prove this a very wise decision.

When the SM1 astronaut crew came to Goddard to practice inserting both WFPC2 and COSTAR, the actual flight units, into the new HFMS, they had difficulty sliding them into the replica spacecraft. It turned out that they had been practicing that job in the underwater, neutral buoyancy training facility at Marshall with mockups that differed slightly from the final flight designs of the two new instruments. The mockups slid straight in. However, both flight units had small protrusions on their exteriors—the head of a bolt on COSTAR, insulation bumping over a small bracket on WFPC2—that got in the way. In the simulations in the large water tank the astronauts had been training with incorrectly designed mockups. Without their practice sessions using the HFMS, they would not have discovered these deviations until they got to orbit. At that point it would, at best, have caused a pause in the spacewalks, and a waste of precious EVA time, while the Project team in Houston tried to figure out why WFPC2 and COSTAR were not sliding smoothly into Hubble's aft shroud. In the end the astronauts in orbit were able to maneuver both instruments to work around those unexpected bumps, because they had "discovered" the problems as they attempted to insert the flight instruments into the HFMS in the big clean room at Goddard.

A far more threatening issue arose when the test engineers hooked up to the VEST the first of the two new gyroscope packages that had been built to fly on SM1. As they turned it on to test it, a fuse blew in the VEST power system. Then the second gyroscope package was installed. It also blew a fuse. The VEST power system was identical to that within Hubble in orbit. Something was terribly wrong. Without these tests in "Cepi's Sandbox," we would not have known that the fuses then flying on Hubble were under-rated; they could not carry as much electrical current as the new gyroscopes or other new pieces of equipment required. If they had blown during servicing when the astronauts installed the new boxes, or a bit later,

Figure 4.4. The High Fidelity Mechanical Simulator (HFMS) located in the world's largest Class 10,000 clean room at Goddard. It was designed to be an exact duplicate of the aft end of the Hubble spacecraft in orbit. Each servicing mission payload element was fully tested in the HFMS to assure that it would fit smoothly into Hubble. The shuttle EVA astronauts also used it to train for their tasks in removing and installing scientific instruments and other components. (Credit: NASA, GSFC.)

after Hubble had been deployed back into its orbit, it would most likely have been the end of the Hubble mission. Hubble would have been dead in the water. So, Cepi was right in pressing for test facilities on the ground that exactly duplicated what was in orbit. The lack of those facilities could easily have resulted in another deeply embarrassing Hubble disaster for NASA.

Dan Goldin took over as NASA Administrator on 1992 April 1. At that time, we were far along into preparations for SM1. This mission was undoubtedly the most complex and challenging of the Shuttle era up to that time. As a "rescue" mission for Hubble, it was also more highly visible than usual. Hubble's spherical aberration had left NASA's reputation seriously damaged. The Agency was not in the best of health in the early 1990s. The common wisdom was that its future depended on the

Figure 4.5. The Vehicle Electrical Simulator for Testing (VEST) in the big clean room at Goddard. This is a full-scale duplicate of the Hubble spacecraft's equipment bays, containing computers, data recorders, batteries, and myriad other electronics boxes. Every new payload component was hooked up to the VEST and was functionally tested. Each component could be commanded from the Space Telescope Operations Control Center (STOCC) just as it would be in orbit. (Credit: NASA, GSFC.)

success of SM1. All of this was dumped into Goldin's lap the day he took the reins. He reportedly was extremely worried about the current state of affairs.

Among those of us far down the NASA food chain who hadn't yet encountered Dan, gossip had it that those who had worked under him were of the opinion that he was a temperamental boss, possessing a short fuse and a brutal manner of expression when he was not pleased. All of us on the Hubble Project were highly sensitive to that possibility and were on our guard.

Over the years I only encountered Goldin a few times and never personally witnessed his temperament. However, high-level people at Headquarters and at the NASA Field Centers, including Goddard and Hubble Project management, did encounter his famous management style. Wes Huntress, the Director of the Solar

System Exploration Division and subsequently the Associate Administrator for Space Science at HQ under Goldin put it this way (Huntress 2003):

"When Dan Goldin came on board, it became clear immediately that he was going to be a challenge to work for. I had had some experience with people like Dan. I mean, Dan is a certain personality, and I'd had experience with somebody like him before, which has stood me in good stead, because I didn't make the initial mistakes a lot of the other people at NASA Headquarters made with him. I knew not to make them. And I got along well with Dan. I developed my own means to deal with him. Dan's a very intimidating person, and the way in which he works with his direct reports is to intimidate them, put the fear of God in them. Well, I never let him know I was scared, ever. Never let him know I was off balance. So that kept him off balance. So that's the way you deal with him. I love dan, and he is a great visionary, but he's difficult to deal with."

Responsibility for the success of satellite missions and flight hardware resided primarily with the NASA field centers at that time. Headquarters had little involvement or responsibility in the five or so readiness reviews that typically were conducted at Goddard to assess the progress of robotic space mission development. The first Hubble servicing mission didn't fit that paradigm very well. So, to ameliorate his worries, Goldin ordered up a sequence of independent reviews of the work we were doing in preparation for SM1–18 reviews in all.

Most of these reviews were held between April of 1992 and March of 1993. Each review relied on experts from other parts of NASA, industry or academia to provide input to Goldin. Each typically lasted 2–3 days but required a week of preparation by many of the key Hubble Project technical staff beforehand. In all, that added up to the equivalent of several months of actual time and several person-years of labor being devoted to the review process during that period. While our engineers were preparing good review presentations, they were not totally focused on the technical work at hand. This was a worry to Joe Rothenberg, but as was his style, he was determined not to complain, but to extract value for his Project from these reviews.

One important problem was getting the astronauts who would actually fly on the mission assigned quickly. Astronaut Greg Harbaugh had been working with the Hubble Project in making plans for SM1. But Greg had been assigned only as an advisor to the Project. He was not assigned to fly on SM1. The actual flight crew needed to get involved in the technical details as early as possible. We wanted their names. It was a very complex mission; they would require a lot of training. Only the spacewalking crew could give us accurate input on how they planned to work, how the ballet dances called EVAs should be choreographed. We had to get on with the work of designing and building scores of tools for their individual use in individual tasks on Hubble. It would make a difference in tool design, for example, if a particular member of the crew were left-handed versus right-handed. A tool design might have to be modified if it didn't have the right "feel" to the astronaut who would be using it.

In prior shuttle missions, the final selection of the flight crew had been done about one year before the launch date. We needed to have the crew on board much sooner than that. This problem was brought to the attention of one of the independent review committees chaired by General Tom Stafford. Stafford was a former astronaut, famed for his flights on Gemini, Apollo, and the joint Apollo-Soyuz mission with the Russians. At the end of the review, Stafford's committee gave a private briefing of their findings to Joe and Cepi. He said, "It sounds like you guys need to have your astronaut crew assigned very soon." "Yes, like tomorrow," Joe replied. So, in Joe's presence Stafford called Dan Goldin and told him that the Hubble Project needed to have their flight crew assigned as soon as possible, otherwise they were going to fail. His review team had validated this, and it was the Project's number one need. A week later we had the names of all the SM1 astronauts. The crew came on board in July of 1992, fully a year and a half before launch.

Another major issue was the number of spacewalks that would be allocated to SM1. In prior missions the number had been **one**. The Shuttle Program had no experience doing more EVAs in a mission than that. To complete the long list of servicing tasks we had lined up for SM1, we believed we needed at least four EVAs. At another of the independent reviews, chaired this time by Joe Shea, Rothenberg broached the issue. He didn't overtly plead for more EVAs. He simply listed the tasks needing to be done, which of those could be fit into one spacewalk, which could be done in two spacewalks, and which would be left undone when the shuttle returned to the ground if we were limited to one or two. Shea was a pioneer aerospace engineer and had led the Apollo Program at NASA Headquarters. At the time of the review he was an adjunct professor at MIT. Once again, a review chairman came to our rescue. Shea reported to Goldin that there were a lot of really important things to be accomplished on this mission and they couldn't be done in only one or two spacewalks. A short time later we were authorized to plan four EVAs. That number was ultimately expanded to five.

During the summer and fall of 1993 Dan Goldin continued to fret about whether NASA would be able to pull off this complicated mission. He mused about whether SM1 should be divided into two missions (at almost double the cost of one), each to be simpler and, in his mind, more likely to succeed than the all-or-nothing version we then had in place. He began pressing the Project with his concerns and talked about yet another independent review committee to look into the subject. At that point Joe Rothenberg had had enough. Joe called Goldin and, in effect, threw his badge on the table. "Dan, we can either have a review of an alternative mission, which will distract the Team, or we can keep the Team focused on ensuring the success of the planned mission. We can't do both." That ended the discussion of SM1a and SM1b.

After WFPC2, COSTAR, and each of the other payload components were thoroughly tested and verified to be flight ready at Goddard, they continued to be kept under the carefully controlled environmental conditions in the clean room until it was time to ship them to KSC for final integration onto the flight carriers and into Endeavour's payload bay. The plan was to wait until the last possible moment to transport them to the Cape, so that they would remain totally protected for as long as possible. Transporting them to KSC was relatively straightforward. Each unit would ride in an environmentally controlled van with an air-cushioned suspension to assure a soft and gentle trip. The Project began shipping the instruments and other delicate flight hardware to KSC in this manner in August of 1993. However, it was far more challenging to transport the flight carriers that would be mounted in Endeavour's payload bay to provide structural support for the SM1 payload during launch and in orbit.

The carriers were very large. The Flight Support System (FSS), a horseshoe shaped structure about 15 feet in diameter, carried a large "lazy Susan" platform to which Hubble would be latched while the astronauts were working on it. The Solar Array Carrier (SAC) had the primary job of holding the two new ESA arrays, each rolled up like window shades about 12 feet long. The Orbital Replacement Unit Carrier (ORCU) was built onto a U-shaped ESA Spacelab Pallet, about 8 feet long and 14 feet wide, designed to fit snuggly into the payload bay. It would carry the WFPC2, COSTAR, and numerous other replacement units and spare parts to be inserted into Hubble, all in their protective enclosures.

At midnight one night in late July of 1993 a caravan of behemoths made its way out of the West Gate of the Goddard Space Flight Center. Escorted by Maryland State Police, it slowly wended its way around the Washington Beltway and up Interstate 95 toward Baltimore. The caravan was transporting three huge, ultra-clean shipping containers within which were the large carriers.

The largest of the carriers, the ORUC, was being carried in a shipping container left over from NASA's Long Duration Exposure Facility (LDEF) shuttle mission. The container was 18 feet wide and 15.5 feet high. It was being hauled on two double-drop lowboy flatbed trailers that had been welded together side-by-side for a total width of about 16 feet. The other two carriers were also being carried in enormous containers that also required double-drop lowboy trailers. This was an "extra wide load" if ever there was one. And at total heights approaching 16 feet, they would just barely clear the numerous bridges and overpasses along the route. This is why the caravan moved along these major highways during the hours after midnight when traffic was very light.

At the front of the caravan was a "pole truck." It was simply a pickup to which had been attached a 16 foot-tall bamboo pole. A typical overpass had a clearance of 16 feet 6 inches. The tallest container reached 15 feet 6 inches off the ground. So, it was reasoned that, if the bamboo pole cleared an overpass, the caravan's cargo also would clear it. Bruce Milam was the Hubble manager responsible for getting all the carriers transported safely to KSC. He was riding in an SUV just behind the pole truck. Behind Bruce's SUV followed the three heavy-duty trucks hauling the three shipping containers. At overpasses where it looked like the clearance was going to be tight, Bruce stopped the caravan, stood in the road with a flashlight and used hand signals to guide each truck slowly under the overpass or bridge. Each container had a built in environmental control system to provide a clean room environment for the flight hardware. However, to be safe, the caravan also included a tube truck, filled with zero-humidity compressed air that could purge each container with ultra-clean, dry air for up to 25 days if necessary. Behind the tube truck came various other vehicles transporting support staff.

The 36 mile trip from Goddard to the Defense Logistics Agency (DLA) Terminal at Curtis Bay at the mouth of the Patapsco River, where it flows into the Chesapeake Bay south of downtown Baltimore, took four hours. Awaiting the caravan and its critical cargo was one of the barges that NASA had been using for years to transport shuttle external fuel tanks to KSC from the plant where they were manufactured at the Michoud Assembly Facility near New Orleans. It was enormous—80 feet wide and 300 feet long with an 80 foot high enclosure for carrying cargo.

The barge had all the comforts of home for its crew and passengers. However, it could not propel itself. That required two tugboats, a large one at the front to pull the barge through the water at the end of a thick steel cable, and a smaller river tug at the rear to provide extra control while they were navigating through narrow shipping channels. In the open sea only the large tug was needed.

Typically the large tug pulled the steel cable at a distance of 1200–2000 feet, far enough in front of the barge so that the cable would sometimes disappear from view below the water's surface. The crew had to be constantly on the lookout for submarines. On several occasions in the past the conning towers of submarines had inadvertently caught the cable of a tugboat pulling a barge, causing the tug to capsize and sink. This hadn't happened to one of the NASA external tank barges, but was nevertheless a safety threat of which they had to be aware.

It took the remainder of that day to load the three carriers on board the barge and to tie them down securely to the vessel with cables and turnbuckles. Bruce Milam divided his team into three shifts so that someone would be on watch at all times until the voyage was completed. At 7:30 the next morning they departed the port, much to the chagrin of Baltimore rush hour commuters whose travel was interrupted when a drawbridge on the Baltimore Beltway, I-695, had to be opened to allow the 80 foot high vessel to pass.

The trip began uneventfully, until the barge and its two tugboats approached the waters of the shipping channel off a peninsula called the Northern Neck of Virginia. There the winds picked up strongly, waves began to stack up until the small river tug at the stern of the barge began to take on significant water washing over its side. The captain of the larger tug allowed the smaller boat to leave to seek shelter. This left him in a difficult situation. The 80 foot-tall cargo enclosure acted like a sail. It, together with the "skeg" structure under the stern, caused the barge to continually veer toward the upwind side of the channel. Every 2–3 min, the tug captain had to steer downwind to pull the barge back toward the center of the channel. This exhausting cycle repeated continually all through the night.

Bruce's wakeup call the following morning was the sound of a loud crash. He leapt out of his bunk and rushed to the porthole in time to see that the barge had just sideswiped a sea buoy and had broken all the buoy's glass lenses. The captain later told him that he was totally exhausted from maneuvering the tug all night and had cut the path by the buoy "a little too close." Thankfully, no harm had befallen the crew or the Hubble cargo as a result.

Departing the Chesapeake, the tug and barge turned south and proceeded down the coast. The seas and winds were calm until, between Jacksonville and Daytona Beach, Florida, thunderstorms and "beam" seas hit them—high waves were hitting at right angles to the sides of the two vessels. Continuously for 12 hours they rolled right and left. Some of the team became seasick. The radio antenna of the tugboat was struck by lightning and emitted green smoke. Fortunately, the tug's crew had some spare parts; they were able to restore communications.

At 1:30 in the morning, they arrived at the first sea buoy marking the channel into Port Canaveral. They stopped in Port Canaveral to rest for a few hours, then proceeded up the Banana River to the KSC turning basin in front of the Media Center. There they docked, unloaded the shipping containers from the barge and moved them in another convoy to a high bay building that had been reserved at KSC for work on the Hubble payload. Putting all the pieces of the SM1 payload together awaited the arrival of the air-ride vans carrying WFPC2, COSTAR, and the other spacecraft components.

The Shuttle team at KSC had developed an ingenious process to load bulky, heavy payloads like ours into the payload bay of an orbiter standing vertically on its launch pad. The Hubble payload—instruments, spacecraft components, protective enclosures and carriers—weighed in at about 24,000 lbs. To transport all that equipment to the launch pad required a payload canister, a long box whose interior duplicated the dimensions and layout of the shuttle payload bay—15 feet across and 60 feet long. The canister had long clamshell doors that were duplicates of the payload bay doors on the shuttle orbiter. The payload was installed into its canister in KSC's Payload Processing Facility in exactly the same way, with the same layout and attachments, as it would be mounted into Endeavour.

The canister plus its Hubble cargo weighed about 135,000 lbs. It required a lot of industrial strength lifting with heavy-duty cranes. First, it got rotated to the vertical position. Then, on October 26, it was transported at about 5 mph to the launch pad on an enormous, self-propelled 48 wheel truck. Endeavour was waiting for it there.

The launch pad contained an enormous service structure that looked as though it had been constructed from a giant-size erector set (Figure 4.6). One part of the service structure was fixed—a tall narrow tower from which workers and the astronaut crew could gain access to the orbiter. The larger part of the structure was movable. It could be rotated to bring it into close proximity to the orbiter's payload bay, allowing access for payload installation. Centered on the rotating service structure was a tall compartment, the payload changeout room (PCR). A massive crane hoisted the canister holding our payload up (Figure 4.7) so that it's clamshell doors fit snuggly into the doorway of the PCR, sealed there by a large, flexible gasket to prevent outside air from leaking in. The PCR doors were then opened, followed by the canister doors, exposing the Hubble components to the clean environment inside. Strong, heavy cantilevered shelves were then extended into the canister. On those the three Hubble carriers could be extracted into the PCR. Afterward, the doors of the canister were closed, the tightly sealed doors of the PCR were closed, and the canister was removed from the pad to be used later for some other shuttle payload.

Figure 4.6. The launch pad with its service structure. The tall tower at the center is fixed in place. It provides access to the orbiter by workers and the astronaut crew. The wider structure to the left rotates to provide direct access to the shuttle's payload bay. Centered on the rotating structure is the white payload changeout room into which the complete Hubble payload would be mounted prior to its insertion into the payload bay. In this view the shuttle stack is being transported to the pad on the heavy crawler. (Credit: NASA/Michael Soluri.)

Figure 4.7. The payload canister containing the entire Hubble payload, protective enclosures and carriers being hoisted into the payload changeout room on the pad. (Credit: NASA/Michael Soluri.)

The next task was to move the rotating service structure into position, so that the PCR was aligned and tightly sealed with those same gaskets against the body of Endeavour. Once again the PCR doors swung open, followed by the orbiter's payload bay doors. Just as the Hubble carriers had been unfastened and pulled out of the payload canister, the reverse process allowed them to be inserted and fastened inside the payload bay itself. In this way, the shuttle team placed Hubble's payload precisely into the correct locations within the payload bay, while it was still being protected by the PCR from the wind, moisture, and sand of the beachfront environment outside—or so we thought.

I'm a firm believer in Murphy's law: If anything can go wrong, it will. My mantra is a corollary to Murphy's law: Nothing is ever simple. This certainly proved true after we had loaded the Hubble SM1 payload into Endeavour's payload bay on Pad 39-A at KSC.

On Sunday, October 31, I got a phone call at home from Joe Rothenberg. "Dave, I'm really sorry to disturb you at home over the weekend, but we've had a problem at KSC, and I need you to get down here ASAP to have a look and tell me what you think." I was on the first flight out of Baltimore to Orlando the next morning. Rothenberg, Cepi, and I attended a 1 pm meeting at KSC where the problem was discussed. There had been a strong windstorm the prior Friday night. Somehow a large amount of sand got into the PCR and all over the Hubble equipment. The tightly sealed and bagged protective enclosures had apparently kept the instruments and other boxes clean. But there were sand particles all over the solar arrays.

Later that afternoon I went with Rothenberg and Cepi to Pad 39-A to see the problem first hand. Standing there in full bunny suit regalia, Joe asked me, "What do you think?" "How big are the particles?" I asked. "About 200 µm seems to be the average." I told him, "This could be bad. That's about the size of the smallest entrance apertures of the two spectrographs on Hubble. If sand particles made their way into Endeavour's payload bay, they could migrate into Hubble's aft shroud while it was open during the EVAs. I can't predict where the particles might end up, but there are certainly very bad places where they could do damage to Hubble—like at those entrance apertures, for example." That's all Rothenberg needed to know.

At a follow-up meeting the next morning, the KSC managers discussed options for removing the Hubble payload from the PCR so that the latter could be cleaned. Some of the sand apparently was left over from sandblasting that had occurred during earlier refurbishment of Pad A. The pad had been vacuum cleaned afterward, but that had been limited by the lack of reach of the vacuuming tools. Some of the sand had remained in the upper reaches of the PCR, but then got blown by the high winds back down onto our payload. Some of the sand was also off the beach. The gaskets sealing the PCR from the outside world were leaky.

Cepollina, speaking for the entire Hubble team, responded harshly. "You guys [the KSC shuttle team] have lost your credibility. The situation is fraught with the temptation for you to say that the PCR is clean, when in fact it might not be. The potential for in-orbit contamination of Hubble is serious. We insist that our payload be removed from the PCR, placed back into the payload canister and returned to the processing facility where we can thoroughly inspect and clean it. We also want to change pads from 39-A to 39-B, after you guys have cleaned it to our satisfaction." The KSC leaders agreed. Endeavour was moved from 39-A to 39-B. The Hubble payload was transported to 39-B on November 14, and was transferred into the orbiter's payload bay on November 19. We were still on track for launch during the first few days of December.

References

Huntress, W. T. Jr 2003, Johnson Space Center Oral History Project, NASA History Office, January 9, 27, et seq

Leckrone, D. S. 1991, ApJ, 377, L37

AAS | IOP Astronomy

Life With Hubble
An insider's view of the world's most famous telescope
David S Leckrone

Chapter 5

To the Rescue: The First Servicing Mission

It was a crystal clear night at the Kennedy Space Center (KSC). A full moon lit up the western sky. A chilly breeze blew in from the Atlantic and across the Turning Pond in front of the KSC Media Center. I was at the Media Center to serve as the scientific public spokesman for the Hubble Project, participating in briefings and answering reporters' questions. A few minutes before 2:30 am several of us stepped outside to look for the Hubble Space Telescope passing overhead. Hubble was launched due east out of Cape Canaveral on 1990 April 24. The latitude of the Cape is about 28.5°N. So, the plane of Hubble's orbit, centered on the center of the Earth, is inclined 28.5° to the plane of the equator. Occasionally the rotation of the Earth, the revolution of Hubble around the Earth, and the slow drift or precession of Hubble's orbit around the globe all conspired to bring the spacecraft back over its original launch site, passing directly overhead on a due-easterly course over KSC. It was on such occasions that shuttle missions to service Hubble could be launched. To catch up with Hubble, the shuttle had to be launched into the same orbital plane as the telescope and a bit behind it. So, here we were, shivering in the chill of the morning of 1993 December 2, a few hours before dawn, watching Hubble pass overhead and anticipating that Shuttle Endeavour would rocket away from Pad 39B in about 2 hr to go catch up with the observatory.

In its payload bay Endeavour held precious cargo for its trip to Hubble: the two new instruments designed to correct the telescope's spherical aberration, WFPC2 and COSTAR; a new set of solar arrays from the European Space Agency (ESA) intended to correct the tendency of the original ESA arrays to flex and jitter; two new sets of gyroscopes to replace failed units on the spacecraft; a computer co-processor to augment the capabilities of the original spacecraft computer; two replacement magnetometers needed to measure Hubble's orientation relative to the Earth's magnetic field; new electronics boxes to provide operational commands and electrical power to the gyroscopes and the solar arrays; a cabling system to enhance electronic redundancy in the Goddard High Resolution Spectrograph (GHRS) and give it a longer life; and new fuses to replace the original units whose reliability had been of concern. This set of payload hardware illustrated the overarching goals of the mission—to restore and improve scientific performance, to replace failed spacecraft components, to upgrade spacecraft capabilities, and to demonstrate the concept of shuttle-based servicing to enable a long and productive lifetime for the Hubble Space Telescope.

Kennedy Space Center was a place of deep contrasts. It was as though "Star Fleet Headquarters" had been built within "Jurassic Park," though there were no dinosaurs. KSC was located on the Merritt Island Wildlife Refuge near Titusville, Florida. The setting was primeval with wetlands, swamps, jungles, sea, and sky. It was home to American alligators, loggerhead and green turtles, osprey and bald eagles, bobcats and Florida panthers, and the West Indian manatee. The parking lots at the Visitors' Center abutted dense jungle. Signs there warned visitors about the poisonous snakes. Co-existing with the gators and the coral snakes were the remarkable technicians, engineers, and managers who worked at Kennedy, their offices, shops, and laboratories nestled a short distance from the Vertical Assembly Building, the Shuttle Processing Facility, the Launch Control Center and a plethora of launch pads, both active and historical. And then there was Kennedy Space Center the "theme park." During the day some areas were overrun with tourists and tour buses. The Visitors' Center charged a steep fee to get visitors onto tours that transported them to special locations within KSC. There was a strong sense of fantasy about the place; this was the closest ordinary people in their daily lives would ever come to "Star Trek."

There I was standing outside at the Press Site looking at a *real* space ship sitting on a *real* launch pad, the *real* Shuttle with its huge rust-colored external tank (ET), loaded with liquid hydrogen and liquid oxygen, and its white Solid Rocket Boosters (SRBs). Powerful spotlights beaming upward into the dark night brilliantly illuminated the scene (Figure 5.1). Endeavour itself was glistening white, its wings outlined in a thin line of black tiles. Delicate little plumes of condensed water vapor and evaporated liquid oxygen misted down the side of the orbiter. A sharply pointed rod required to arrest lightning strikes capped off the gantry, the tallest tower on the pad. The gantry was bejeweled in orange and white lights, glittering like a Christmas tree. A long white arm extended a circular cap over the top of the external tank, sucking vented oxygen gas away from its hull. It was also decorated with lights. The external tank was eerily luminescent with a bright spotlight reflecting off it. To the left of the rocket stack was an ordinary-looking water tower. Just before the main engines and SRBs ignited, it would release hundreds of thousands of gallons of water onto the mobile launcher platform to suppress the intense sound waves that would otherwise be reflected back upward, potentially damaging the orbiter and its payload. A single red blinking light on its top warned errant planes away. A small orange flame emitted at treetop level from a flare stack reminded me of the flames I had seen in oil fields in my youth, burning off methane. In this case the flame was burning excess hydrogen gas, vented from the external tank.

Later that morning, if STS-61, the first shuttle servicing mission to Hubble, successfully got off the pad, then suddenly it would all become extraordinarily real,

Figure 5.1. Launch pad 39A at night prior to a shuttle mission. This is the view from the Media Center looking across the turning pond. (Credit: Wes Meltzer.)

the pinnacle of this hierarchy of nature and human endeavor. Dreams would be converted to a sensory overload of fire and thunder. The fantasy would fade away. The ordinary work-a-day world at KSC would be supplanted for a few minutes by experiences more wondrous than any to be found at Disney World.

The huge countdown clock display sitting on the grounds of the press site indicated T minus 1 hr, 20 min and counting. At that moment the astronaut crew was being assisted into Endeavour's cabin and strapped into their seats. They were shouldering two heavy burdens: to complete the process of redemption of the Hubble Space Telescope and to give a boost to support for the construction of the International Space Station (ISS). The former required working in the harsh environment of low Earth orbit, handling large unwieldy structures, constrained by bulky spacesuits, using specially designed tools and following a complex and carefully planned choreography. So did the latter.

If the STS-61 crew couldn't service Hubble with relative ease, Congress would certainly question NASA's ability to piece together a large, massive, functioning space station. In June of 1993, only six months before STS-61 was scheduled to fly, an amendment to delete space station funding from NASA's authorization bill in the House of Representatives failed to pass by only one vote. So, although the focus of the world's attention that December morning was on a dangerous and complex mission to rescue Hubble, there was even more at stake for NASA, the survival of the human spaceflight program and possibly the survival of the Agency as a whole.

This was a big mission with big objectives, big challenges, and big risks. Seven astronauts were willing to put their lives on the line to make Hubble whole. It was difficult to assess the probability that they might die attempting to do so. In selling the Space Shuttle Program to Congress, NASA had originally estimated the probability of loss of vehicle and crew in a catastrophic accident as 1/100,000—one thousandth of one percent. But Challenger and its seven-person crew had been lost eight years before after only 24 successful shuttle launches. Of the 58 missions prior to STS-61, one had ended in disaster—a very small sample, but sufficient to make us all worry. Each astronaut had concluded that the service they were about to perform for humankind in repairing Hubble was worth that exposure to serious personal danger.

The crew had been training for 17 months for this night and for the 11 days and nights to follow. Now it was time to go fly.

The countdown process began with a "call to stations" for the ground control team 72 hours before the launch window opened. The launch window was the interval of time during which it became possible, following the laws of celestial mechanics, for Endeavour to be launched into an orbit from which it would be able, after a series of rocket engine burns, to catch up to Hubble some 360 miles overhead. On the morning of 1993 December 2 that window opened at 4:26 am and closed at 5:38 am. The launch team had 72 min to get the shuttle off the pad.

The countdown was heavily scripted with built-in periods when the clock was running down and built-in holds when the clock was purposely stopped. During each period, a list of specific tasks and tests had to be completed to bring the entire shuttle and ground support systems steadily up to full readiness for launch. Up to the moment Endeavour, its external fuel tank, and the two solid rocket boosters lifted off the pad the entire countdown and launch process was under the control of the KSC Launch Director, Bob Sieck, and his team at Kennedy. At the moment the shuttle cleared the tower, however, responsibility for the flight would be handed over to Flight Director Milt Heflin and his Mission Control team at the Johnson Space Center in Houston.

The last built-in hold was at $T - 9$ min. At this point the launch team refined its calculations of the opening and closing times of the launch window—precise to one thousandth of a second. The flight data recorders were turned on, and a poll was taken of all the responsible shuttle and payload managers and engineers seated at the launch control consoles—"go or no go?" This included the Hubble Project Manager, Joe Rothenberg, sitting at the "Payloads" console within the cavernous Launch Control Center. Prior to that moment, Joe had checked with his Project Management Team leaders to get their nod that all elements of the observatory—in orbit, within Endeavour's payload bay, and at the Space Telescope Operations Control Center (STOCC) at Goddard—were ready for this critical mission to commence. So, when the call came from Bob Sieck, Joe replied with a brisk "Go!"

The countdown resumed, "$T - 9$ min and counting." It is at this point that my pulse began to race, shortness of breath set in and my mental state became a mix of fear and exhilaration. As the final 9 min were counted down, critical events were checked off in rapid succession. The computer-controlled automatic ground launch sequencer program took over. At $T - 7$ min 30 s the Orbiter access arm through which the crew had walked to climb on board Endeavour was retracted. In an emergency it could be moved back into place in a matter of seconds. At $T - 5$ min the three auxiliary power units (APUs) were started. The APUs provided the power needed to drive the orbiter's hydraulic system. They were critical for moving Endeavour's three main engines to provide thrust in the right direction, for moving the aerodynamic control surfaces (elevons, rudder/speed brake, and body flap), and for deploying the landing gear at the end of the mission. So, they had to work! It was the failure of an APU, discovered around this point in the countdown that delayed the original launch of Hubble on STS-31 in 1990. At $T - 3$ min 55 s the three main engines were moved around on their gimbals and the aerodynamic control surfaces were moved back and forth to assure that they were working properly. At $T - 2$ min Endeavour's Commander Richard Covey and his crew closed and locked the visors on their helmets. At $T - 50$ s Endeavour switched from external to internal electrical power. At $T - 31$ s the ground launch sequencer turned launch implementation over to the internal computers on board Endeavour—"auto sequence start." From this point on, the shuttle was in control of all its critical functions. On its own it could complete the launch sequence, or it could call a halt to the countdown because it had detected an anomaly. The human ground controllers could also intervene manually if they detected that something was amiss. At $T - 16$ s that ordinary-looking water tower next to the pad rapidly began to dump 300,000 gallons of water onto the pad's surface, to suppress by half the roaring 145 dB of sound generated when the main engines and SRBs fired up. The three main engines started in sequence, engine 3 at $T - 6.48$ s, 2 at $T - 6.36$ s, and 1 at $T - 6.06$ s. The stack of orbiter, plus external tank, plus solid rocket boosters rocked forward momentarily and then sprang back fully upright. At $T - 0$ the explosive nuts on four massive bolts holding down each SRB blew apart, freeing the solid-fuel rockets to ignite and propel a total mass of about 4.5 million pounds upward off the pad. If a last second anomaly arose when the main engines ignited at $T - 6$ s, it was still relatively straightforward to turn them off and stop the launch sequence. However, once the SRBs ignited at $T - 0$, there was no turning back. The astronauts were leaving the surface of the Earth no matter what.

I was standing near the bank of the Turning Pond along with hundreds of reporters and NASA people. Video cameras were cranking away and scores of still cameras stood on their tripods at the ready. Behind us about a half dozen major TV news outlets had small huts or trailers in place from which they covered the launch. Their famous space news reporters and other staff stood on the roofs to get a good look. The Press Site was actually the best location from which to watch the launch, about 3 miles from the launch pad, closer than any other visitors' viewing location. Everyone listened intently as KSC public affairs spokesman Bruce Buckingham called out the final seconds of the countdown over the PA system:

$T - 15, ..., 11, 10, 9$, we are go for main engine start

At main engine start, $T - 6.6$ s, enormous quantities of water began flowing onto the mobile launcher, reaching a peak rate of about 15,000 gallons per second. All the while, the burning propellant mixture that produced thrust as it was ejected from the engines also deposited enormous amounts of heat onto the pad. The result was

Figure 5.2. Liftoff of Space Shuttle Endeavour on STS-61, the first servicing mission to the Hubble Space Telescope. The white clouds are steam from the 300,000 gallons of water rapidly dumped onto the pad to suppress the acoustic energy that might otherwise damage the pad and the shuttle. (Credit: NASA, KSC.)

the rapid production of a huge cloud of white steam, looking much like a roiling cumulus cloud, rising up until, for a few moments, it obscured the pad. Through the dense cloud I continued to see the ethereal glow of the main engine exhaust.

..., 5, 4, 3, 2, 1, ...

At $T - 0$ the solids ignited.

and we have liftoff, liftoff of the Space Shuttle Endeavour on an ambitious mission to service the Hubble Space Telescope! (Figure 5.2).

It was as though the Sun had risen two and a half hours prematurely. The exhaust plumes were so intensely bright that they momentarily overwhelmed my dark-adapted vision and left lingering residual bright spots on my retinas. As Endeavour gained altitude the entire sky lit up with an eerie yellow–green–gray glow. It was bright enough to read a newspaper by.

Like thunder and lightning, the sound of the launch reached us well after the sight, some 14 s after. When the sound finally arrived I felt it as much as I heard it— a strong subterranean bass percussive impact on my torso, overlaid with a loud staccato noise. It took my breath away.

As the spaceship gained altitude and distance it came into the sunlight beaming over the horizon. Endeavour became a bright point, trailing the exhaust of its rocket engines. It looked like a comet. At about $T + 2$ min, forty miles high and 50 miles down range, the two SRBs separated from the vehicle, creating another visual astronomical metaphor. It looked like a triple star system. After about 3 min all had disappeared from view.

At this point my job, along with the rest of the Hubble Management Team, was to get to JSC in Houston with all deliberate speed. We would pick up the mission timeline there. Cepi's majordomo, Jimmy Barcus, had arranged for two private NASA planes to fly us to Ellington Air Force Base near JSC. He had organized a team of people to collect our luggage and preload it on the planes, so that we could depart quickly. Jimmy always reminded me of someone straight out of "Hogan's Heros"—extremely resourceful and efficient in arranging all the details of complicated logistics, so that the management and technical teams could focus on their Hubble work. Getting all of us to Houston quickly was business as usual for him.

Forty-five minutes after Endeavour had left the pad, we were "wheels up" off the runway from the Shuttle Landing Facility, the same runway on which the shuttle would land about 12 days hence. As a kid I had dreamt of being an astronaut. That didn't work out. But rolling down Runway 15/33 at the Kennedy Space Center gave me goose bumps. It was in a very small way a shared experience with the men and women who actually did fly rocket ships into space and later came back safely to that very place.

It was sadly ironic that, after everything Frank Cepollina had done over two decades to prepare NASA and the Hubble team for this mission, 2 days after the launch and prior to the first spacewalk his daughter, Paula, died at age 28 from complications related to juvenile diabetes. Jimmy Barcus was still at KSC at the time and so was Cepi's wife, Ann. Barcus took her to Orlando and put her on a flight back to Washington. At the same time Cepi left Houston, and met Ann at Dulles airport.

During the primary phase of the mission, the Goddard Project management team was scheduled to be on duty through the night, roughly from 11:30 pm to 12:30 pm the following day, both in the customer support room (CSR) in Building 30 at JSC and in the STOCC at Goddard. It was during these periods over five days that the EVA astronauts would be out in the payload bay of Endeavour doing their hands-on work on Hubble. Our shift schedule would allow those of us with senior leadership responsibilities to monitor closely what was going on in orbit during the EVAs and to make any decisions necessary if contingencies arose. So, under normal circumstances, Cepi would be sitting with the rest of us at a console in the CSR during each spacewalk. He also had a large group of specialist engineers from his servicing team present to resolve any technical problems that might arise.

Now, at the worst possible time, tragedy had unfolded in Cepi's personal life. Back in Maryland, during each night's management shift, he sat in the STOCC at Goddard with a headset on, or on the telephone, talking to his team back in Houston, doing the work from a distance for which he had been preparing. During the daytime he made funeral arrangements for his daughter.

Endeavour orbited 340 miles above the Earth, below and behind Hubble. The orbiter was traveling faster than Hubble and catching up at a rapid clip. For the first two days after launch the seven-person crew had concentrated on getting some sleep, adapting to weightlessness, inspecting the payload and the robot arm (the Canadian RMS), and preparing spacesuits and tools for the coming repair work on Hubble. But now, on Flight Day 3, it was time to rendezvous—time to catch up to the observatory and capture it. Commander Dick Covey and Pilot Ken Bowersox ("Sox") went through a series of thruster burns to adjust the height and velocity of their spaceship to match exactly that of Hubble.

Astronomer–astronaut and Mission Specialist Jeff Hoffman grabbed a pair of high-powered binoculars and began searching the sky for Hubble. He spotted it as a distant point of light, growing gradually brighter as the shuttle closed in on it. After many minutes had gone by, Hoffman could begin to make out its shape; it started to look like a real object. Then he called out to his colleagues, "Hey guys, I can see the solar arrays and the one on the right is bent over; it looks mangled." These arrays were to be replaced on the second spacewalk during the mission. They had been causing problems ever since Hubble's deployment in 1990, warping and vibrating due to rapid changes in temperature twice each orbit as they moved from darkness to daylight and vice versa. But when Hoffman reported what he saw to us on the ground, some of us were a bit surprised. We hadn't expected those thermal stresses to be strong enough to actually break an array. We had practiced for a contingency where one of the solar arrays wouldn't completely roll up and would have to be jettisoned overboard by an astronaut during an EVA. Now it looked like that preparation might be called into play.

At approximately 2:35 am (EST) on the morning of 1993 December 4, Covey and Bowersox fired the Reaction Control System (RCS) thrusters for the terminal initiation (TI) burn to place Endeavour on its final intercept trajectory to Hubble. At that point Covey took over manual control and maneuvered the orbiter to within 30 feet of the free-flying observatory. Although both Endeavour and Hubble were orbiting the Earth at about 18,000 miles per hour, Covey had nulled out the relative motion between them so that, as seen by the astronauts, Hubble appeared to be standing still, floating over the payload bay. Mission Specialist and Swiss astronomer Claude Nicollier guided the 50 foot long robot arm to grab hold of the grappling fixture mounted on the side of Hubble. He then pulled Hubble down aft-end first and locked it firmly into three clamps on the Flight Support System platform at the back end of the orbiter's payload bay. Hubble, captured and berthed, was no longer a free-flying spacecraft. Commander Covey radioed to the ground, "We have a handshake with Mr Hubble's telescope." All of us on the Hubble team at JSC had been following these events on TV monitors in the customer support room; a euphoric cheer went up when we heard those words.

There was much for the astronaut crew to do to prepare for the first spacewalk the next day. But they couldn't help taking a moment to enjoy the spectacular beauty of Hubble. Its exterior was covered with aluminized insulation that reflected light brilliantly, like a silvery mirror. That silver color coupled with the gold iridescence of the solar arrays made the spacecraft a breathtaking sight in itself. Beyond that, Hubble reflected the hues of its surroundings—the Earth below, blue oceans, orange and brown deserts, green forests, the brilliant reds and yellows of the setting and rising Sun, and the velvet black of the sky. Hubble was constantly changing color. Its mirrored surface reflected sunlight intensely. The shuttle crew, especially the remote arm operators, Claude Nicollier and his backup Ken Bowersox, often were forced to don sunglasses as they looked through the back windows of the cabin toward the blinding gleam of the spacecraft.

As the seven astronauts floated in their cramped quarters three hundred sixty miles above the Earth, a much larger team below—hundreds of engineers, scientists, and managers, divided between Texas and Maryland—was working intensively around the clock directing and supporting the mission. Even though *Hubble* was fastened down firmly in the orbiter's payload bay, it was still an active spacecraft, powered up, accepting commands from the ground and telemetering data back. The engineers in the STOCC at Goddard were still operating it. During the spacewalks to come over the next five days, they would turn the electrical power off on individual boxes before the spacewalkers went to work replacing them. Afterward, they would restore power so that the new units could be tested. All these procedures had been choreographed and practiced many times in advance. Coordination between the STOCC and Mission Control at Johnson could not have been any closer.

In Building 30 at JSC, the historic building from which all the Apollo missions to the moon had been directed, the STS-61 Flight Directors and their large teams of supporting engineers were working in multiple shifts in the large MCC—the Mission Control Center. They were responsible for the well being of the astronauts, the proper functioning of Endeavour, and the successful accomplishment of the shuttle's mission. Only they could communicate directly with the orbiter; in fact that job was usually confined to a single person—the Capsule Communicator or CAPCOM—to keep the lines of communication clear and simple. Their work environment was a scene straight out of the movie "Apollo 13," except with more modern equipment. Long banks of computer consoles and video displays, arranged in several rows, were populated with very professional looking men and women, wearing headsets with microphones, talking quietly among themselves over the communication links. "Payloads, this is EVA on Flight Ops." That was the typical way a Johnson engineer in the MCC responsible for monitoring extra-vehicular activities would use the communication link dedicated to flight operations to hale her counterpart who was responsible for the payload and for coordination with the Hubble payload team from Goddard sitting down the hall in the customer support room (CSR).

For the purposes of this mission the Shuttle Program thought of the Hubble Program as its "customer." Thus, we were ensconced in the smaller CSR. I sat with the Hubble Project Team—Project Manager Joe Rothenberg, Goddard Center Director John Klineberg, and other HST managers and engineers from Goddard and from our contractors. The NASA Headquarters contingent sat at a separate table, including Program Scientist Ed Weiler, his boss, Deputy Associate Administrator Al Diaz, and Hubble Program Manager Ken Ledbetter.

The CSR had its own banks of computer consoles and video displays. Engineers who were experts in various Hubble systems worked there, monitoring the highly detailed flight timeline as various events in orbit unfolded and closely watching the telemetry that indicated how well the spacecraft was working—or not. Any problem would rapidly be detected and the management team would be informed. To keep the line of communications between the MCC and the CSR clear and authoritative, a single individual on each shift, one of the HST engineers, called the Servicing Mission Manager, was designated to be our point of contact over the communication links.

Rothenberg sat at a table close to the engineers. That kept him, and the rest of us within earshot, aware of what was going on. We all could listen to "air-to-ground," the radio channel carrying the voices of the astronauts at work during spacewalks. Sometimes we were lucky enough to be able to watch them on video transmissions, though that was intermittent and had a habit of cutting out at the most critical moments.

Typically problems—big or minor—would arise each day. It was Rothenberg's job to engage his team in discussions about to how to respond to such contingencies. That was the main reason we were there. We had been planning for all the contingencies—anything that might go wrong—that any of us could think of, and practicing how we would respond to them over and over again for many months as an extended Johnson/Goddard/Headquarters team. These exercises in the practice of anxiety were called Joint Integrated Simulations (JISs). They engaged the entire group that would be actively working on the actual mission, including the astronauts working in the shuttle simulator, in highly realistic dress rehearsals. We were as well prepared to handle problems on the fly, as we knew how to be. We had little doubt that bad things would happen during the flight that we hadn't thought of in advance. In such situations we relied on the fact that our team was made up of smart and experienced people. We had learned to be good decision makers.

I was thrilled to be sitting in that historic place, on what was surely an historic mission in the history of space flight, deeply embedded within a large team of the world's smartest space cadets. It was, every day, an edge-of-your-seat experience—a heady blend of worry and elation.

Most of us from the Hubble Project at Goddard and from our numerous contractors had been working together as a "badgeless team," day after day, year after year for a decade or more. We were a pretty well-oiled machine, fully prepared for the challenging technical job that lay before us. But there was a subtle, psychological divide separating us from the NASA Headquarters group. Their job, sitting within the Washington, DC beltway, was to set overarching NASA science policy and manage a large budget that encompassed Hubble and a lot of other NASA flight programs. They managed scientific peer reviews every few years out of which came the selection of new instruments to fly on the observatory. And they provided an interface with the Executive Branch, Congress and the Public.

Our job as a Project was to implement the entire Hubble effort within the constraints of budget, schedule, and policy dictated by Headquarters. Headquarters had minimal involvement in the technical or scientific details. They stood apart and a bit aloof—not really a part of our otherwise cohesive team. So, the question—always unstated—in the back of some minds was, "Who really is in charge here?" Ed Weiler opined on several occasions that, "These missions are not Goddard missions, they are NASA missions, so NASA Headquarters should be in charge." But no one at Headquarters had the experience or knowledge to take on such a role, anymore than the Secretary of the Navy would have the capability to take personal charge of the Pacific Fleet during a major battle. This ambiguity of authority and responsibility would continue to fester and cause conflict between the Hubble Project and NASA Headquarters for years to come.

At 10:46 pm EST on 1993 December 4—2 days, 18 hr, 19 min into the flight (mission elapsed time or MET)—Jeff Hoffman and Story Musgrave opened the airlock hatch and emerged into Endeavour's payload bay. Both had awakened early. They couldn't wait to get started. So, EVA 1 began an hour earlier than planned. The view of Hubble with Earth in the background was glorious. But time was fleeting and they needed to get on with the job. Nominally, they had six hours to complete their work. That was a conservative number. Their air supply, the capacity of the lithium hydroxide canisters in their spacesuits to absorb the carbon dioxide they exhaled, and their other expendables could last 2–3 hr longer, depending on the individual. However, Flight Directors were always stingy in allowing EVA "overtime," as they wanted to assure adequate margins in the expendables to protect the astronauts in case of an emergency.

Story was lucky to be there. He was the Payload Commander, the lead EVA astronaut for the mission. He was a very smart man, with six academic degrees in areas as diverse as medicine (an M.D.), math and statistics, computer programming, chemistry, physiology and biophysics, and literature. But Story had an unusual, somewhat quirky personality that some at JSC and within NASA found off-putting. When I first met him, I especially recall his handshake—soft and light. It made me think that he protected his hands, perhaps as a surgeon would. It was astonishing, then, that he did serious damage to his hands during training for the mission.

Engineers at Goddard and Johnson had been mulling over what orientations Endeavour should maintain in orbit so as to protect the Hubble telescope optics from bright sunlight. One idea was to keep the underside of the orbiter pointing toward the Sun with the payload bay exposed either to the bright Earth or to dark space at all times. However, this would allow the payload bay to get very cold. To test how effectively the crew could work under those conditions, Story suited up and carried some tools into a thermal vacuum chamber to practice. After a while his hands felt cold, then they had no feeling at all. Foolishly he, a physician, ignored these warning signs because he wanted to complete the test. His hands were severely frostbitten. It was bad. His skin turned black.

Musgrave's carelessness angered a lot of people, some of who had axes to grind against him anyway. They tried to get him removed from the STS-61 crew. For Story, however, this episode had a happy ending. He flew to Alaska to seek treatment from a leading frostbite specialist. His hands recovered. Dick Covey, the mission Commander, and the rest of the crew came to his defense and persuaded management that he was too important to mission success to be removed. As it turned out, that ultra-cold orbiter attitude was never adopted for the flight. Other orientations were found that would protect Hubble and keep the payload bay at a tolerable temperature.

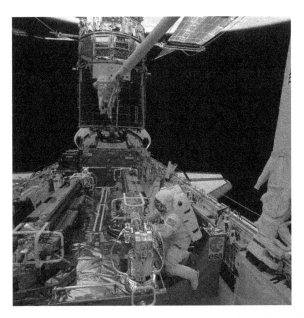

Figure 5.3. Astronauts Musgrave and Hoffman prepare to go to work on EVA 1. (Credit: NASA.)

So here Musgrave was near the end of Flight Day 3, following Hoffman into the vacuum. After completing a series of housekeeping chores to prepare for the tasks at hand (Figure 5.3), they moved to two large doors in Hubble's aft shroud. Hoffman was attached to the end of the remote manipulator arm, being guided by Claude Nicollier inside the cabin. Musgrave was moving about freely using handholds and tethers to assure he didn't float away. They unlatched and opened the two doors. Their first job was to replace two remote sensing units—two boxes, each of which contained a pair of gyroscopes essential to keeping Hubble properly oriented and precisely pointed on the sky. As planned, Hoffman supported Musgrave as the latter carefully maneuvered inside the spacecraft, behind the three large, fragile Fixed Head Star Trackers, to reach, disconnect, and replace the old gyro units. If he bumped one of the star trackers, he could damage it and that would raise big problems later for operating the telescope. He had practiced his choreography many times before and executed it flawlessly. When he was finished Hubble once again had six working gyros—as good as new.

On the ground in the CSR everyone was excited as the first EVA was underway. A cacophony of loud conversations filled the room. Pete Spidalieri, the lead Hubble engineer in the room, sitting at his console attempting to talk to his shuttle team counterpart in the MCC, had had enough. He stood and shouted in a very loud, stern, commanding voice, "Quiet in the CSR. Quiet in the CSR!" I couldn't decide whether to be embarrassed, shamed, or amused by this very conscientious young Goddard engineer shouting at the assembled group of very senior managers and scientists to *"Shut Up!"* as though we were unruly school children. But maintaining discipline in that room, given the seriousness of the job we all were responsible for, was Pete's job as a leader. And he did it well. The room quickly became hushed.

When we launched Hubble in 1990 it was full of gas, very clean air. In the vacuum of space that air dissipated rapidly. Nevertheless, we had to be certain that it was all gone, that a very high vacuum had been achieved, before we could turn on the scientific instruments in orbit. Some of them were operated at a high voltage—up to 12,000 V. If a little residual air was present when their high voltage came on, they might arc like a tiny lightning bolt and short circuit, causing irreparable damage. As a precaution, two of the instruments—the Faint Object Spectrograph and the Faint Object Camera—had built-in pressure gauges. These were sensitive enough to detect local gas pressures down to a millionth of the air pressure at the surface of the Earth. Serendipitously, they could also detect the presence of human beings in spacesuits, sensing the tiny amounts of water vapor their suits vented.

In the CSR we watched the video as Musgrave and Hoffman opened the aft shroud doors and began their work inside to replace the gyroscope packages. One of the engineers on console saw the telemetry signals from the two pressure sensors rise and fall as the astronauts moved about. He printed a hard copy plot of pressure versus time and showed it to some of us. The astronomers sitting nearby, Peter Stockman, Ed Cheng, and I chuckled quietly when we saw the plot. It was rather awesome to see our instruments detect something so non-astronomical as intruding astronauts. But on a serious note we all were a tiny bit paranoid about anything that might contaminate Hubble's delicate parts, including water vapor. Nothing could be done about it, of course. And the risks were infinitesimal. Still we were always prone to worry.

I thought Ed Weiler might like to be "in" on the unexpected observation. He was sitting at the Headquarters table chatting with Trish Pengra. Trish was a contractor employee, trained as a TV producer, who provided most of the knowledge and brainpower behind the public outreach initiative that Ed led. She had a good nose for news and could sense when a Hubble science story would produce good press coverage. Beyond that, Ed often consulted her. She was a good sounding board for him.

I approached their table and said, "Ed, I thought you might like to see this," holding out the pressure graph. "What is it?" he snapped. I explained what the graph showed. "Well what do you expect me to do about it? You're interrupting my conversation!" "I realize that," I replied in an anodyne tone. "Sorry." I was startled. Ed and I had over many years shared a friendly, collegial relationship. His words and tone were out of character.

Al Diaz watched this exchange, his eyebrows raised. A little later he asked Ed to join him on the portico in front of the building—a good chance for Ed to take a cigarette break. I later learned that he had directed Ed to go back to his hotel and get some sleep. Perhaps that was it, a simple case of Ed being stressed and sleep-deprived.

I went back to work, following the progress of EVA 1. A serious problem was brewing.

After finishing the installation of the new gyroscopes, Musgrave had moved on to prepare for the next tasks—installing new electronic control boxes for the gyroscopes, changing out those weak fuses and readying the solar array carrier for the next day's work. He left it to Hoffman to close the two large doors they had opened earlier to get access to the gyro packages. That should have been a simple, one-person job. There were four latches up and down the doors. Jeff had to rotate a handle that engaged all four latches. He then would tighten a bolt that spanned each latch to secure it in place. He rotated the handle and then asked Nicollier to maneuver him down to the bottom of the doors. He tightened the latch bolt there. Next he rode up to the top of the doors. But that top latch wasn't engaged properly. The top of one of the doors was sticking out. "Let's try it the other way," Jeff thought to himself. He undid the bottom latch and started the process again at the top of the doors. But then the bottom latch didn't engage; the bottom of the door was protruding outward. The door seemed to be warped.

These doors had last been opened years before on the ground in one-g at room temperature. Perhaps, Hoffman thought, the harsh thermal environment or the lack of weight in microgravity had deformed what had originally been a straight door. In any event he needed Musgrave's help; this was a two-person job. They needed to push in the top and the bottom of the door and bolt the two latches shut simultaneously. Fixed at the end of the remote arm, Jeff could readily apply force to his end of the door. But Story was a tethered free-flier. He had to grab hold of the massive telescope, push in the door, and secure the latch bolt all at the same time. It couldn't be done with only two hands.

On the ground this sequence of events caused a furor. The Hubble engineers struggled to come up with several approaches for the astronauts to try. None of them worked. Minutes went by—then an hour of precious EVA time. One engineer asked me, "If we can't close all of the latches, which would be better to leave open, the top of the door or the bottom?" There was really no good answer to this question. Either way, this would be devastating for Hubble's ability to do its scientific work. The interior of Hubble, the telescope and instruments, would be exposed to unwanted sunlight or light from the bright Earth. Thermal control, keeping the interior of the spacecraft at a steady, benign room temperature, would be very difficult given this new pathway for heat to leak out into space. Intuitively, I "winged it" and answered, "If you absolutely must leave one latch open, make it the top one." This may well not have been the correct answer. Only detailed stray light and thermal control computer analyses could yield the best answer, if indeed there was a best answer. But there was no time for that. Something had to be done—and soon!

Then Musgrave had a bright idea. On Shuttle missions they regularly fly a contingency tool, a payload restraint device or "Come Along." It's similar to a tool anyone could buy in a hardware store. It's a small hoist with a ratchet and handle. You pull the handle to wind up a cable and the ratchet holds the cable tight, so that it doesn't unwind as you continue to pull. "Come Alongs" are used in construction to pull structures together and hold them in place until they can be permanently fastened. In this case, the shuttle's "Come Along" might work to pull the misaligned doors together and hold them in place while Story bolted the latch shut.

Back at Mission Control and in the CSR this idea produced skepticism and worry. That tool could exert thousands of pounds of force, if it were over-tightened.

The astronauts might well break the doors in the process of trying to close them. After several minutes of heated discussion, it was up to Hubble Project Manager Joe Rothenberg and Flight Director Milt Heflin to make the call. They elected to give it a try, cautioning Musgrave and Hoffman to proceed deliberately, using as little force as possible. They did. And it worked. The aft shroud doors were closed and latched. The rest of EVA 1 went smoothly. It had lasted almost 8 hr—far from the desired 6 hr. But the first day of working on Hubble had been successful. From that moment on, NASA knew for sure that it could do this kind of big-time maintenance and construction work in orbit.

Hoffman and Musgrave returned to the shuttle cabin. Kathy Thornton (better known as "KT") and Tom Akers, who would be the spacewalkers the next day on EVA 2, helped them out of their suits and began preparing themselves for their own upcoming excursion into the vacuum. A command was sent to Hubble from the STOCC at Goddard to roll up the floppy old solar arrays. Thornton and Akers would be removing them in EVA 2 and replacing them with ESAs new and improved version. It surprised no one that the mangled array wing would not roll up properly. It got stuck at the point where its support structure was bent. The ESA solar array manager who was with us in the CRS pleaded with Joe Rothenberg to request that the astronauts manually crank the array to complete its roll up at the beginning of EVA 2, so that it could be returned to the ground. The ESA engineers back at British Aerospace in England would really love to have a close look at it to diagnose why it had failed so badly. But Rothenberg would have none of that. He was worried that an astronaut working so close to a broken and potentially sharp metal structure was at risk of injury or worse from a puncture in their space suit. The need to get the array back home wasn't truly essential; it would be nice to do, but was not required. Also, contingency plans had been developed and rehearsed to dispose of one or both old array wings, if any problems were encountered in rolling them up and stowing them. Joe made the call in consultation with Milt Heflin. EVA 2 would begin with that array being dumped overboard.

At 10:29 pm EST on December 5—MET 3 days, 18 hr, 2 min—Thornton and Akers floated out of Endeavour's airlock and began setting up the payload bay for the day's work. Situated on the end of the remote arm, KT attached a handle to the partially rolled up broken wing and grabbed hold. Tom disconnected the array from Hubble. Claude Nicollier drove KT up as high as he could above and away from the shuttle (Figure 5.4). As the Sun rose, she simply let go of the handle, leaving the array floating above the payload bay. Pilot Ken Bowersox fired the shuttle's maneuvering thrusters and slowly backed the entire vehicle away from the array. As the two separated, the thruster exhaust hit the wing and it began flapping as though it were a giant bird. From the perspective of the TV cameras on Endeavour capturing this image, it looked like the "bird" was flying away from the orbiter, though in fact the opposite was true.

In the CSR we watched on the monitors, totally awestruck by this incredible scene. Later, at a press briefing, I described it as a mythical "Wagnerian" image, that would forever be "burned into my brain." It brought to mind the opera, "Die Walküre" (The Valkyries) from Richard Wagner's Ring Cycle. Here was a woman,

Figure 5.4. Astronaut "KT" Thornton prepares to release the damaged solar array into the void. (Credit: NASA.)

a "Valkyrie," possessing extraordinary powers, flying high above the Earth, releasing a great bird toward the rising Sun. If ever space flight could be described as "Romantic," with a capital "R", this was the moment.

Inside the cabin the crew watched, transfixed, as the solar panel flapped its wings, somersaulted over and over and disappeared into the distance. It would be caught up in Earth's thin upper atmosphere and ultimately de-orbit and burn up. The cabin crew was awakened from this reverie by a voice on the radio. Akers impatiently asked, "Isn't somebody supposed to be reading me the procedures? We have work to do."

Over the next six hours Akers and Thornton mounted a new solar array onto Hubble's side where the damaged wing had resided and plugged its electrical cable into a receptacle. They then removed the second, fully rolled up old panel from the opposite side of Hubble's body, stored it on the solar array carrier, mounted its replacement and plugged it in. A quick test from the STOCC showed that the two new arrays were functioning well. Hubble would have over 4500 W of reliable electrical power at its disposal for years to come, hopefully without the serious jitter problems created by the first generation arrays.

After a day or two we had settled into a rather arduous routine. For 13 hours on the EVA shift we worked with the shuttle mission team during the spacewalks. Then followed a shift handover meeting at which the team, led by Rothenberg, discussed all the day's problems and concerns. We created action plans for the group of people working on the next shift—the planning shift—so that they could get us prepared for the next day, the next EVA. After that the JSC Flight Director, together with a few of us from Headquarters and from Goddard, briefed the press about the events of the EVA and any issues that had arisen. Then it was back to our hotels for a quick supper, a few hours sleep, and "breakfast" (at 9 or 10 at night). Back at JSC the cycle would begin again with an 11:30 pm handover meeting to learn what the planning shift had accomplished while we were away.

Joe Rothenberg didn't follow this routine very faithfully. He couldn't sleep. He was the guy in charge of Hubble, and had a hard time tearing himself away from the action, day or night.

The public and media interest in the mission was vast, global and non-stop. Typically every morning Ed Weiler would be interviewed on the major TV network morning programs, The Today Show, Good Morning America, and CBS This Morning. He was an effective spokesman for Hubble, speaking in language "Joe Sixpack" might understand. Stories appeared every evening on the national network newscasts, usually featuring Weiler and several astronauts not directly involved in the mission.

The Public Affairs people asked me to handle some of the "lesser" interview requests, often in the wee small hours of the morning. That was fun. People across America and in other countries were so intense and excited about what was going on every day during the mission. I loved being able to reach out to them. I enjoyed interviews with veteran space correspondent Bill Harwood and Sharyl Attkisson on *CBS News Up To The Minute* and with Tom Donovan on *NBC Nightside*. The only difficult part of talking in real time on national television was avoiding declaring victory before victory was actually in hand. One key measure of "victory" would be the successful completion of most of the tasks scheduled for the five spacewalks and the safe return of the shuttle crew to the ground. But we still couldn't be certain of ultimate mission success until several more weeks had passed and we had aligned, focused and tried out the corrective optics in WFPC2 and COSTAR. It was very difficult to be so patient for so long.

Astronomers all over the world were awaiting the third spacewalk with high anticipation. EVA 3 was where the rubber would meet the road in so far as correcting Hubble's flawed optical performance was concerned. Which space telescope would we end up with, one that continued to underperform woefully, or one that realized the expectations and dreams of a whole generation of scientists?

On December 6 at 10:34 pm EST—MET 4 days, 18 hr, 8 min—Musgrave and Hoffman exited the airlock and began the work of installing WFPC2 with its specially shaped corrective mirrors.

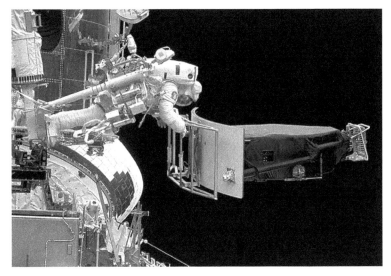

Figure 5.5. Astronaut Jeff Hoffman moves the original WFPC to a holding fixture on the side of the payload bay during EVA 3. It was stored there temporarily while WFPC2 was removed from its protective enclosure and inserted into Hubble through an opening in the side of the aft shroud. (Credit: NASA.)

The original WFPC was not enclosed within the spacecraft like all the other instruments. To acquire crisp, clean pictures with low background noise its CCD light detectors had to operate at a very cold temperature, about −88°C (−126°F). To get so cold without the help of any kind of internal refrigerator, both the original WFPC and WFPC2 had to dump the heat they produced out of the instrument and outside the spacecraft into empty space.

Years before, the designers at JPL had been given a special dispensation. The Hubble spacecraft would be designed with a rectangular hole in its side, just the right size to allow WFPC to be inserted but with a large curved sheet of metal attached to its tail, extending outside of the spacecraft to serve as a radiator. Cooling pipes carried heat from the CCD's inside the cameras to the external radiator, where it radiated away into the cold darkness. Baffles around the edge of the radiator, coupled with baffles around the edge of the rectangular hole, sealed the instrument so that no stray light would make its way from the outside into the aft shroud of the spacecraft.

Now Hoffman and Musgrave had to attach a handhold to the external radiator of the original WFPC, disconnect electrical cables, unbolt the camera from its latches and slowly, carefully slide it out of the Aft Shroud. Recalling the problems they had at Goddard during practice runs inserting WFPC2 into the High Fidelity Mechanical Simulator, the two astronauts slid WFPC back into Hubble again to practice, observing how that felt before trying the procedure for keeps with WFPC2. Claude Nicollier carried Hoffman, holding the old camera, on the remote arm to the sill of the payload bay where he stored it temporarily on a holding fixture they had placed there earlier (Figure 5.5). Then the two spacewalkers opened the WFPC2

Figure 5.6. Astronauts Jeff Hoffman and Story Musgrave prepare to remove WFPC2 from its protective enclosure and insert it into the rectangular opening seen here in the side of Hubble's aft shroud. Note the original WFPC now attached to the sill of the payload bay to the right, out of the way as the astronauts install WFPC2. (Credit: NASA.)

carrier box, attached the handhold to the new camera's radiator panel, unbolted the instrument, and with Nicollier's assistance lifted it out.

The JPL engineers had designed WFPC2 to be inserted sideways through Hubble's aft shroud (Figure 5.6). To allow the camera to see the sky, to capture the telescope's optical beam and relay it into the instrument, a flat pickoff mirror was attached to its front, tilted at an angle of about 45° (see Figure 3.1). If that mirror got bumped out of alignment, the camera wouldn't be able to view the sky; it would become useless. "Fragile, handle with care," was the operative instruction to the astronauts. A metal cage enclosed the pickoff mirror to protect it from accidental bumps. While Hoffman held WFPC2 against his body to keep it stable, free-floater Musgrave detached the cage and pulled it straight out taking great care not to wiggle it and hit the mirror.

In the CSR the whole team was watching this unfold on the TV monitors. Thirty or so people were all holding their collective breath, hoping that the delicate removal of the protective cage wouldn't end in disaster. At the critical moment, just as Musgrave was beginning to pull the cage away from WFPC2, we lost the video feed

Figure 5.7. Astronauts Hoffman and Musgrave ride the RMS arm to the top of the telescope to replace two failed magnetometers on Hubble. (Credit: NASA.)

from orbit. We couldn't see what was happening! A chorus of epithets filled the room.

But Musgrave and Hoffman, who had practiced that step many times on the ground, did it perfectly. Hoffman and Nicollier lifted the new instrument, with its now exposed pickoff mirror, up to the empty cavity where a short time before the old, handicapped WFPC had resided. Musgrave firmly placed his feet in a foot restraint he had mounted on the side of Hubble so that he could apply gentle force to help Hoffman slide the camera onto its guide rails and into the interior of the spacecraft. It slid in effortlessly. Hoffman tightened a single, critical bolt holding the massive, 600 lb box firmly in place. About a half hour later, the ground controllers in the STOCC reported that the new camera was alive electrically and functioning properly. But it would be weeks later before we would learn whether the corrective optics were working well.

The two spacewalkers completed their EVA with a ride on the remote arm to the top of Hubble near the aperture door (Figure 5.7). They needed to replace two failed magnetometers, devices that determined the spacecraft's orientation relative to the

Earth's magnetic field—something like a magnetic compass. Musgrave and Hoffman were as far away from the Shuttle as they could be while still being attached to it. The view was spectacular.

The original magnetometers were not expected ever to fail. So, they could not be removed easily. Instead, Hubble engineers designed new units that could be attached on top of the old ones. While Hoffman and Musgrave were mounting those, they noticed some flaking paint and two loose side panels on the old devices. On the ground engineers began brainstorming how to cobble together protective covers that could be used to enclose the magnetometers and keep the environment around the telescope free of floating particles of paint and other contaminants. If time permitted on the final spacewalk two days later, those would be installed.

Hoffman and Musgrave re-entered the airlock after a spacewalk that had lasted 6 hr and 47 min. As Hoffman later reported (Hoffman 2009), he was elated with their success and with the promise that the new WFPC2 held for astronomy. Musgrave seemed depressed, however. "I feel terrible," he said. "What are you talking about?" Jeff asked incredulously. Story replied, "The placement of the foot restraint wasn't right. I couldn't reach the way I wanted to reach." Hoffman was amazed. "Story, it was a success. We did it—we installed WFPC2!" "Yes, but my plan didn't work right." In Hoffman's own words, "That's just Story. He had planned to do it a certain way and it didn't work quite that way. He was just really upset, not that we didn't celebrate that night; everybody felt really good. I think Story got over it pretty soon. I was surprised because I was so happy. The fact that Story wasn't elated, or at least that there was this other thing that was getting in the way, got my attention. I thought it was strange."

Back in the CSR we were bouncing off the walls with joy—handshaking, back slapping, and a few hugs punctuated the moment. The feeling of exhilaration was intense. Ed Weiler later told reporters, "This is the first day I've been really excited. This is the first day for astronomers."

KT Thornton was a remarkable person, one of a small cadre of female astronauts who thrived in the male-dominated world of human spaceflight. She was an experienced spacewalker, a PhD nuclear physicist, and a mom to three girls. She had been known to bake cookies for school functions at 2 am, while working intensively at JSC in training for her next flight during the day. During the week she functioned as a "single" mom in Houston while her husband, Steve, resided in Charlottesville, working as a physics professor at the University of Virginia. Steve flew back to Houston every weekend to be with his family.

Thornton's EVA partner, Tom Akers was equally impressive in his own way. He graduated as valedictorian of his high school class in the small town of Eminence, Missouri, in 1969. He took a summer job as a Park Ranger in the national forest adjacent to his town, while studying Applied Mathematics at the University of Missouri at Rolla. When he was only 24, Tom became principal of his old high school in Eminence. But he longed for the bigger world. He enlisted in the U.S. Air Force in 1979 and became a test pilot/engineer, working on a number of weapons programs before being selected for the NASA astronaut program in 1987. He brought to the Hubble repair mission spacewalking experience on two previous shuttle flights.

Now it was Thornton's and Akers' job to complete the mandate to NASA voiced by Riccardo Giacconi over three years before, that a repair of Hubble's poor vision should restore the entirety of Hubble science capabilities, not just those that, incidentally, produced beautiful pictures. They had to successfully install COSTAR.

It was now 5 days, 17 hr, 46 min since launch (December 7 at 10:13 pm). The airlock hatch opened. Akers and Thornton stepped into the void. Right away KT began having problems communicating with the crew inside Endeavour's cabin, the IVA (intra-vehicular activities) team. During a spacewalk, the two EVA astronauts don't work in isolation. The five astronauts inside the cabin provide critical support. In addition to operating the remote arm, they follow a step-by-step checklist detailing every move the EVA team is to perform and read this out to them over an intercom. This is important; obviously the two people in spacesuits can't carry a thick book of instructions with them around the payload bay. They need the support of the IVA crew. Without a good intercom link this becomes impossible. After a few minutes spent in trying debug the communications problem, and consulting with Mission Control, the decision was taken to continue with the spacewalk, with all communications relayed to KT by Tom Akers.

Their first big task was to remove the High Speed Photometer (HSP) from its axial bay inside Hubble's aft shroud, to make room for COSTAR. With KT riding the remote arm and Tom free-floating with safety tethers attached, the two proceeded to open two tall aft shroud doors. Claude drove the remote arm to bring KT within her own arms' reach of two handles on the side of the HSP. She slowly pulled the big box out, with Tom watching and giving her directions. He had a better view than she did, as the phone-booth-size instrument blocked her vision. Together they practiced re-inserting it to get a feel for the critical step of inserting COSTAR a few minutes later. They then moved the HSP to its holding fixture on the sill at the side of the payload bay and anchored it there temporarily, while they removed COSTAR from its protective enclosure.

As I watched this unfold from my vantage point in the CSR, I wondered what Bob Bless and his HSP science team were feeling at that moment. The development of the HSP had been an "experiment" in spaceflight project management. Principal Investigator Bless was a professor of astronomy at the University of Wisconsin in Madison. He originally proposed the HSP in 1978 as an inexpensive, breadbox-sized module that he hoped could fly "piggy-back" within one of the much larger new instruments. The instrument concept was modest and scientifically narrowly focused. But that science was potentially very important. For example, the HSP could follow what happened to matter—gas, dust, stars—in the final seconds as they spiraled into and were consumed by a black hole. Exciting stuff!

For the Hubble Project this design idea was a non-starter. All the instruments had to be self-contained within their own enclosure and had to meet a strict set of interface requirements with the spacecraft. Major spaceflight scientific instruments were usually designed and manufactured by large aerospace contractors, such as Ball Aerospace or Lockheed-Martin, or by NASA centers, such as JPL or Goddard. We were curious to find out if a scientifically worthy, well-designed instrument, suitable to fly on a major space astronomy mission like Hubble, could be produced at low cost in-house by an academic group in a university setting. That was the "experiment." So, Bless' "breadbox" was built into the same kind of telephone-booth-size box as Hubble's other, industrial grade axial instruments—FOS, GHRS, and FOC.

Now the HSP had to be sacrificed for the greater good of scientific progress. For 15 years the Wisconsin team had put as much blood, sweat, and tears into their little instrument as everyone else. So, this was surely a sad day for them. But we had done everything we could to give them a lot of observing time on the telescope prior to this mission, to get as much science out of the HSP as they could, limited as we were by spherical aberration.

Nicollier maneuvered the remote arm to allow KT to remove COSTAR from its protective enclosure. He then placed her at the open doorway of the aft shroud (Figure 5.8). Tom Akers helped align the big box onto its guide rails, so that KT could slide it into its latches inside Hubble.

Years later, Mission Commander Dick Covey described the scene (Covey 2007). "That was, to me, one of the great moments in spacewalking. COSTAR weighed about 800 pounds, and Kathy Thornton was the one that was moving that guy

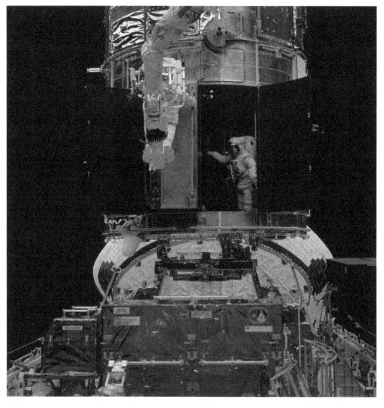

Figure 5.8. RMS arm operator Claude Nicollier drives KT Thornton, carrying COSTAR, to the front of the Hubble axial instrument bay where it will be mounted on guide rails, slid into position behind the telescope, and locked into place. Tom Akers stands at the ready inside the aft shroud to assist KT in lining up the instrument onto its guide rails. (Credit: NASA.)

around. She was the one that had to manually hold it and take it into the telescope. Kathy was the smallest of our spacewalkers and maybe one of the smallest spacewalkers ever, and 800 pounds was nothing to that woman. She did wonderfully with it. It just shows that zero gravity is a great equalizer for people of all sizes, and it was kind of neat seeing her get to do that."

After COSTAR was inserted, engineers at Goddard commanded its tower, its adjustable optical bench, to deploy. That was the structure that held all the coin-size mirrors in place that were designed to correct Hubble's bad eyesight. If something went wrong and the tower got hung up part way through its deployment it would be useless and might block light from getting into the three instruments it was intended to help. This had been a worry of astronomers when the idea of COSTAR was first brought up in 1990. To ameliorate everyone's worries, COSTAR's designers at Ball Aerospace added a crank that an astronaut on an EVA could use to manually retract the tower (see Figure 3.6). As it turned out, the tower deployed properly. KT and Tom didn't have to use the manual safety mode. We were all relieved.

The final job for the day for Tom and KT was the installation of a supplementary computer called a Co-Processor. At the time of Hubble's launch in 1990 the original spacecraft computer, dubbed the DF-224, was already badly out of date, having been designed in the 1970s. Its six memory cards provided a total of 48 kilobytes of data storage capacity. (For comparison, the iPhone that I carry in my pocket as I am writing this in 2019 provides 64 gigabytes of data storage, or more than 1.3 million times the capacity of that original computer.) The Co-Processor had one megabyte of memory and could operate as a stand-alone computer, independent of the DF-224 in case the latter failed entirely. The installation was uneventful, but what followed was a storm within NASA—not a "hurricane," just a brief squall. But we were startled and bemused by its sudden onset.

After the Co-Processor was installed, Hubble engineers in the STOCC followed normal procedures and ran aliveness and functional tests to verify that it worked properly. They loaded data into its memory banks and then read the memory back out. Over the course of several hours the new computer seemed to be having problems. The data read from its memory kept changing, sometimes with 1s arbitrarily changing to 0s and vice versa (bit flips), sometimes dropping out entirely. Something clearly was wrong.

At the Flight Director's press briefing after the shift had ended, Joe Rothenberg reported on the apparent problem with the Co-Processor. He concluded, "So, even if we have to disconnect the Co-Processor and not use it at all, I still think we won't be disappointed with the overall mission."

Word of the anomaly quickly spread to NASA Headquarters in Washington. When Dan Goldin heard about it he reportedly became livid. Rothenberg later told us that Goldin had called the Center Director's Office at Goddard, shouting into the phone something like, "You idiots! How could you have put a piece of crap like that on the Hubble Space Telescope?" In the midst of a mission that had gone nearly perfectly so far, this outburst seemed really out of place. It was hard to take at that moment.

Through the rest of that day and into the evening, our engineers back in the STOCC puzzled over the anomalous data they had been receiving from the

Co-Processor. After a while, they began to notice a pattern, a correlation between the times when they saw spurious readouts from the computer memory and the way the shuttle's antenna was oriented toward the Tracking and Data Relay Satellites up above in geosynchronous orbit. While it was in the payload bay, Hubble sent data to the ground through Endeavour's communications channel, via the TDRSS satellites. Our engineers proved that the problem was not an internal failure of the Co-Processor, but simply a problem with shuttle communications due to imprecise pointing of the high gain antenna toward the communications satellites. The Co-Processor actually was working well.

The fifth and final spacewalk was something of a "mop-up" operation. Musgrave and Hoffman exited the airlock at 10:14 pm EST on December 8 (MET 6 days, 18 hr, 6 min). Musgrave moved to his position, riding on the end of the remote arm. Hoffman was the free-floater for the day. Their tasks included installing an electronics box containing the circuits that drove the movements of Hubble's solar arrays, the Solar Array Drive Electronics or SADE. There was a simple set of cables to be installed on the Goddard High Resolution Spectrograph, a "cross-strapping" kit, to provide redundancy in its electrical power system. Once again they were to ride the remote arm up high to the top of the telescope to install the jury-rigged covers over the magnetometers that Nicollier and Bowersox had fabricated from spare pieces of multi-layer insulation. Those covers would contain any paint chips that broke off the sides of the magnetometers and prevent them from contaminating other parts of Hubble. Lastly, they needed to monitor the opening of the new solar arrays and prepare Endeavour's payload bay for the trip home. All those tasks sounded routine, but in the end they provided more excitement than any of us had bargained for.

For redundancy there were two identical SADE boxes inside an equipment bay about halfway up the length of the spacecraft. Since they were considered unlikely to fail, their design wasn't amenable to EVA servicing. The main problem was that, to remove a box, a space-suited astronaut had to disconnect eight small electrical connectors that were plugged in and held in place by pairs of tiny screws. Removing those connectors proved difficult for Musgrave and Hoffman, struggling with bulky gloves and with the power tools provided for the task that weren't very well suited to the job in the vacuum of space. But with considerable effort they finally managed to get all the connectors unfastened and pulled away from the SADE box. However, another surprise was in store. The 16 screws were not properly held in place or captivated by the connector plugs at the ends of their respective cables. The screws could become loose and float around freely. Indeed some of the screws started unthreading themselves and drifting away. That was a bad situation. If a vagrant loose screw found its way into Hubble, or into a mechanism in the payload bay, one of those that closed the shuttle's doors for example, it could potentially cause serious, even mission-threatening damage.

Story saw one screw floating nearby, grabbed it and put it into his trash bag. Then he saw another, and another. As the bag began to fill with screws freely drifting around inside, he found it difficult to keep them all contained. When he opened the bag to put another screw in, others would float out. He could barely keep up with the swarm of screws. At one point Story grabbed for a screw but didn't catch it. Instead he hit it, propelling it downward toward the floor of the payload bay 25 feet or so below. In microgravity and vacuum there was nothing to slow down that screw. Its momentum simply carried it along at a constant velocity—Newton's first law of motion, and all that.

As the free-floater, Hoffman could cover more distance than Musgrave, attached as the latter was to the arm. Jeff held onto the arm with one hand and reached for the screw with the other. He missed. Bowersox, operating the RMS inside the cabin, said, "Hold on Jeff, I'll move you down on the arm." Unfortunately, the maximum velocity the arm could achieve equaled the velocity of that screw. So, it remained out of Hoffman's reach as he rode the arm downward. Then, Bowersox had a brilliant idea. On the fly, he quickly changed a setting in the computer program that controlled the RMS, essentially tricking it to move faster. In that way Hoffman caught up with that screw, depriving it of its newly found freedom.

In the end, all the screws were rounded up and safely herded into Story's trash bag. For the rest of the mission that episode came to be known as "the great screw chase."

Musgrave and Hoffman's hands-on EVA skills also came in handy when the engineers in the STOCC back at Goddard sent the commands to Hubble to take the first step in deploying the solar arrays—to move the two array masts from their folded positions at the spacecraft's side to their extended positions perpendicular to Hubble's body. One of the masts simply wouldn't obey the command. It sat there motionless. On a free-flying robotic spacecraft, the failure of a major mechanism to deploy an appendage might well mean mission failure. In our case, with these two men in spacesuits at the ready in Endeavour's payload bay, it became a simple matter of asking them to manually crank the solar array mast down into place. No problem.

The remainder of EVA 5 went smoothly. Once again Jeff and Story greatly enjoyed their ride on the RMS up to the magnetometers at the top of the telescope, 45 feet or so above the payload bay. There they encased the magnetometers in their jury-rigged enclosures. The view up there was magnificent. Jeff could briefly let go of the remote arm, though still loosely tethered to it, and feel the exhilaration of being a separate, freely orbiting satellite of the Earth, unconnected to anything.

After 7 hr, 21 min Musgrave and Hoffman returned to the airlock. Theirs was the final spacewalk of the mission. The crew had accomplished everything they had set out to do. They had proven beyond doubt that humans in space were fully capable of performing complex, intricate tasks working with large structures, such as would be required in assembling a space station. The physical work of repairing Hubble was complete and had gone very well. It remained to be seen if the telescope's vision had been fully corrected.

A few minutes before midnight on December 9 the umbilical line between Endeavour and Hubble was disconnected and the observatory went on internal power. Hours before, the solar arrays had successfully been rolled open and moved to face the Sun so that they could charge Hubble's batteries. Some flaky data from one side of a data interface unit, one of four DIU's on board, gave us pause. We asked to delay deploying the telescope briefly while the engineers in the STOCC did some troubleshooting. If one side of the DIU failed, a test run from the STOCC gave us confidence that the redundant second side could be brought successfully on line. Besides, there was no recourse. There was no spare DIU available in the payload. So, Joe Rothenberg gave Milt Heflin a "go" to deploy the telescope.

Claude Nicollier grabbed hold of Hubble's grappling fixture with the robot arm. The three latches holding the spacecraft down on the Flight Support System platform were opened. Claude lifted Hubble up above the payload bay. The telescope's aperture door was opened. At 5:26 on the morning of December 10, 8 days and 59 min after the flight had begun, Nicollier released the RMS's grasp on Hubble. He did that gently, carefully, so as not to impart any motion to the giant observatory. From the ground we watched it hanging there in space, resplendent in brilliant shades of silver and gold, on its own, free of the shuttle. Covey and Bowersox fired Endeavour's small maneuvering jets and piloted the big bird slowly away from the Hubble Space Telescope. As we watched, it slowly grew smaller and faded into the distance.

The customer support room and the Mission Control Center simultaneously burst into boisterous celebration. We had done it. This whole, enormous, brilliant team had pulled off the most complicated shuttle mission up to that time, though we hadn't forgotten that a safe landing back on Earth was still ahead for the shuttle crew. For the most part the EVA tasks had gone smoothly. The astronauts were virtuoso repair people. At that morning's press briefing, I suggested to the media that the seven members of the crew be awarded AAA ratings as space mechanics. A day later the American Automobile Association did exactly that.

The shuttle mission management team from the MCC joined us in the CSR. Hands were shaken; backs were patted; words of congratulations were exchanged. Some hugs were shared. Trish Pengra presented Joe Rothenberg with a small bottle of Crown Royale Canadian whiskey—his favorite. Someone had smuggled cigars and illicit champagne into the CSR. We couldn't light the cigars up, but we clinched them in our teeth. Scores of plastic glasses were filled with champagne and hoisted in toasts to this team, to the mission, to the Hubble Space Telescope and to the fervent hope for a safe return of the shuttle and its brave crew to the ground. Endeavour did indeed land safely on Runway 33 at the Kennedy Space Center at half past midnight on December 13.

Several days later I was back in Maryland, dividing my time between Goddard and the STScI. The reporters who had been covering the mission closely at KSC and JSC were now impatient to hear how everything had worked out; most especially of course whether the optical corrections had been successful. No one was more eager than I to find out. But we had to keep counseling patience. After all, it might take weeks to test the new imaging capabilities and make adjustments in WFPC2 and COSTAR to get the best quality we could.

I had vivid and unpleasant memories of the first time we looked at the original images of stars taken with Hubble in 1990, with the media in the room, and were embarrassed by what we all witnessed. This time we needed to allow ourselves some space, some breathing room to assess what SM1 had accomplished and to do our best to make things work correctly before we had to face the press and the public.

It was also true that Ed Weiler and the NASA Headquarters public affairs group were almost paranoid about preventing the press from putting out anything that even hinted at negative stories about Hubble. We held weekly press telecons, but carefully coordinated in advance what we would and would not talk about. For months prior to SM1 we astronomers had been working on a list of really interesting and beautiful celestial targets that we could observe with WFPC2 and the other COSTAR-improved instruments to show off what Hubble could now do. We had done this also after the observatory's deployment in 1990, though some Hubble veterans thought it ill advised given the spherical aberration embarrassment. We talked in generalities to the media about these SM1 "Early Release Observations," that they might expect to see sometime in the future. But it was verboten to divulge the identities of the specific targets. The reasoning was that, if we decided at the last minute to change targets, or if a particular observation didn't work out as well as we had hoped, we didn't want the press to report that in some way as a negative outcome.

The checkout process went more smoothly than we had ever imagined it could. We purposely waited a week or two to allow any residual gaseous contamination within the newly launched instruments and electronics, or any contamination left behind by the shuttle to dissipate. We couldn't turn on the high voltage light sensors in FOS, GHRS, and FOC until we were certain that this out-gassing process had run its course and there was no longer any risk of damage from electric arcing. So, bringing COSTAR on-line and putting it through some tests had to wait a while. However, there were no such limitations on using WFPC2. It did not contain any sources of high voltage.

It was shortly after midnight on 1993 December 18, eight days after Hubble had been deployed from Endeavour and only five days after the shuttle had landed back at KSC. The "first light" images taken with WFPC2 were scheduled to be read down from the spacecraft. The operations room at the STScI was filled with astronomers, anxiously waiting to get the first glimpse of a test image that would reveal whether or not the spherical aberration correction had worked. Unlike the "first light" debacle of 1990, no press were invited. Only a lone videographer was present to capture the moment. The result was a now oft-replayed video showing WFPC2 Principal Investigator John Trauger and his science team, STScI optical scientist Chris Burrows and Ed Weiler gathered around a monitor as the first image of a small cluster of stars scrolled onto the screen. It was clear that the image quality was good. The group erupted into whoops. Trauger shouted, "Hey, hey, hey." There was relieved laughter, handshakes all around. Weiler famously mouthed, "Holy shit!" A few minutes later Weiler took Trauger aside for a quiet conversation. "Is it really good? Does it really look like the correction worked?" he asked. Trauger reassured him. "It's just what we were looking for."

Over the next few days, the team sharpened the WFPC2 images even more with a tiny adjustment of the telescope's secondary mirror focus position and a small tilt of the instrument's pickoff mirror. Now we were ready to start taking the first dazzling astronomical pictures to show the world.

Over the next week a similar process played out with COSTAR. Its pairs of small mirrors were tilted out into the telescope's light beam. Spacecraft operators turned on the Faint Object Camera's high voltage power so that it could obtain star images through COSTAR's corrective optics. The images were superb.

Now we knew what we had. Hubble was no longer visually challenged. Weiler's public outreach team, together with the STScI public outreach office and astronomers from each instrument science team—began to plan the Early Release Observations (ERO's) in detail and with relish. These would represent our first attempt to show how the universe was supposed to look through a correctly working Hubble Space Telescope.

Each instrument team had submitted a long list of potential ERO's. At regular meetings of the outreach team during the months leading up to the mission, we sat around as a group and discussed the targets. We wanted to select the best two or three from each instrument and evaluate whether there were any technical issues with them, whether they were in a part of the sky that Hubble could observe that time of year, for example. We had three main goals: to demonstrate to the public that Hubble's vision had been successfully corrected; to show what that meant in terms of producing some images that would demonstrate how breathtaking and beautiful the universe is when at last you could see it clearly; and to enable some "quick look" science that would draw the attention of the astronomical community.

Once the technical details of observing each of the candidates had been worked out, with some targets being eliminated by this process, we needed to discuss which targets best met our criteria. Trish Pengra was our most authoritative expert on shaping NASA's message for public consumption. She had a great eye for selecting which images and which astronomical story lines were most newsworthy and would garner the greatest favorable attention. Her input played a major role in guiding the final selection of ERO targets that would be released to the outside world.

With the ERO's and some other demonstration images in hand, we now had the evidence to show the world that the Hubble team, the one that had refused to be defeated, had successfully brought the Hubble Space Telescope up to the standards of performance that we all had counted on going back to the time of Chairman Spitzer's "Little Black Book." We were a few years late, but humanity could now see the universe clearly for the first time, as originally promised. However, we had to wait a bit longer to shout out the news. The Senior Senator from Maryland, on whose support in Congress we strongly depended, wanted to be the first to give the good news to the American public. We were more than happy to wait a week or two to accommodate her schedule and give her that opportunity.

At a press briefing at Goddard on 1994 January 13 Senator Barbara Mikulski held aloft an image of the central region of galaxy M100, showing views before and after the correction of spherical aberration by WFPC2 (Figure 5.9). "The trouble with Hubble is over!" she gleefully proclaimed.

A few moments later Jim Crocker showed a before-and-after image of a star observed with the Faint Object Camera (Figure 5.10). The insertion of COSTAR's mirror pairs had transformed the image from a broad smear with a tiny bright central core to a tightly focused, bright bundle of light. The image core now had

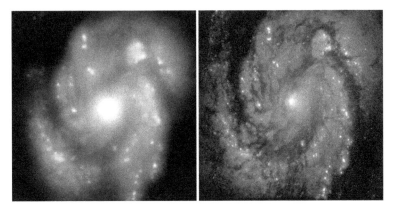

Figure 5.9. The two images of the central region of spiral galaxy M100 shown by Senator Mikulski at the Hubble press briefing on 1994 January 13. The image at the left was obtained with the original WFPC shortly before the launch of SM1. The image at the right was obtained by WFPC2, corrected for spherical aberration, soon after the conclusion of SM1. (Credit: NASA, ESA, the WFPC2 Team, STScI 1994-01.)

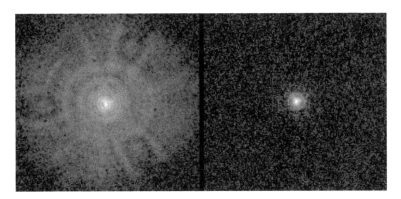

Figure 5.10. Images of the bright star Melnick 34 obtained with the ESA Faint Object Camera without COSTAR to correct spherical aberration (left) and fully corrected with COSTAR (right). Jim Crocker showed this comparison at the Hubble press briefing on 1994 January 13. (Credit: NASA, ESA, the COSTAR Team, STScI 1994-08.)

concentrated within it about 80% of the starlight instead of the original 15%. Crocker announced to the assembled media, "The quality of Hubble images is now the best that engineering can achieve and the laws of physics allow."

References

Covey, R. O. 2007, Johnson Space Center Oral History Project, NASA History Office, March 28, 15
Hoffman, J. 2009, Johnson Space Center Oral History Project, NASA History Office, November 12, 30

AAS | IOP Astronomy

Life With Hubble
An insider's view of the world's most famous telescope
David S Leckrone

Chapter 6

Transfiguration: Life Without Spherical Aberration

It was mid-January of 1994, 1 month after completion of the Hubble Servicing Mission, SM1. The big winter meeting of the American Astronomical Society was underway in Crystal City, Virginia, just across the Potomac from Washington. I was standing in a large foyer outside one of the auditoriums where new scientific results were being presented in plenary talks to the several thousand assembled astronomers. I was caught off guard when a strong pair of arms enclosed me from behind in a massive bear hug and I was lifted a few inches off the ground. Turning, I saw a beaming John Bahcall.

John was one of the founding fathers of the Hubble Space Telescope. He had championed it from its infancy to the scientific community and to NASA. During the 1970s he had spent more time than anyone else on Capitol Hill, patiently explaining to key members of Congress why the Large Space Telescope was critically important to the advancement of human knowledge and why it should be funded by American taxpayers. John said, "What is at stake is not just a piece of stellar technology but our commitment to the most fundamental human quest: understanding the cosmos." Once, when he was asked in an interview why the funds needed for the telescope shouldn't instead be devoted to cancer research, he said that both were extremely important and that there were ample funds in the federal budget to support both. John's tireless effort had been seminal in making the dream of the Hubble Space Telescope a reality. Now he and the rest of the astronomical community were poised to see those decades of effort bear fruit at last. On that day in January he was a happy fellow.

Bahcall could be very emotional at times. Three years before at a meeting of the Hubble Science Working Group at Goddard, he left many of us shaking our heads as he railed angrily against NASA's decision to release early observations from this badly flawed telescope to the public; it was just too embarrassing for those like him who had advocated Hubble's virtues in public for so many years. He punctuated his outburst by storming out of the room, loudly slamming the door in the process.

Now things were very different. John had every justification for being elated. At this meeting he was to be presented the prestigious Dannie Heineman Prize in Astrophysics, jointly awarded by the American Astronomical Society and the American Physical Society, primarily for his studies of neutrinos, nearly massless subatomic particles, coming from the Sun. John was an active researcher with the HST, focusing on the properties of the gas lying between galaxies, the intergalactic medium, as well as quasars, the quasi-stellar radio sources. It's hard to imagine a research topic in astrophysics more far afield from Hubble than solar neutrinos. But John was a polymath who followed his curiosity no matter where it led him. And today there was icing on John's cake. This was the first conference where scientific results from the optically corrected Hubble Space Telescope were to be reported. At last, after all these years of leading the charge in support of Hubble and then bearing the frustration of the telescope's spherical aberration, John could finally take pride publically in what had been and would be accomplished with the telescope.

This meeting of the Society was a landmark in the history of astronomy for another reason also. Here, the first conclusive observational evidence in support of the Big Bang theory for the origin of the universe was to be unveiled.

John Mather, my colleague at Goddard, presented an invited lecture revealing the final results from his instrument on NASA's Cosmic Background Explorer (COBE) satellite. John was the Lead Scientist on that mission and the Principal Investigator of the FIRAS (Far-Infrared Absolute Spectrophotometer) instrument designed to measure the spectrum of the cosmic microwave background radiation. He showed a single graph of measured data, a plot of the brightness of the cosmic background radiation over a range of microwave frequencies. With exquisite precision the curve exactly fit the theoretical emission from a perfect radiator, a black body with a temperature of 2.725°C above absolute zero (physicists refer to this as 2.725 K or Kelvin).

Previous attempts to measure the cosmic microwave background from balloons or rockets had yielded noisy data and puzzling results. Now we had the answer with clarity and certainty—the universe did indeed originate in a Big Bang.

The Big Bang theory made a very specific prediction about the existence and intensity of the relic radiation it left behind. As measured at the Earth, after a passage through the expanding universe that began when the universe was only about 300,000 yr old, the radiation should be that of a perfect black body radiator at a temperature of around 3 K. Mather and his team had verified that prediction with breathtaking observational precision. Ideally this is how science is supposed to work. And this was science that addressed the most fundamental of questions—where did we, and everything else in the universe, come from? The audience stood as one and cheered. I had never seen such a display before. Scientists tend to be more reserved. But this was cause for unbuttoned jubilation. Twelve years later Mather and fellow COBE Principal Investigator George Smoot would be awarded the Nobel Prize in Physics for this discovery.

A second major scientific milestone was the revelation that the Hubble Space Telescope had been successfully repaired—its faulty eyesight fully corrected. The servicing mission flown a month before had been 100% successful. We had withheld the details of Hubble's superb new performance until Senator Mikulski could take the lead in announcing it to the public at the press conference at Goddard on Thursday, January 13. We had to accommodate her busy schedule and keep the results under wraps, while at the same time encouraging the Hubble science teams to get ready to give scientific papers showing their new Hubble observations at this AAS meeting that had started two days before. For a couple of days the Hubble astronomers attending the meeting were supposed to bite their tongues and not share what was the greatest news of their professional careers with their close friends and colleagues. This required a lot of self-restraint, and for the most part they did a good job of it. Now, the pleasure of revealing how clearly we could see the universe through the eyes of Hubble was especially sweet. The word "transfiguration," a complete change of form or appearance into a more beautiful or spiritual state, seemed apt to describe the exquisite quality of the new Hubble data.

Beginning the afternoon after Ms Mikulski had exclaimed to the media that *"the trouble with Hubble is over"*, the wraps came off. We held a media briefing at the conference to reveal the Early Release Observations (EROs) taken over the prior three weeks. The Hubble astronomers presented those same observations to their colleagues in more technical poster papers displayed among hundreds of other papers in an enormous room reserved for displays and exhibits. Hubble observations now fit gracefully into the huge show without any of our previous reticence. The EROs had been designed primarily to showcase Hubble's wonderful resolution. The new images were crystal clear, sharp, colorful, and evocatively beautiful. In addition to the central region of the beautiful, face-on spiral galaxy M100 shown by Senator Mikulski at the press briefing (Figure 5.9), there was exquisite detail in the WFPC2 image of the Orion Nebula (Figure 6.1) and in the eruptive binary pair of stars, eta Carinae (Figure 6.2).

One of my personal favorites was a Faint Object Camera + COSTAR image of the densely packed core of the globular star cluster, 47 Tucanae. Globular clusters are like great spherical lanterns in the sky, ablaze with the light of hundreds of thousands or millions of stars—big and little; hot and cool; blue, yellow, red. Observed from the ground, their central regions are so densely packed with stars that they usually appear as burned out, amorphous blobs of light in astronomical images (Figure 6.3). Now, at last with Hubble, astronomers could resolve each individual star and precisely measure its brightness and color (Figure 6.4). This gives a very direct measure of how old the cluster is and how stars of various masses have evolved as they have consumed their nuclear fuel over their lifetime—a laboratory for the study of stellar evolution by our remarkably keen-eyed telescope.

John Mather had used COBE to observe the first light from the birth of the universe. With Hubble we now had the means to probe in great detail how the universe developed and evolved in the eons that followed. It seemed that everywhere we looked on the sky with the Hubble Space Telescope, we now saw wondrous

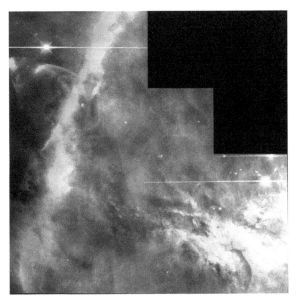

Figure 6.1. SM1 Early Release Observation of a portion of the Great Nebula in Orion taken with WFPC2. The nebula is a giant cloud of gas and dust, illuminated by a small number of massive, hot stars. It is a stellar nursery in which new stars are regularly being formed. Astronomer C. Robert O'Dell first discovered the small round or elliptical nodules in this and other Hubble images. He called them "Proplyds," or protoplanetary disks, and indeed they contain young stars surrounded by disks of gas and dust from which young planets are undoubtedly being formed. (Credit: NASA, ESA, C. R. O'Dell, STScI 1994-10.)

things. For the first time in history we humans had a clear view of the universe. We collectively stood in awe of the beauty and truth that it revealed.

A few days later, near the close of the AAS meeting, Robert Smith, the professional historian, who had been allowed to work "inside" the Hubble project to document the observatory's development from the beginning, approached me. Smith made an astute observation. The enormous and expensive Superconducting Super Collider particle accelerator project had been canceled by Congress just three months before. Robert said, "The difference between the Hubble Space Telescope and the SSC was that you [the HST program] explained the importance of the science you planned to do with modesty and humor. The public understood and supported you. The SSC team did a very poor job of that. The average man on the street had no understanding of why it was important to find the Higgs boson. So, they failed, and now you guys have succeeded spectacularly." He was right. Ed Weiler and our HST Public Outreach Team were responsible for that and deserved full credit for maintaining enthusiastic public support for Hubble until its optical problem could be corrected.

In the following five chapters I'm going to tell some remarkable stories of scientific campaigns undertaken by astronomers to exploit the sharp vision of the Hubble telescope on the heels of SM1: the startling, serendipitous, once-in-a-millennium spectacle of the 21 fragments of Comet Shoemaker/Levy 9 crashing into the atmosphere of Jupiter; the first indisputable proof that a super-massive

Figure 6.2. SM1 Early Release Observation of the highly unstable binary star system eta Carinae, observed with the Planetary Camera mode of WFPC2, fully corrected for spherical aberration. The system dramatically erupts from time to time, ejecting the material seen here, and on at least one occasion in the nineteenth century, becoming the second brightest star in the sky. (Credit: NASA, ESA, J. Hester, STScI 1994-09.)

Figure 6.3. The globular star cluster 47 Tucanae observed from the ground. The core of the cluster is so densely packed with stars, that it appears here as an unresolved blob of light. (Credit: AURA/UKSTU/AAO, Digitized Sky Survey, STScI 2013-25.)

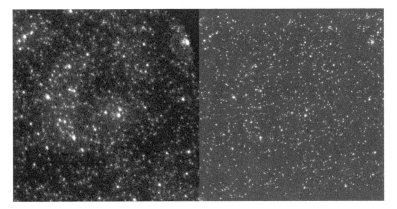

Figure 6.4. Images of the central part of the core of 47 Tucanae observed with the ESA Faint Object Camera. The image at the left was taken without COSTAR and shows the smeared out, overlapping, aberrated images of a multitude of stars. The image at the right was taken with COSTAR's corrective mirrors inserted into the telescope's light beam and the spherical aberration fully corrected. (Credit: NASA, ESA, R. Jedrzejewski, STScI 1994-11.)

black hole lies at the center of almost every galaxy and is the hungry monster that powers quasars; the precise measurement of the rate of expansion and the age of the universe; and the first clear view of the universe as it appeared long ago and far away. The science was revolutionary and it took Hubble to do it.

These are also compelling stories of human struggles, struggles to wrestle the truth out of Mother Nature and social struggles within a community of professional scientists. People often typecast scientists as unemotional, rational, coldly logical beings, something like Star Trek's Mr Spock. But scientists are not Vulcans. They are *Homo sapiens*. They are extensively trained to carefully observe nature and to rationally interpret what they observe on the basis of what they know about physics and mathematics. On the other hand they are often proud, passionate, quarrelsome, prejudiced, and possessive people. Those qualities do provide an engine of sorts for scientific progress. Their propensity to argue, challenge each other, and defend as strongly as they can their own points of view frequently leads to a clearer understanding and sometimes to a grudging acceptance of truth.

Life With Hubble
An insider's view of the world's most famous telescope
David S Leckrone

Chapter 7

Once in a Thousand Years: Preparing for Comet S/L-9

It was a clump of ice, rock, and dust, the nucleus of a comet, probably a mile across, a primal relic left over from the formation of the solar system 4.6 billion years ago. Over the eons its orbit around the Sun had settled into a location spanning the region between the asteroid belt beyond Mars and the orbit of Jupiter, the Sun's largest planet. Sometime, probably in the 1920s or 1930s, the comet came a little too close to Jupiter and was swept into a highly elongated orbit around the massive body. It took the comet about two years to complete one circuit around the planet. Caught in a gravitational tug of war between the Sun's gravity and Jupiter's, the comet's orbit was unstable and changed over time.

In July of 1992 it reached a point of closest approach, only 25,000 miles above the cloud tops of the planet. Here the force of Jupiter's gravity became irresistible, and what had once been a single comet nucleus fractured into about ten to twelve individual pieces, each of which became a smaller comet in its own right. As time went on, these fragments continued to break apart into smaller pieces, ultimately becoming a train of about 25 individual mini-comets. A few of them were very tenuous and eventually faded away. The remaining fragments would only pass by Jupiter once more in their lifetime, their orbit taking them too close to the massive planet, whose gravity would inevitably pull them into a collision course and destroy them. It would be a spectacular astronomical event that might occur once in a thousand years. One small group of dedicated scientists would make sure it didn't go unseen.

On 1993 March 24 astronomers Eugene and Carolyn Shoemaker and their frequent collaborator, amateur astronomer David Levy, were doing what they had done so many times before: searching for slowly moving objects in the sky—asteroids and comets in our solar system. Once a month the Shoemakers traveled to Palomar Mountain in southern California to observe with the 18 inch Schmidt telescope at the Palomar Observatory. Schmidt telescopes were designed to take undistorted wide-angle pictures. They were efficient machines for surveying large swaths of sky in a short time. The 18 inch Schmidt was the oldest telescope on the mountain, dating back to 1936. Still, it had performed well for the Shoemakers, who used it to discover 32 comets and hundreds of asteroids over the years.

Gene Shoemaker had designed an effective tool to make it easy to spot comets and asteroids. After pointing the telescope toward the desired location in the sky, he would take two exposures, separated in time by about 45 min on specially sensitized photographic film. In that time, a comet or asteroid would be seen to move slightly from one exposure to the other, while the background stars would, of course, remain fixed. This small change in position of an object in the picture is akin to the small displacement of images projected on the screen in a 3D movie with each image observed by only one eye through 3D glasses. When the two images are combined and processed in the viewer's brain, the result is an illusion of depth and dimensionality. Gene Shoemaker's stereomicroscope allowed him to observe the two pictures simultaneously. Each image was seen by only one eye. When the brain processed the two, any moving object, displaced by a small angle, would appear to "float" above the flat background of distant stars.

On this particular night, the team of comet hunters was battling clouds, an approaching storm, and a shortage of film, much of their stock having been accidentally ruined when someone had opened the box and exposed it to light. Nevertheless, they had a few pieces of usable film and an occasional break in the clouds through which they could collect a small number of pictures before the storm moved in. One of their fields contained a burned out image of ultra-bright Jupiter. As Carolyn Shoemaker scanned the two pictures with the stereomicroscope, she passed by a puzzling object. She later said that at first glimpse it could have been an edge-on galaxy. But that wasn't possible, because it appeared to "float" above the flat background. It had to be in the solar system, and it had to be moving. She went back to have another look. The object looked like a long bar of light with multiple tails. "It looks like a squashed comet," she exclaimed.

Astronomers, both amateur and professional, who discover transient phenomena in the sky, like comets, asteroids, or supernovae (exploding stars that brighten rapidly), follow a well-known protocol. As soon as they are reasonably confident that their observations are sound, they report them to the Central Bureau for Astronomical Telegrams of the International Astronomical Union (IAU). This allows an IAU Circular to be distributed rapidly to the entire astronomical community around the world, so that other astronomers can quickly get to work at their telescopes, both to confirm the phenomenon and to perform follow-up studies. The Director of the Central Bureau and Editor of the IAU Circulars at that time was Brian Marsden of the Smithsonian Astrophysical Observatory in Cambridge, Massachusetts. Marsden was also a leading expert on the calculation of the orbits of comets and asteroids within the solar system.

The Shoemakers and David Levy emailed news of their discovery to Marsden. He suggested that they contact a couple of other observers who could home in on the object with larger telescopes. In particular they wanted Jim Scotti, who was then observing with the 36 inch Spacewatch camera at Kitt Peak Observatory near

Tucson, Arizona to have a look. Marsden reported the outcome in IAU Circular No. 5725, on 1993 March 26, stating that Scotti had confirmed the appearance of the comet. Marsden named the object Comet Shoemaker–Levy (1993e). It would later be known as P (for periodic)/Shoemaker–Levy 9 or S/L-9 for short.

Over the next few days, observations of the comet came in from other observers. The most significant and evocative of these was from Jane Luu and David Jewitt, who used the 88 inch (2.2 m) University of Hawaii telescope on Mauna Kea and a large format CCD light-sensing detector to take a spectacular image of the comet. They wrote, "We resolved 17 separate sub-nuclei strung out like pearls on a string."

As the weeks went by following the discovery, astronomers from around the world made many measurements of the changing position of the center of the "string of pearls" as it moved along its orbit. But what was its orbit and what relation, if any, did it bear to nearby Jupiter? Brian Marsden used these incoming reports to continuously refine his calculation of the orbital parameters. Near the end of April, he was convinced. The comet train was in orbit around Jupiter. Moreover, his calculations, and those of other experts, showed that, on 1992 July 7, the comet (or its original parent body) had passed very close to the planet, close enough that the tidal forces of Jupiter's immense gravity must have torn it apart. But Marsden drew an even more startling conclusion from his calculations. The "string of pearls" was on its last pass around Jupiter.

The point of closest approach to Jupiter during the current orbit was computed to occur at less than Jupiter's radius. The twenty or more mini-comets that had been left after the parent comet had broken apart during its prior closest approach in July of 1992 were destined to crash through the cloud tops and penetrate the atmosphere of Jupiter in mid-July of 1994. The comet fragments would not survive their next pass around Jupiter.

The prospects of actually witnessing, for the first time in recorded history, a major collision of bodies in the solar system, at a predicted specific time 14 months in the future, with plenty of time to prepare observing campaigns at observatories around the world, galvanized the international community of astronomers. This was a thrilling and unique opportunity to view, on a relatively small scale, how the solar system was made.

Collisions between planets and asteroids in the early epochs of the formation of the solar system had played a major role in shaping the worlds we observe today. It was the process of two celestial bodies colliding into one another that tore a giant chunk from the early Earth and created our Moon, for example. Several periods of major collisions of asteroids and comets onto the young, nascent Earth around four billion years ago had undoubtedly impeded the formation of the earliest microbial life forms, perhaps inducing several episodes of mass extinction before life finally became firmly established. The dinosaurs were likely eradicated by an asteroid strike 65 million years ago. The largest recorded impact on the Earth's surface took place two billion years ago, in what is now South Africa, resulting in a crater 380 km (236 miles) in diameter. That's an expanse larger than the distance between Seattle and Portland. At the same time, impacts from comets or asteroids weren't always apocalyptic events. It is likely that such strikes on the early Earth delivered most of the water to our future home planet.

Clearly, the impacts of the 21 "pearls on a string" of S/L-9 would be an event of considerably smaller scale than those that took place in the early solar system. Still, it would give scientists a once-in-a-millennium opportunity to observe the collision process as it was happening. No one could predict with certainty what, if anything would be seen during the impacts. But potentially there was knowledge to be gained both about the comet's physical nature—its structure and makeup—and about the gas giant Jupiter's atmosphere. Mother Nature would be conducting an experiment in real time, and solar system experts across the globe wanted to be ready to collect and interpret whatever data they could.

At the Space Telescope Science Institute (STScI) in Baltimore, astronomer Hal Weaver was following the unfolding story of S/L-9. In the decade since he had completed his PhD in physics at Johns Hopkins, Hal had established himself as one of the world's leading authorities on comets, particularly on observing comets from above the atmosphere.

In 1986, Hal Weaver went to work on Hubble, helping to develop its capabilities to track the motion of objects in the solar system. Precise tracking was necessary so that Hubble's cameras and spectrographs could stay locked on planets, satellites, asteroids, and comets while collecting data.

Hal was the right person at the right place when news of the Shoemaker and Levy's discovery arrived. He phoned Gene Shoemaker to have a more detailed discussion about the fragmented comet and then arranged to have Hubble take images of it with the Wide Field and Planetary Camera (WFPC). Those were acquired on 1993 July 1.

Though Hubble still suffered from spherical aberration and the associated blurred images at the time, Hal's first pictures were nevertheless sharper than any obtained from the ground. They showed a sharp central point of light near the center of each fragment. This was not the nucleus, which was too small to be seen even by Hubble, but the atmosphere or "coma" of dust and gas that surrounded each nucleus and glowed by reflected or scattered sunlight. No one knew how big each fragment was, or how big the original parent body had been. But from the WFPC pictures, Hal could estimate a rough upper limit: the fragments were probably no bigger than 3 miles (5 km) across.

Later in July in the torrid heat of Tucson, Arizona, Weaver attended an ad hoc meeting of astronomers from around the world who were excited by the possibilities of what might happen a year later. They needed to discuss what was known and not known about this comet. They needed to plan and coordinate the various observing programs that each of them might undertake. What became clear at the conference was how little was really known about S/L-9.

Estimates of how much energy would be released in Jupiter's atmosphere during the impacts varied widely. But those present seemed to think a conservative estimate would be the equivalent of tens of millions of megatons of TNT—thousands of times the energy equivalent of the Earth's entire arsenal of nuclear weapons at the height of the cold war.

How rigid were the fragments? Would they hold together as they approached closer to Jupiter, or would they crumble further into clouds of small particles? In that case most probably nothing would be seen from Earth as fragments collided with Jupiter. A scenario in which the impacts would simply fizzle, presented significant risks to the Director of the STScI and to the Hubble Project. If a lot of precious observing time on Hubble were allocated to observing the S/L-9 impacts and nothing very interesting was seen, the wasted resources could produce a scandal within the astronomical community and potentially another black eye for NASA following the spherical aberration debacle.

Despite the risks, the opportunity to see an event so rare was impossible to pass up. The overall probability of a comet being captured into orbit by Jupiter's gravity, remaining in orbit given all the pushes and pulls of other forces acting on it, passing close enough to Jupiter to be pulled apart into many small pieces, and then actually colliding with the planet, was very small—an event that might happen once in a thousand years. Humanity was remarkably lucky to be well prepared to observe the collisions with the remarkable tools of modern astronomy led by the Hubble Space Telescope. A servicing mission to correct Hubble's bad eyesight was planned for December of 1993. The timing of the mission was, of course, coincidental to the appearance of S/L-9. But, if it were successful, astronomers would have the great fortune to focus humanity's sharpest vision ever on this cosmic event. The scientific potential was too great to ignore, so that both NASA and the National Science Foundation reprogrammed $750,000 each to support research on the comet.

In the early 1990s, Heidi Hammel was part of an emerging wave of brilliant young women scientists who were steadily breaking down the ancient social barriers that inhibited the full and equal participation of women in the male-dominated physical sciences—especially physics and astronomy—in the United States. Heidi got her undergraduate degree in Planetary Science at MIT in 1982. She studied at the University of Hawaii for her PhD in Physics and Astronomy, which was awarded in 1988. Her dissertation research was based on observations of Neptune and Uranus obtained at the University's 2.2 m (88 inch) telescope atop the extinct Mauna Kea volcano. Afterward, she took a post-doctoral position at the Jet Propulsion Laboratory, where she worked on planning the encounter of Voyager 2 with Neptune in August of 1989.

During the "cruise phase" of the Voyager mission leading up to the flyby, Heidi and her colleague, Candy Hansen, obtained images of the distant planet, observed its rotation and cloud features, and generally did the scientific preparations that laid the groundwork for Voyager's short stay in Neptune's vicinity.

Among astronomers, Neptune and Uranus traditionally had the reputation of being "dull" places, appearing through a telescope to be featureless and generally uninteresting. However, Heidi's observations for her PhD thesis showed that this was far from the truth. Neptune's atmosphere did have clouds and dark spots that changed with time, proving that it was a far more dynamic place than people had ever imagined. Heidi and her colleagues discovered Neptune's Great Dark Spot, a furious storm the size of the planet Earth, a tempest with winds traveling in excess of

1500 miles per hour, the fastest winds ever recorded in the solar system. The furthest planet in our solar system, Heidi showed, was far from dull.

As Voyager 2 sped closer to Neptune in 1989, Brad Smith, the leader of the Voyager imaging team, approached Heidi with a request. The new images to be obtained with the spacecraft needed to be placed in context. The mission needed a set of Earth-based observations to which the Voyager images could be compared to provide ground truth. The team needed someone with deep knowledge of the planet, and experience observing it, to obtain images of Neptune with a large mountaintop telescope before, during, and after the encounter. Heidi was that expert. So, she trudged back to Mauna Kea, observing from there, essentially alone, as all the excitement at JPL erupted when the spacecraft hurtled by Neptune.

Having done so much work with Voyager over many months as it cruised toward Neptune, she missed the emotional payoff—the thrill of being there among the close-knit team of engineers and scientists at the moment of their success. Thousands of miles away on a mountaintop, Heidi couldn't experience the intoxicating rush that the rest of her teammates enjoyed when Voyager flew only 3000 miles above the north pole of Neptune on 1989 August 25. She had gone to Mauna Kea because it was important to support the team. But in return she was left feeling isolated and disappointed. Five years later, Mother Nature would provide Heidi with karmic justice as the 21 fragments of S/L-9 crashed into Jupiter.

In 1990, Heidi returned to MIT as a Principal Research Scientist in the Department of Earth, Atmospheric, and Planetary Science. One afternoon in May of 1993, a graduate student in her department, Joe Harrington, came running down the hall exclaiming, "Have you heard? A comet is going to crash into Jupiter!" Heidi simply shrugged her shoulders and said, "So what? Jupiter is huge. Comets are tiny. Nothing's going to happen."

Not everyone felt the same way. In 1993 September the newly appointed STScI Director, Bob Williams, issued a special call to the community, inviting astronomers from around the world to submit scientific proposals for participation in the comet campaign. It took courage on Williams' part to do this, to allocate about 100 orbits of precious Hubble observing time to a program that might produce only negative results. If he were wrong, senior members of the astronomical community would certainly come down on his head for wasting so much Hubble time. But his scientific judgment demanded that the risk be taken for such a rare opportunity offered up by Mother Nature.

The Principal Investigators on the best proposals would be selected to serve on a small Science Observing Team (SOT). It would be the job of the SOT to formulate the detailed observing strategy and reach a consensus on scientific priorities for the Hubble S/L-9 campaign.

Back at MIT, Tim Dowling, a theorist whose specialty was modeling the atmospheres of Earth and other planets, grew excited about what might happen to Jupiter's atmosphere when the fragments of S/L-9 hit. His calculations indicated that, following a big impact, giant ring-shaped waves would spread out across large expanses of the atmosphere, like the ripples in a pond after you've thrown a stone into it. The waves might be observed and followed as they propagated, perhaps giving insight into the properties of the atmosphere. Not being an astronomer, Tim approached his colleague Heidi Hammel and asked her if she would write a proposal to look for these expanding rings around an impact site in images taken with Hubble. Heidi had misgivings about this. Her intuition said that, at best, one might only observe small, white cloud puffs resulting from a collision and little more. However, she had written one unsuccessful Hubble proposal in the past to observe Neptune, so she at least was familiar with the proposal process; and she was intrigued by Dowling's model predictions. She agreed to give it a shot.

The proposals were due in mid-November and the final selection of SOT members would be announced in late December. Sometime in early January of 1994 Heidi received a phone call from Bob Williams. He told her that her Hubble proposal had been accepted along with several others that required imaging observations. He wanted her to combine the proposals into a single coordinated imaging program. And he wanted her to be the imaging team leader. This was a shock to Heidi because, as she explained, "I had never used Hubble at that point, I had never written a paper about Jupiter, although I had looked at Jupiter through a telescope. So, I felt uniquely unqualified to do this. But you can't say *no*. You have to say *yes*."

The competition to use Hubble to study S/L-9 produced a first-rate Science Observing Team. Hal Weaver's observations would focus Hubble's new camera, WFPC2, on the string of comet fragments as it drew ever closer to the giant planet. Heidi Hammel and her imaging team planned to use WFPC2 to follow the clouds and wind patterns in Jupiter's atmosphere and to look for waves that might ripple through the atmosphere after each collision. Bob West from JPL planned to study any haze produced in the planet's stratosphere as a result of the impacts. John Clarke from the University of Michigan was a specialist in using Hubble's cameras at ultraviolet wavelengths to study the auroras, the northern and southern lights, of the gas giant planets in our solar system. He wanted to see if the impacts had any discernable effects on Jupiter's auroras. Keith Noll and Melissa McGrath, who both worked as research scientists at the STScI, were spectroscopists experienced in using the observatory's two spectrographs, the Faint Object Spectrograph (FOS) and the Goddard High Resolution Spectrograph (GHRS). They planned to measure the changes in the chemical composition of the impact sites and also to look for effects in the giant magnetic field surrounding the planet. Joining the group as a theoretical expert on the properties of Jupiter's atmosphere was Andy Ingersoll from Caltech. With this talented group, all the scientific bases would be covered when (or if) observable phenomena were seen the following July.

The highly successful first servicing mission in December of 1993 had indeed cleared up Hubble's poor eyesight. The excitement about using this new treasure to watch the S/L-9 impacts became intense. One of the first scientific observations with the newly refurbished telescope in January of 1994 was a panoramic sequence of images of the S/L-9 string of pearls made by Hal Weaver with the newly installed WFPC2. It was extraordinary! All of the fragments, twenty of them in this view, precisely aligned in the direction of their motion toward Jupiter, stood out in sharp relief, beautiful ... and doomed.

The brightest of them had clearly changed dramatically since the murky Hubble image taken the prior July. It had split into two pieces. But would other fragments, or perhaps all of them, disintegrate further as they traveled closer to Jupiter and were tugged more forcefully by its gravity? Would anything be left to see in July of 1994?

At a meeting of the American Astronomical Society in Philadelphia in January of 1991, Charlie Pellerin, the head of Space Science missions at NASA Headquarters, and Hubble Program Scientist Ed Weiler's boss, called a small group of Hubble team members together. He wanted to plot a strategy for counteracting the negative publicity that had followed after the public announcement of spherical aberration the previous June. Charlie asked Ed to lead a new, long-term Hubble outreach team. Even the optically flawed telescope was producing some interesting science. NASA needed to let the world know about that—no hype, just the facts clearly presented so that the general public could understand what Hubble was accomplishing despite its handicap. The core membership of the new team included public outreach people from NASA Headquarters, Goddard, and the STScI. As the Deputy Senior Project Scientist I was asked by the Hubble Project at Goddard to serve as its public spokesman and representative on the Weiler team.

It was a challenge. Early on, Weiler's team had to contend with the public's disappointment after spherical aberration was discovered. Later it had to deal with the public's stratospheric expectations after the success of the first servicing mission to repair Hubble's optics in December of 1993. It was against this backdrop that the media's interest began to mount steadily in anticipation of S/L-9's encounter with Jupiter the following July. What would Hubble see? How spectacular might it be?

At the weekly meetings of our outreach team, Weiler was very cautious about the impacts. He was convinced that S/L-9 would be unspectacular and, if we over-hyped the potential of the July encounter, it would leave NASA deeply embarrassed once again. He directed the team to downplay the comet story. Nobody wanted to be involved in another "Comet Kohoutek."

Czech astronomer Lubos Kohoutek discovered Comet Kohoutek in March of 1973. It appeared exceptionally bright given its distance from the Sun at the time of its discovery. Astronomers inferred that it must be a large comet. Moreover, they were fooled by its orbit, which appeared at first to be hyperbolic—that is, the orbit of a comet originating very far from the Sun and making its first visit to the inner solar system. Consequently, it was thought that it must be rich in ice and other volatile materials that would evaporate off the comet in large quantities due to the heat of the Sun as it made its closest approach in late 1973. Thus, Comet Kohoutek held the promise of putting on a spectacular display for observers on Earth and in the Skylab space station in orbit. The press designated it as the "comet of the century."

NASA organized a formal effort called "Operation Kohoutek" to coordinate worldwide observing programs, as well as observations from space. As it turned out, Comet Kohoutek made a respectable showing in that it was bright enough to be seen with the naked eye. But it was by no means as spectacular as originally anticipated, no more so than numerous other comets. The press and public considered it a "dud" and the NASA publicity leading up to its apparition as it rounded the Sun proved embarrassing.

The rest of Ed's team argued with him that we should be prepared for success, that the S/L-9 impacts might be visible, even compelling, and might lead to important science. NASA would look bad if it appeared that we had not taken the opportunity seriously. This internal debate proved moot, however. The press latched strongly onto the story on its own without any prodding from NASA. Public interest built exponentially as the date of the impacts approached. Stories appeared in Time, Newsweek, on TV national news, and in all the major newspapers.

Hal Weaver continued to follow the evolution of the S/L-9 train of mini-comets during the months, days, and hours leading up to the impacts (Figure 7.1). The string of pearls changed markedly as time went on. Astronomers labeled the fragments with the letters "A" through "W" with A leading the pack and W at the rear. The letters "I" and "O" were omitted to avoid confusion with the numerals 1 and 0. Fragments J and M were seen in some of the early ground-based images in 1993, but had faded from view by the time Hal took his first Hubble images in July of 1993. The brightest and presumably the largest fragments were E, G, H, K, L, Q, R, S, and W. If any of the mini-comets were to make a big, visible splash in Jupiter's atmosphere, surely these would be the ones. By the time July of 1994 rolled around, the train of fragments extended to well over a million miles. Because of the enormous length of the string of pearls, it would take 5.5 days for all the fragments to crash into Jupiter.

Some of the fragments had broken apart and their pieces were beginning to spread out. In May both ground-based and Hubble images had shown that one of the biggest ones, fragment G, had a companion, indicating that G had begun to come apart. It was disquieting to all of us. These mini-comets looked fragile. The "$64 question" was, would anything at all be seen on July 16 and during the 5.5 days thereafter?

As we entered 1994 July, anticipation of the S/L-9 Jovian encounter grew as hot as the temperature outdoors in the Maryland summer. We were preparing for the

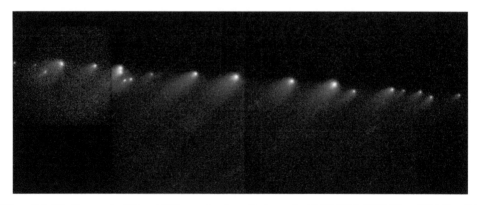

Figure 7.1. The fragments of Comet P/Shoemaker–Levy 9 observed with Hubble's WFPC2 on 1994 May 17. (Credit: NASA, H. Weaver, T. Smith, STScI 1994-21.)

media to descend on us. Detailed observing plans began to gel for Hubble, the Galileo spacecraft then approaching Jupiter, and many ground-based telescopes. The members of the Science Observing Team had taken up residence at the STScI, moving into temporary offices, installing their equipment, and checking out the computer codes they would use to analyze the observations as they came down from Hubble.

Heidi Hammel and her student assistants, Joe Harrington and Jennifer Mills, installed a new, powerful computer to display the comet images they expected to obtain. The plan was that those images would be processed and put into good form for scientific study by the image-processing software that the STScI normally used for WFPC2 observations. However, the turnaround time between reading the data down from Hubble to the ground and producing the final data files that an astronomer could interpret was usually about three days.

"Three days! That is simply unacceptable!" Heidi railed. Three days was a lifetime. If the imagery that Hubble collected were poor, the delay would only make the media regard it as a bigger failure. But if Hubble were actually able to capture some major fireworks on Jupiter, then requiring a three day lag time would severely blunt that success. If the first images indicated some kind of adjustment required of Hubble, then the entire comet campaign could be derailed. The images needed to be seen in minutes, not days.

Fortunately, John Clarke had already developed calibration software for his ultraviolet images that was much faster. Heidi put Jennifer Mills to work adapting Clarke's software for visible-light images. Jennifer got the turnaround time down to about 10 min! Without that effort there would have been no images to interpret for science and nothing to show the public for days on end. Young Jennifer Mills became a hero to the imaging team.

Unfortunately, another development would demand heroics of a different sort from a Hubble team member back at Goddard, a contribution so vital that failure would have doomed the entire S/L-9 observing project.

It was July 7, nine days before the calculated time when the first comet fragment would crash into Jupiter and only four days before a critical pre-impact Hubble observation of all 21 comet fragments was scheduled. Larry Dunham and a group of operations engineers were seated at consoles in the area of the Space Telescope Operations Control Center (STOCC) at Goddard Space Flight Center devoted to testing software and new procedures before they were uploaded into Hubble's spacecraft computer, the DF-224.

Dunham was the leader of the team of Lockheed engineers operating Hubble from the STOCC. In recent weeks the DF-224 had been experiencing intermittent "hiccups," having trouble properly interpreting commands. On July 5 the STOCC team had purposely put Hubble into a light version of safemode called Inertial Hold—sort of an induced coma—so that they could try to diagnose and solve the problem. They found a faulty location on one of the computer's memory boards and were trying to work around it by patching in some new memory from the computer Co-Processor, a piece of hardware that had been inserted into Hubble by the astronauts in the first servicing mission the previous December.

The test of the new setup was going well. The team was about ready to declare the problem solved and bring Hubble back up to full operation when a technician who had been on console in the mission operations room next door rushed in, exclaiming:

"The door closed! We just entered Zero Gyro Sunpoint."

Dunham and his colleagues were shocked. "What the hell? What's going on now?" he muttered. "Now we have a gyro failure?" They checked the telemetry from the spacecraft and found two of the four operating gyros flagged as "bad," yet in other telemetry all the gyros looked normal.

Hubble had placed itself into a deep state of safemode, the Zero Gyro Sunpoint (ZGSP) mode. The spacecraft had never gone into ZGSP before, except during an early test explicitly of that mode. The telescope's aperture door closed, all the gyros that guided the pointing of the spacecraft were taken off-line, and the solar arrays were automatically oriented face-on to the Sun to maximize electrical power. Hubble basically had put itself into a protective fetal position to assure its own survival and then "called home," requesting help from the engineers in the STOCC.

How bad could the failure in Hubble be? The nightmare scenario was that there had been some kind of hardware failure, perhaps a computer or gyro failure, on the spacecraft. As far as the comet observing campaign was concerned, Hubble was now dead in the water.

The news couldn't have come at a worse time. This was supposed to be the moment of Hubble's crowning success. Mother Nature was giving us the perfect opportunity for a highly visible demonstration to the world of what Hubble could do now that its vision had been restored during SM1. This was the year of the Shoemaker–Levy 9 comet. The whole world knew about it. The press was in frenzy about it. Hubble was to be the key to unlocking its secrets—but only if the Hubble Space Telescope were functional.

News of the Hubble emergency spread quickly up and down the line at Goddard, NASA Headquarters, and the STScI. John Campbell, the Hubble Project Manager, took the phone call from the STOCC. "John, we're in safemode, and this one looks bad," said Ann Merwarth, the manager responsible for operations. "It's Zero Gyro Sunpoint. We have never been in it before."

Campbell sighed deeply and then replied, "Okay. Get your folks to pull all the data together and look it over. Let's schedule a meeting later this afternoon to see where we're at and what we should do."

Campbell followed his normal protocol of going up his chain of command to let Goddard senior managers know about Hubble's status and then on to the Hubble Program Manager at NASA Headquarters. Soon thereafter he got a call from Dan Weedman, the Director of the Astrophysics Division at NASA Headquarters. Weedman was an academic scientist, new to NASA, having come recently from his faculty position at Penn State. He had been apprised of the bad news and wanted to get the details directly from Campbell.

"Right now, Hubble is down," Campbell told him. "We don't know why, but we'll do everything we can to get this figured out ASAP. We all know the comet is coming soon."

Weedman replied, "Yeah, the whole community is watching. This is a really unique opportunity for science. That said it's more important to do this right than to do it fast. Don't rush anything or cut corners without following the proper protocols. Hubble's welfare comes first. If we miss the comet, we miss the comet."

On 1994 July 7 in the hours following Hubble's entry into ZGSP, the prospects for using the observatory to acquire exquisitely clear visual images and unique spectroscopic observations of the S/L-9 fragments crashing into Jupiter were fading fast. The comet was expected to strike the planet on the evening of July 16. Time was not on our side.

Later that afternoon, at the 4:00 pm meeting in the STOCC, Campbell and other Hubble senior managers pressed the operations engineers on the possibility that they had not fully diagnosed what was wrong with the DF-224 computer when they first put it into Inertial Hold. If they hadn't actually found the root cause of the computer hiccups, if something more serious was wrong with the computer, then the fix they had just tested wasn't going to solve the problem. It would have been premature to bring the spacecraft back up to normal operations until other possibilities had been exhausted.

This sounded like criticism to Brian Vreeland and he took it personally. Vreeland was a junior engineer who had been transferred from the main Lockheed plant in Sunnyvale, California to Goddard the year before specifically because of his expertize with the DF-224 computer, a 1970s era computer that was still in use on many spacecraft. Only a few people remained in the industry that could even program it anymore, it was so out of date. It had 64 kB of memory, the same as an old-fashioned Apple IIe. Vreeland was at the center of the earlier discussions in the STOCC about what was wrong with the computer and how to resolve the commanding problem. Now, management was starting to question whether he, Dunham, and the other operations engineers really understood the issue. Maybe there was something deeper going on.

After the meeting Vreeland sought the solitude of a back room to think through what had happened. He was determined to find the explanation and to do it quickly. He knew there wasn't enough time to solve the problem in the normal way, with a slow and laborious step-by-step forensic analysis of all the possibilities—hardware and software—and still be able to observe the comet. For him it was a matter of pride. He pondered the problem into the evening and resumed the next morning.

At 10 am the STOCC Conference Room was packed with senior managers and engineers for a meeting to continue discussions about the mysterious ZGSP safemode entry. About halfway through the meeting Brian Vreeland came tearing up the hall to the doorway. He couldn't get into the room; even the "standing room only" walls and corners were occupied. He waved his arms frantically to get Larry Dunham's attention. Dunham walked out of the conference room and asked Vreeland, "What's going on?"

"I think I have it! I think I understand what happened," Vreeland exclaimed. ***"THERE's NOTHING WRONG WITH HUBBLE!"***

The two of them walked into the mission operations room and sat at a table. Vreeland began to walk his boss through his reasoning. He was excited, couldn't wait to share what he had discovered:

"It's possible of course that we had a random hardware failure. That would have been quite a coincidence to have that happen only two and a half days after another safemode entry. But bad things happen, right?" Dunham nodded, as one with much experience dealing with "bad things." Larry had years of prior experience leading the operations team in solving technical problems that threatened Hubble and in pulling the spacecraft out of numerous safemodes.

"If it was a random hardware failure, then for sure we've lost the comet. But I wanted to check out the possibility that it wasn't a coincidence, that the Inertial Hold safemode on the 5th and the Zero Gyro safemode on the 7th were somehow related. I knew I was grasping at straws. But I started with the timing. How much time elapsed between the first safemode entry and the second? It was a bit more than 48 hr. But that didn't mean a whole lot to me at first. So, I thought, what if I took this number and look at it in terms of a counter?"

"And we have lots of counters in the computer, right?" Dunham interjected. "As I recall they're mostly 'good counters.' They check a component, and if it's still operating correctly, they increase the count by one."

"That's right," Vreeland responded. "We have hundreds of them." He continued, "I started going through the arithmetic. The system checks each component 40 times every second. If the component is still 'good,' the system increments its counter by one. Each counter can count up to 37,777,777."

Looking a little puzzled, Dunham interrupted, "That's an octal number right, the kind the old DF-224 is most comfortable using?"

"You've got it," Vreeland replied, getting impatient to get on with his story. "It's an efficient way to express a total number of counts in ordinary base 10 arithmetic of 8,388,607. The system could check that a component was still 'good' that many times, before its 'good counter' maxed out, or rolled over."

Dunham sat up in his chair, suddenly keenly attentive. He was a supervisor, not an expert on the spacecraft computer. That was Vreeland's job. "So, let me see if I understand. This is like a very old truck that has just hit 999,999 miles. Its odometer only has six wheels, so it can't display a larger number. The next number that comes up is 000,000. Even though the truck has been driven a million miles, its odometer says it hasn't yet been driven one mile. It has overflowed—it's giving false information."

"Yep," Vreeland said. "But wait, it gets better! If you count up to the maximum, 8,388,607 counts, at a rate of 40 counts every second, it takes 209,715 s for the counter to overflow or roll over. Guess how many days that adds up to?"

"I can sense you can't wait to tell me. Go ahead," Dunham replied.

"It's 2.427 days. And guess how much time ran off the clock between the first safemode event, Inertial Hold, and the second one, Zero Gyro." Vreeland couldn't hold back, and the answer poured out, "2.427 days!"

We had a counter roll over. But which counter? There are so many of them. And why did this one max out?

Dunham added, "Yeah, all the counters are read out in telemetry, but normally the guys sitting at the consoles don't monitor them. They're never supposed to overflow. A 'good counter' resets whenever the component it's monitoring is used; that's usually many times a day."

Vreeland continued, "So, I went back and had a look at what was happening to the counters at the exact moment when Hubble put itself into ZGSP. It was the gyro 'good counter' that had rolled over. And that made sense. The next octal number that came up as the counter overflowed was 40000000. The 4 stood for a minus sign. The computer read that as minus zero. Tilt! It choked and sent erroneous commands to take the gyros off-line and head into safemode."

"And now I'm with you," Dunham said. "We forgot about all of this when we put Hubble into Inertial Hold a couple of days ago. We made a mistake in not remembering that the gyro 'good counter' can only count for so long before it overflows. Hubble just kept pointing to the same spot in the sky. The gyros were never used. The 'good counter' was never reset. And it just kept counting and counting and counting, until it couldn't count anymore, 2.427 days later. There's really nothing wrong with Hubble! The Zero Gyro entry was a false alarm!"

Dunham was immensely relieved. "You've nailed it, Brian. Great work! Now we just have to persuade NASA management that we're good to recover and go back to normal ops. Shoemaker–Levy, here we come!"

Vreeland presented his findings to the operations team and to senior project management. Campbell sat at that briefing with total relief written all over his face. "Okay, let's just recover and go," was his direction to the team.

Word quickly propagated up and down the line in NASA and to the anxious astronomers awaiting S/L-9 that nothing was wrong with Hubble. It had been a computer glitch that was easily fixed. It seemed the whole world breathed a huge sigh of relief. At NASA Headquarters Dan Weedman was so thankful, he sent Brian Vreeland a personal letter of gratitude.

A new observing schedule was generated by the STScI, transmitted to Goddard and uploaded to the spacecraft. Hubble resumed normal operations on July 9, only two days before Hal Weaver's next scheduled pre-impact WFPC2 images of the comet train and one week prior to the impact of the first fragment onto Jupiter. A close call, but disaster had been averted because of the imaginative problem-solving skill of digital detective Brian Vreeland. From that time on, the spacecraft controllers always scheduled Hubble to be re-pointed at least once per day, so that the gyro "good counter" would never overflow again.

Life With Hubble
An insider's view of the world's most famous telescope
David S Leckrone

Chapter 8

When Worlds Collide: The Comet Campaign

D-Day for the assault of the fragments of Comet S/L-9 on Jupiter was 1994 July 16. At JPL, Paul Chodas and Don Yeomans had used the best orbit solution for each individual fragment to calculate the expected time and location of each impact. They predicted that fragment A would impact on Saturday, July 16 at Universal Time $19^h59^m40^s$ (about 8:00 pm Greenwich Mean Time, or 4:00 pm Eastern Daylight Savings Time). They predicted that the impact would happen out of sight of observers on Earth on the backside of the giant planet. Jupiter rotates faster than any other planet in the solar system. Its "day" lasts about 9.9 hours. If Chodas and Yeomans got the numbers precisely right, it would be about 13 min after impact before the A impact site would rotate into view for Hubble and for other observatories on Earth. But there was uncertainty in their calculations on the order of a few minutes, so the Earth-based observers needed to be vigilant, not knowing the exact moment when something might be seen.

Chodas and Yeomans also predicted that each fragment would be traveling at a blazing speed of about 137,000 miles per hour when it hit. Since the sizes and masses of the fragments were highly uncertain, and widely debated, it was impossible to predict accurately how much energy each fragment would release into Jupiter's atmosphere. But a reasonable estimate was somewhere in the range of between 100,000 and a few million equivalent megatons of TNT, or about 100 million times the energy released by the atomic bombs dropped on Hiroshima and Nagasaki. What effects would each impact produce on Jupiter while the world watched? We didn't know, but we were about to find out.

The Hubble Project team at Goddard felt major pressure to get its part of the action right. After all, the 21 impacts were transient events happening in real time. If there were any more operational glitches during the upcoming week, there would be no "re-dos." The stakes both for science and for NASA's public image were very high. The Space Telescope Operations Control Center (STOCC) operations team held a "Flight Readiness Review" on Thursday, July 14 to go over all the preparations in detail one last time.

The imaging team began observing Jupiter with Hubble around 8:00 am on Friday, July 15. The intent was to see precisely what Jupiter looked like just before the first comet fragment hit. The first data from the Hubble observations was downloaded at 4:50 that afternoon. Heidi Hammel's team was the first group to look at this set of images. "Oh…My…God!" Hammel exclaimed. "This is bad! We have to change this. And we have less than one day!"

The images taken through the red filter were badly overexposed. Apparently, a computer setting that controlled the sensitivity of the light detectors in the WFPC2 cameras had been left in the wrong position. This meant that the commands that had already been loaded into the instrument computer onboard Hubble to take the first impact observations the next afternoon were wrong.

Hammel was very aware that Hubble observations were always scheduled at least a week in advance. Last minute changes were usually not permitted. There was always the risk of making a mistake and leaving things worse rather than better, or even putting the telescope at risk. "I'll go talk to Doxsey," she said.

Astronomer Roger Doxsey was the head of Hubble science operations at the Space Telescope Science Institute (STScI). The entire Hubble team revered him as one of the few people alive who might actually know everything there was to know about the Hubble Space Telescope. He was as dedicated as anyone could be and a brilliant problem solver.

Stepping into Roger Doxsey's office, Heidi blurted out with a note of worry, "Roger, we've got a serious problem. We've screwed up royally. The red images of Jupiter that just came down are badly overexposed; they're unusable."

Doxsey pondered that for a few seconds and asked, "Do you know what went wrong?"

"The gain settings on the WFPC2 CCDs were left in the wrong position. It's a simple fix, just throwing a digital switch. But it's a crisis! Tomorrow we've scheduled the observations of the impact of fragment A. The press will be watching. The world will be watching. And the color images are going to look really bad."

"Let me make a call to our guys in Planning and Scheduling. We'll see what we can do," Doxsey said as he picked up the phone.

Within two hours, Hammel got the news. She had a green light to change the computer commands. Doxsey told her, "We can schedule a real-time communications link to Hubble in about an hour from now. If you can give us the revised commands that quickly, we'll send them up."

"No problem," she replied, greatly relieved.

The Planning and Scheduling Team at the STScI and the STOCC operations team at Goddard had waived the normal rules. They knew that in this case, the risks were worth taking. After all, this would be the first impact, and so the one most highly anticipated by the scientific community and the public. This particular impact would never happen again. Hubble couldn't miss it, and the data had to be perfect. The team did what had to be done in real time to reload the correct settings in the instrument. Another big sigh emanated from everyone involved; we had dodged another bullet at the last minute.

The estimated time for the impact of Fragment A was approximately 4:00 pm on Saturday, July 16. Hubble's WFPC2 camera was set up to snap a series of eight images through color filters designed to isolate near-ultraviolet, violet, blue, red, and far-red light. The four far-red filters were specifically designed to detect the emission or absorption of light by molecules of methane, CH_4, a well-known abundant constituent of Jupiter's atmosphere. The first image exposure was timed to begin at 4:13 pm. The extra 13 min was the amount of time needed for Jupiter to rotate the A impact site around the morning limb so that it would just begin to be visible to observers on Earth. Then, one Hubble orbit (or about 95 min) later, the WFPC2 would take a second sequence of images. In those additional 1.6 hr, Jupiter's rotation would carry the impact site fully into view.

The usual afternoon meeting of the Science Observing Team (SOT) in the Institute boardroom was canceled for July 16. There was little more to discuss and nothing more to do, until the first observations came in. Two days before, in the July 14 issue of *Nature* magazine, astronomer Paul Weissman had published an article entitled, "Comet Shoemaker–Levy 9: The Big Fizzle Is Coming" (Weissman 1994). Now the moment of truth was close at hand.

Around 5:30 pm, I encountered Melissa McGrath virtually skipping down the hall on the second floor of the STScI. She told me that a message had been posted from the Calar Alto Observatory, a joint German-Spanish observatory in southern Spain, reporting a sighting at infrared wavelengths of a bright plume emerging above the southwestern limb of Jupiter. It had been observed at about 4:18 pm, Eastern Daylight Time. A bit later, a set of deep-infrared observations was reported from the European Southern Observatory in La Silla, Chile. For the Hubble scientists who had to wait three more hours until their first observations could be read down from the spacecraft, these emails produced a mixed reaction.

The good news was that the first impact of one of the smaller fragments had produced an observable response in Jupiter's atmosphere. However, it had been observed in the infrared at wavelengths corresponding more to heat than light. The team wondered: could this be a purely infrared phenomenon produced by friction heating the incoming fragment? Would anything be seen in visible light? If not, surely the media and public would wonder why Hubble saw nothing, while the impact was easily seen by ground-based telescopes—potentially another bad day for Hubble.

At 7:30 pm a large group of reporters assembled in the Institute's auditorium for a NASA press briefing. The panel presenting the briefing consisted of Gene and Carolyn Shoemaker and David Levy. Carolyn described how the comet had been discovered "on a dark and stormy night" in March of 1993. David rhapsodized about the upcoming comet campaign as an illustration of how exciting and fun science can be to everyone—laypeople as well as professional astronomers. He emphasized as he had on several other occasions, that he doubted the public in general, or even amateur astronomers with small telescopes, would be able to directly observe any evidence of the impacts on Jupiter; it would take professionals observing with large telescopes to see anything.

Gene's remarks focused on what we might expect to see when a comet fragment hits Jupiter. His colleagues, Paul Hassig, David Rodie, and Andy Ingersoll, had used a theoretical computer model to create a simulation of the response of Jupiter's atmosphere to a 1 km wide fragment with the density of water ice, coming in at a 45° angle, at 134,000 miles per hour (216,000 km hr^{-1}), and penetrating to a depth of about 20 miles (32 km) below the cloud tops. At that point, heated to a temperature as high as 40,000 K, the fragment would blow apart. Gene showed a video of the simulation. The fragment generated a shockwave in the atmosphere that rapidly heated the surrounding gases, creating a blazing hot fireball. Most of the material, a mix of comet debris and Jovian air, expanded into a hot bubble that rose buoyantly back up the evacuated tube that had been dug out by the incoming fragment. Rising above the cloud tops, the fireball created a plume, very much like the mushroom cloud produced in the explosion of an atomic bomb. It reached an altitude of about 620 miles (1000 km) in 5 min or so. As the top of the plume cooled, it flattened out and sank back down onto Jupiter's stratosphere. In this simulation, the flattened plume spread out into a "pancake" about 1240 miles (2000 km) across. The model predicted a tremendous display. Would that play out in reality?

Shortly after 7:30 pm we all crowded around a monitor in the STScI Operations Center. Heidi seated herself directly in front of it, so that she could get a good look at her first WFPC2 observations. Slowly, the first image scrolled onto the screen, a black-and-white picture of the southwest quadrant of Jupiter's disk, showing the bright limb or edge against the darkness of space. It showed nothing else. *Aha!* Heidi thought to herself. *Just as I expected.*

Then a second image came up. Suspended just above Jupiter's limb was a small white dot. Heidi pointed to it and asked no one in particular, "Could that be a satellite?" She was referring to any of the four large, bright Galilean satellites: Io, Europa, Callisto, or Ganymede. Hal Weaver said, "I'll check." He went to a nearby bookcase and pulled out the 1994 edition of "The Astronomical Almanac." This ancient reference produced jointly each year by the navies of the United States and the United Kingdom has, in some form, provided accurate positions of planets, satellites, and stars to navigators and astronomers since 1766. Now it was going to guide a team of modern space explorers in interpreting the first image they saw of a comet's collision with Jupiter. Hal and Melissa McGrath poured over the Almanac's table of the times of rising and setting of the Galilean satellites for each day of the year. "No! It's not a moon," they proclaimed confidently.

The next image to come up on the screen showed nothing at the position of the white dot—only the blackness of space beyond the limb. Image number four, however, was a revelation. The white dot was back, but it had spread out into an extended fuzzy patch. The subsequent five pictures showed it clearly growing into a fully formed plume, rising thousands of kilometers above Jupiter, spreading out horizontally, and then settling back down onto into a flattened "pancake" above the cloud tops (Figure 8.1). Each image showed the plume as being separated from Jupiter's limb by a dark band. That was the shadow of the planet cast onto the

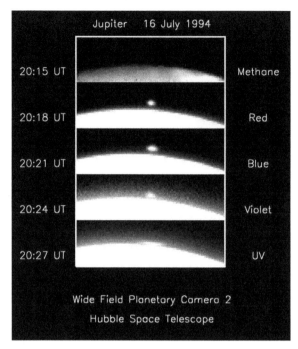

Figure 8.1. Time sequence of WFPC2 images showing the evolution of the plume created by the impact of Fragment A into Jupiter's atmosphere. This was our first view of an impact event. The plume rises and spreads out into a flattened "pancake" over a period of about 10 min. (Credit: NASA, H. Hammel, the Comet Science Team, STScI 1994-30.)

plume. Jupiter had not yet rotated enough to allow the astronomers to see the plume fully illuminated by sunlight. It was so high, however, that it reached far above Jupiter's horizon and into the sunlight. Remarkably, the observed appearance and evolution of the real plume was a close match to the theoretical simulation being shown to the reporters by Gene Shoemaker at the same time in the auditorium one floor above.

At this point, the imaging team, and the other dozen or so onlookers in the Operations Center, erupted in giddy excitement. For the occasion Melissa McGrath had purchased two bottles of champagne. To guard against any possible disappointment on July 16, the team had drunk toasts with the first bottle the day before. Now it was time to pop the cork in bottle number two for a full-throated celebration of what Mother Nature had just shown them on Jupiter. Melissa had stored the second bottle in a refrigerator on another floor at the STScI. She dashed to the elevator to retrieve it. After putting the bottle in a brown paper bag to hide it, she ran the gauntlet of STScI people and a few stray reporters wandering the hallways. They kept shouting questions at her, "What's happening? What's going on?" She could only say, "I can't tell you yet." But her happy demeanor surely tipped her hand. Arriving back at the Operations Center, Melissa removed the bottle from the bag. But there was one problem. No one had remembered to bring cups. No matter. The

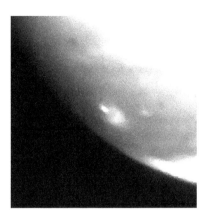

Figure 8.2. Impact zone of Fragment A about 95 min after impact. Jupiter had rotated enough to bring the impact area into view. The image was taken through a filter that passes light at a deep red wavelength, one that is strongly absorbed by methane. As a result, the upper layers of Jupiter's atmosphere look dark, but the impact site looks bright by comparison because it is reflecting, not absorbing sunlight. (Credit: NASA, H. Hammel, the Comet Science Team, STScI 1994-38.)

open bottle was passed around among the science team members, each delighted to take a swig.

Heidi and her team had been lucky. The predicted impact time of fragment A was 4:00 pm. They had scheduled Hubble to begin snapping images at 4:13 pm, expecting that it would take Jupiter 13 min to rotate the impact site to a location where it could be seen from Earth. However, fragment A was late; it impacted about 10 min later than expected. Had it come in "on time," that first image would probably have shown the very final, "pancake" stage of the plume. Instead, the team was able to see the plume grow and develop in all its full glory. And that first image that showed the small white dot had perhaps revealed the initial brilliant glow of the fragment A meteor as it flashed across the Jovian sky.

Still, it was a marvelous feat of mathematics and physics. Yeomans and Chodas had predicted when a body traveling at 137,000 miles per hour would strike another body millions of miles distant, a moving target that was both orbiting the Sun and rotating. And they had been off in their predictions by a mere 10 min.

The next set of images to be displayed on the monitor had been taken one Hubble orbit, or about 95 min, later. In that hour-and-a-half, Jupiter had rotated enough that the A impact site was clearly visible. The methane gas in the planet's atmosphere strongly absorbed light at a particular wavelength in the deep red (890 nm). In an image of Jupiter taken with a filter that passed that color, but no other, most of the planet looked dark. In contrast the new patch left behind in the high stratosphere by fragment A looked very bright. It was plainly visible, very large, and structured (Figure 8.2).

Upstairs in the press briefing, Gene, Carolyn, and David were struggling to answer questions about Hubble observations they had not yet seen. At least they had the two early ground-based reports of bright infrared fireballs seen from Calar Alto and La Silla to fall back on. Something clearly was happening on Jupiter, but what would Hubble's visible-light images show? The briefing was being routed to a TV set downstairs in the Operations Center, so that we could see and hear the questions and answers. Heidi thought to herself, *this is not right!* Even though it normally takes some time to reduce and calibrate the images and put them in a format that would be presentable to the press, Hammel felt that she had to show the media something now, "we had to let them see what we've seen." So, as quickly as they could, her team produced a raw, laser-printer image of the A impact site.

Gene Shoemaker was in the middle of answering a reporter's question. Heidi walked in, carrying a piece of paper. She showed it to Don Savage, the NASA public affairs person who was emceeing the briefing. Then Gene exclaimed, "I think we might have an update from Heidi Hammel!" Heidi said that Hubble had seen the plume, prompting Gene to raise his fist in triumph. She then held up the laser-printer picture that clearly showed that Jupiter had a large, new spot—a "bright streak" with some other "stuff" around it. Earlier in the press briefing a reporter asked Gene what he expected to see from the A impact. His answer was, "a new spot on Jupiter." Now that expectation had become reality.

Heidi left the dais for a few seconds and then returned, holding the partially consumed bottle of champagne. Gene, Carolyn, and David took swigs, all the while laughing. Heidi shook all their hands and raised Gene's arm in victory. Clearly there was great cause for celebration and for anticipation of what would be in store during the remainder of the week. After all, the A fragment was one of the faintest and smallest. What would they see after the larger fragments—G for example—had fallen onto Jupiter?

While watching the first images come up in the Operations Center, Jupiter expert and Imaging Team member Reta Beebe had remained absolutely stoic—stonefaced—staring at the monitor, while the rest of the team was ecstatic and giddy. She later explained what she was thinking. "We are in deep yogurt. That was one of the smaller ones. The big ones are coming later. By the end of the week it's going to be a mess."

At 10:00 pm, a second NASA press briefing took place in the STScI auditorium. This time a pumped up group of astronomers, the principal members of the Science Observing Team, made up the panel. One of the smaller fragments had put on a spectacular show as it plowed into Jupiter's upper atmosphere earlier in the day. It was clear that even bigger fireworks were in store throughout the coming week. So, it was appropriate that all the scientists be present to discuss their plans and expectations.

Heidi showed a much improved version of the A impact site image taken in blue light. Viewed at a large angle near Jupiter's limb, the basic form of the site consisted of an elongated central dark patch, a crescent dark apron extending back to the southeast, the direction from which the A fragment had come, and a hint of a circular dark ring surrounding the central patch. It looked like a huge black eye. This overall structure gave a preview of what would be seen at visible wavelengths at other impact sites throughout the week.

The diameter of the A impact site was comparable to that of the Earth. Our entire planet would fit within it! A relatively small fragment had created a big impact zone. Late in the briefing, a reporter from ABC television asked the panel what this event meant for "the man on the street; what's in it for him?" John Clarke shot back, without missing a beat, "He should be glad he doesn't live on Jupiter!"

John and Heidi expanded on the answer. This was the first time in human history that we had witnessed this kind of large collision event in our solar system. The planets were almost certainly formed and later substantially altered by big impacts. In July of 1994, we Earthlings were very lucky to be able to see and study this kind of process in action. It proved that, even today, we live in a dynamic solar system. These kinds of collisions didn't *just* happen billions or millions of years ago. They're still going on today, albeit less frequently. As Heidi put it, "There are things whizzing around the solar system smashing into other things and generating tremendous explosions. That's a fascinating thing, when you think about it."

Some of the reporters' questions were the same as those on the minds of the scientists. How large was fragment A? How much energy had its explosion released? How deeply had it penetrated? Did it send acoustic waves rippling through the atmosphere? Would it stimulate the formation of a "permanent" cyclonic storm like so many others seen on Jupiter, including the great red spot? What chemical changes in the atmosphere had the impact induced? One particularly interesting question was what was that dark stuff that left such a pronounced "black eye" on Jupiter?

Panelist Bob West said he was very surprised by the black color. He had expected ammonia ice crystals and perhaps water ice crystals to condense out of the plume after it had flattened, cooled, and fallen back onto the cloud tops. Ice crystals like that should be white, not black. He speculated that the black material was made of fine particles of some kind suspended in the stratosphere. But its chemical composition was a mystery. It was early in the observing campaign. Perhaps the remaining twenty impacts would offer further clues.

After the press briefing Heidi needed to go get some sleep, anticipating a busy day ahead. Jennifer Mills remained at the STScI, refining the Fragment A images Hubble had captured. Heidi needed to have some good material to show at the press briefing the following morning. Jennifer was the only one who could do that quickly. Her speedy imaging data reduction software made it possible. She worked deep into the night.

The next morning, Sunday, July 17, the Visitors' Center at Goddard was a beehive, crowded with reporters, TV crews, scientists, NASA public affairs people, and a few hangers-on. Word had spread quickly that the first comet impacts were spectacular. Anyone who had been dubious about potentially wasting time on this story needed no further convincing. Competitive media professionals couldn't afford to miss it. Reporters from across the country and, in a few cases, from around the world, descended on Goddard.

The panelists that morning were Heidi, Gene and Carolyn Shoemaker, and David Levy. By the starting time of the briefing, 10:00 am EDT, three comet fragments had impacted Jupiter: A, B, and C. Gene gave the news that the A and C impacts looked very similar; the fragments must have been similar in size and mass. But the B impact appeared much weaker. In the pre-impact Hubble images of the "string of pearls," fragment B was one of the fainter mini-comets that didn't quite line up with the majority. If you drew an imaginary line through all the brighter fragments, B was a little bit displaced. Possibly B had fallen apart into a swarm of rubble as the comet train evolved in its orbit around Jupiter. It just didn't have enough kinetic energy, enough oomph, to produce the dazzling display exhibited by its larger sisters.

Heidi showed the Hubble images of fragment A taken in blue light (the same image she showed at the previous night's briefing) and in the red methane band. The two images were inverted, like a photographic negative looks inverted in brightness compared to a positive print. The impact zone looked very dark in blue light, contrasted against the bright body of the planet. In the methane band it looked very bright, contrasted against a dark Jupiter. It appeared that the material unleashed by the impact was floating high in the stratosphere, reflecting sunlight upward and making the impact site look bright relative to the background. The sunlight that bypassed the impact material and penetrated more deeply into the atmosphere was being absorbed by the clouds of methane far below, making them look dark.

Heidi also showed the time sequence of WFPC2 images taken as the plume from the fragment A fireball rose above the limb of the planet, expanded horizontally, and collapsed into a flattened "pancake" onto the cloud tops. Gene noted with great satisfaction that the computer model he had shown in his video the previous evening of a typical fragment crashing into Jupiter's atmosphere had "nailed it." Theoretical physics had predicted natural reality very nicely.

Gene estimated the sizes of A and C as "at least one kilometer across, while the larger fragments must be about three km in diameter." Throughout the week the size and power of the successive impacts continued to fool Gene and the other astronomers into believing that S/L-9's parent body and the individual fragments were much larger than they actually proved to be. The scientific consensus after all was said and done was that the parent body must have been between 1.5 and 2 km (0.9–1.2 miles) in diameter. The largest individual fragments were probably between 0.5 and 1.0 km (between 0.3 and 0.6 miles). The awesome displays put on by the impacts became all the more impressive when we realized that the fragments were smaller than expected.

Jupiter is about eleven times larger than Earth in diameter. It could hold about 1400 Earths within its volume. It is about 320 times more massive than Earth. Consequently, it exerts 320 times the gravitational force of Earth on an asteroid or comet that happens to be wandering nearby. In this way Jupiter protects Earth and the other small planets and moons from frequent, catastrophic impacts. It's like a super gravitational "vacuum cleaner," "sucking" up interplanetary debris. On human time scales we are relatively safe. The probability that Earth will be struck by kilometer-size or larger objects any time soon is low. But it is not zero.

On the other hand, public curiosity about the possibility of such a catastrophe actually happening was reflected in a question from a reporter from the Miami Herald. "What effect would the impact of an object the size of fragment A have if it came hurtling into North America?" Gene Shoemaker was, throughout his career, one of the world's preeminent authorities on the formation of impact craters, starting early on with his study of the famous Barringer meteorite crater near Winslow, Arizona. Off the top of his head he answered. "It would produce a crater about 20 km (12 miles) in diameter. If it came in over the Baltimore–Washington area, the impact would kill everyone. It would do enormous local damage. Everything within the crater and everything within the ejecta blanket would be destroyed. Severe damage would extend for hundreds of miles around the impact site. Pulverized material from the incoming asteroid or comet, plus Earth-rock excavated by the impact, would rise high into the stratosphere, generating major climatic changes that would last for years."

Notwithstanding the fact that Gene had in mind a 1 km sized A fragment doing the damage, his answer was qualitatively correct. Even a 0.5 km piece of interplanetary rock speeding into North America at a velocity of at least 10 km s^{-1} (about 16 times the muzzle velocity of an M16 rifle) would do extensive and long-term damage on this regional scale. In contrast, the Chicxulub crater created by the asteroid that wiped out the dinosaurs 65 million years ago is at least 180 km (112 miles) across. So, as large as the S/L-9 fragments were, they were dwarfed by the asteroid that produced the Cretaceous–Tertiary extinction event on Earth.

The full Science Observing Team met at the STScI every afternoon to go over the data that had been acquired during the prior day or so. I sat in on those meetings and took notes. I also discussed with the SOT what aspects of the day's observations were especially newsworthy. Those would go to the top of the list of results to be presented at tomorrow's media briefings.

Afterward, I drove the 25 miles from the Johns Hopkins campus to my home in Columbia, Maryland. I emailed the day's notes and my comments to the public affairs people at Goddard. I ate, slept for three or four hours, and then drove to Goddard to spend the night interpreting the scientific findings to the NASA TV and news team to help them prepare for the next morning's briefing. Needless to say, neither I, nor the scientists on the SOT, nor the NASA news people got much sleep throughout that week. But the experience was wonderful, exhilarating, a true "high," the most fun I had during my 33 years working on Hubble. Adrenaline is an amazing and potent drug. Hubble had done what it had been designed to do, and despite a few hiccups, it had performed absolutely brilliantly. Now there was no time to waste reaping the scientific harvest.

The first of the large fragments (where the size of a fragment was estimated based on the brightness of the corresponding mini-comet in the string of pearls), fragment G, entered Jupiter's atmosphere at 3:34 am (EDT) on Monday morning, July 18. Fragment A had garnered a lot of attention because it was the first to impact and it produced an impressive show despite its relatively small size. But a large fragment like G was expected to yield the most important science, as well as to put on a dazzling display.

At the 8:00 am press conference Monday morning, there had not yet been time to read out the images of the G impact from Hubble and process them to show to the media. That would have to wait until another briefing later that afternoon. However, reports were coming in from ground-based observatories in Hawaii that, at infrared wavelengths, the fireball from G was extremely bright, brighter than Jupiter itself, bright enough to "saturate" or max-out the detectors on the telescopes so that they were temporarily blinded. The impact site was very hot and that great heat translated into intense infrared radiation.

John Clarke was a specialist in studying the gas giant planets in our solar system with orbiting astronomical instruments that are sensitive to ultraviolet light, light with wavelengths ranging from about 100 to 300 nm. He was a leading authority on the auroras of Jupiter, Saturn, Uranus, and Neptune, the equivalent of the northern and southern lights on Earth. Their auroras were most readily observed in ultraviolet light.

When S/L-9 came along, John was keen to use the newly installed WFPC2 on Hubble to observe the impacts on Jupiter in the ultraviolet. At the Monday morning briefing, he showed the first ultraviolet image of impact sites C, A, and E spread across the planet's southern hemisphere (Figure 8.3). In every case the impact site looked much bigger and darker than the corresponding site seen in a violet light image he had taken for comparison just a few minutes before. Obviously, the plumes had flattened out in the Jovian stratosphere over a much wider area than would have been inferred from the visible light images alone. The fine dark particles hanging in Jupiter's upper atmosphere, well above the cloud tops, absorbed ultraviolet light much more strongly than visible light so that even a very small amount of material deposited in a thin layer showed up vividly in John's ultraviolet images. In ultraviolet light, we were seeing the full extent of the region of Jupiter's upper atmosphere affected by the impacts. The features were enormous. Each of them was much larger than the Earth.

At 4:00 pm that Monday, Hammel and Shoemaker showed the assembled media what Hubble had seen during and after the big G impact. Heidi first displayed a sequence of images of the plume glowing brightly in the sunlight as it rose above the limb of the planet, reached its maximum altitude, and then flattened out into a thin, extended pancake resting above the clouds. She estimated the height of the plume as about 2200 km (about 1400 miles). That's about equal to the height of Mt. Everest—stacked over and over, 250 times.

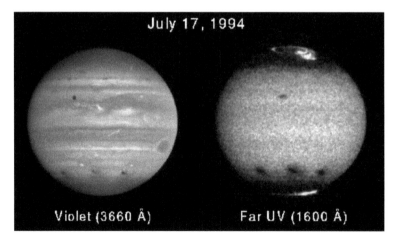

Figure 8.3. Impact zones of Fragments C, A, and E seen in violet light (366 nm) on the left and in far ultraviolet light (160 nm) on the right. The full extent of the impact zones becomes evident when seen in ultraviolet light. The northern and southern auroras can also be seen in the ultraviolet image. (Credit: NASA, J. Clarke, H. Hammel, H. Weaver, M. McGrath, STScI 1994-48.)

As had been the case for the Fragment A impact site (Figure 8.2), a red methane filter image of the G impact site showed it shining brightly in contrast to the darkness of Jupiter's disc below. In an image taken through a green visible light filter, the impact site was huge, beautifully defined, and very dark (Figure 8.4).

Both the plume sequence and the site of the impact seen one Hubble orbit (95 min) later were similar in structure to the fragment A impact two days earlier, but bigger. A wedge-shaped smudge, running from southeast to northwest was undoubtedly the hot, expanding void or "tube" that fragment G bored as it plowed through the atmosphere. At the narrow, northwestern end of the wedge was an intensely dark spot, probably the point at which the fragment exploded, creating a large bubble of extremely hot gas that rose rapidly upward above the cloud tops to create the plume. Centered on the dark spot was a perfectly circular dark ring, about 7500 km (4700 miles) in diameter, later shown to be a wave of some sort expanding outward from the explosion.

To the southeast, once again the Hubble image revealed a big "black eye," a dark semi-circular apron probably composed of a fine soot, the residue of the cooked stew of comet and Jovian material that had blown back out of the evacuated tube along the fragment's entry path. The inner edge of the "black eye" smudge was about 15,450 km (9600 miles) across. The entire Earth would fit easily within it. Heidi noted how very thin that "black eye" was; you could see right through it to the clouds below in Jupiter's atmosphere.

The G impact had been spectacular. A full color image of the G impact site released later in the week became iconic, emblematic of the entire week of the S/L-9 campaign (Figure 8.5).

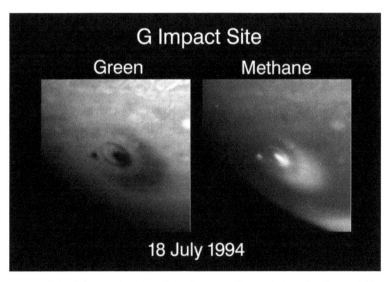

Figure 8.4. Images of the G impact site taken through a green filter and through a deep red filter that passes light emitted or absorbed by molecules of methane. To the left is a spot produced by the impact of the much smaller Fragment D. (Credit: NASA, J. Clarke, STScI 1994-32.)

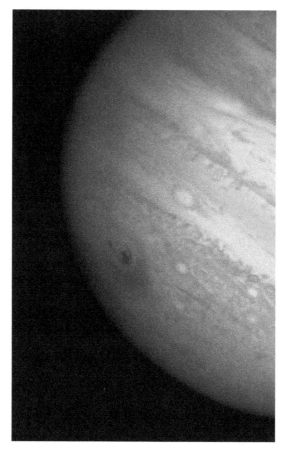

Figure 8.5. The G impact site. The entire Earth would fit within the inner periphery of the dark apron to the southeast. (Credit: NASA, H. Hammel, the Comet Science Team, STScI 1994-33.)

At the first press briefing on July 16, Gene Shoemaker was asked what he expected to see that week. He replied, "New spots on Jupiter." This was a reference to the semi-permanent spots that are always seen in Jupiter's atmosphere, cyclonic storms like the Great Red Spot that can last decades or even centuries. We all started referring to the respective impact sites as "spots"—the A spot, the C spot, the D spot, the G spot, etc. However, when we referred in media briefings or public talks to "Jupiter's G spot," it invariably induced a low level twittering in the audience. Yes, Jupiter had a G spot for a while.

Over the week or so of the S/L-9 observing campaign, the spots, the impact sites, evolved as the winds in Jupiter's upper atmosphere distributed the soot-like material created in the fiery impacts to distant longitudes. Figure 8.6 shows this process over a period of five days for the G, L, and S spots.

The G spot, the first impact site produced by one of S/L-9's larger fragments, was also of intense interest to the Science Observing Team members doing spectroscopy—Keith Noll, Melissa McGrath, and their colleagues. With a limited number of Hubble

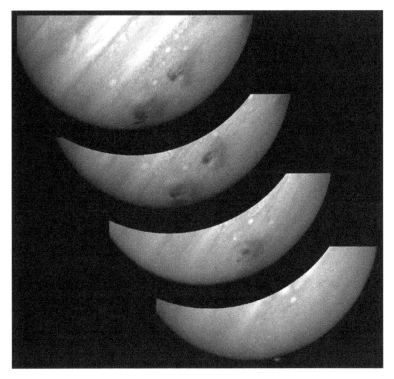

Figure 8.6. Evolution of impact sites over 5 days. At the very bottom the small white dot is the plume produced by Fragment G about 5 min after impact. Second from the bottom is the G site about 1.5 hr after impact. The third image shows how the G site has evolved due to the Jovian winds. It was taken about 3 days after G impact and 1.3 days after the impact of Fragment L seen to the southwest of G. The upper image, taken 5 days after the G impact, shows the continued spread of incinerated material from the G and L sites plus a small impact site from Fragment S almost on top of G. (Credit: NASA, R. Evans, J. Trauger, H. Hammel, the Comet Science Team. STScI 1995-15.)

orbits to use for their observations, they chose to concentrate on the G impact. They wanted to find out what chemical reactions happened when a fragment plunged into Jupiter's atmosphere. What atoms and molecules could be detected that were normally not seen on Jupiter? These could include debris from the comet itself or material that had been dredged up from deep in Jupiter's atmosphere as the fragment penetrated and stirred things up. They might also detect unusual molecules that were formed, at least momentarily, when the mix of comet material and Jovian air was pressure-cooked by the impact into some kind of planetary goulash.

The Faint Object Spectrograph (FOS) and the Goddard High Resolution Spectrograph (GHRS) were one-dimensional instruments. They didn't take 2D pictures. They only needed for the Hubble pointing control system to center their tiny entrance apertures directly over a small area within the impact site so that they could collect as many photons of light as time would allow.

Tiny droplets of water in a rain cloud separate and spread sunlight out into a rainbow of its constituent colors. Similarly spectrographs physically spread out, or disperse, the incoming light—sunlight reflected off Jupiter in this case—into its constituent colors and focus it on an electronic light detector. Spectrographs are the essential tools that allow astrophysicists to study the physical properties—temperature, density, and chemical composition, for example—of objects in the cosmos by analyzing their light. They put the "physics" into "astrophysics."

Every atom, every ion (atoms which have lost some of their electrons) and every molecule in nature absorbs and emits light in a unique pattern at very specific colors or wavelengths, dictated by the laws of quantum mechanics. Patterns of absorption or emission of light are as unique to each atom, ion, or molecule as fingerprints, retinal scans and DNA markers are to a human being. And like CSI experts, astrophysicists use patterns of dark or bright spectral lines or bands to identify the atoms, ions, and molecules that are present at the scene of the "crime"—in this case at the spot where a fragment of a comet has inflicted heavy damage on the upper atmosphere of Jupiter.

The Goddard High Resolution Spectrograph was designed to measure, at any one time, very fine detail over a very narrow band of wavelengths in the spectra of bright astronomical objects. Fainter sources of light were best observed in less detail with the Faint Object Spectrograph. The S/L-9 spectroscopy team took data with both instruments but relied primarily on the FOS. It covered a wider swath of wavelengths in the limited time available, and its lower spectral resolution was sufficient for the science they wanted to do.

At about 5:00 am on Monday morning, July 18, Keith Noll and members of his team gathered in the STScI Operations Center. Fragment G had crashed into Jupiter an hour-and-a-half earlier. The G site was still fresh. Keith selected the location within the G spot to which he wanted Hubble to point, the darkest spot in the center where the fireball presumably detonated. Centering Hubble on that precise location was accomplished with a small, real-time pointing adjustment transmitted up to the spacecraft.

Later that day, Keith and his colleagues had their first chance to look at the G spot spectrum from the morning's observations. Part of what they saw was immediately understood. Another part of the spectrum was baffling.

There was a broad, deep valley or dip in brightness in the spectrum where sunlight had been strongly absorbed by a sequence of molecular bands centered just shortward of 200 nm. This corresponded exactly to the spectrum of ammonia (NH_3) molecules measured in ground-based laboratories. That made sense. Clouds of ammonia ice crystals were seen all the time, concentrated in a layer 15–25 km thick, floating high in Jupiter's troposphere (the troposphere is the deepest layer of the atmosphere where atmospheric convection and weather occur). Presumably these clouds were heated during the impact event, by the incoming meteor, by the exploding fireball, and by the backsplash of the plume as it pancaked down onto the cloud tops. That would cause the ammonia ice crystals to evaporate, producing clouds of ammonia gas that strongly absorbed incoming sunlight over a broad band of wavelengths spanning roughly 180–220 nm.

What Keith's team found baffling was a series of relatively narrow absorption bands, at least sixteen of them in all, spread out between 260 and 300 nm. They

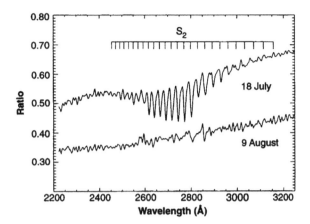

Figure 8.7. Portion of the spectrum of the darkest point in the G spot taken shortly after impact (top) and 3 weeks later (bottom). The "dinosaur teeth" pattern seen on July 18 at 260–290 nm is the characteristic band pattern produced in the laboratory by diatomic molecular sulfur, S_2, now seen for the first time on Jupiter. (Credit: Noll et al. 1995: reproduced with permission © AAAS.)

looked a little like the teeth of a T-Rex fossil (Figure 8.7). The regularity of this structure was characteristic of simple molecules, particularly diatomic (two-atom) molecules. The small spacing or separation between the bands suggested that a heavy molecule produced them. But what was it? It was a mystery.

At the 8:00 am press briefing at Goddard on Tuesday, July 19, Keith showed the shorter wavelength data, the spectrum containing the NH_3 absorption trough. It was easily explained in terms of what we already knew about Jupiter's atmosphere. At that briefing he did not show the longer wavelength observation containing the "dinosaur teeth" pattern of what appeared to be narrow molecular bands. At the moment it defied explanation.

Roger Yelle was an expert on the structure and chemical composition of planetary atmospheres. At the time of S/L-9, he was working at the NASA Ames Research Center in Moffett Field, California. He had come east to the STScI well in advance of the impact events to help the spectroscopy team get ready.

The Science Observing Team had set up a room at the STScI that was their base of operations. They did their work reducing and analyzing the Hubble data there. Yelle was sitting at a table in that room early Tuesday morning, going through a stack of books. It must have been fifteen books tall: all kinds of chemistry reference books, references to laboratory observations of molecular spectra of all kinds, references regarding the chemistry of planetary atmospheres—every source he could lay hands on that might help him solve the puzzle of the "dinosaur teeth" near 280 nm. Heidi heard him mutter in frustration, "I've seen this spectrum; I know what this is; I've seen it before; I just can't remember what it is!" He was leafing through those books when, all of a sudden he shouted, "I've got it!" "What?" Heidi asked. "It's pure sulfur!" he exclaimed. He showed her the book; she looked at the laboratory spectrum in one of the illustrations and nodded, readily agreeing. The sequence of molecular bands seen in the G spot spectrum was exactly like a textbook representation of the spectrum of pure diatomic molecular sulfur, S_2.

The discovery of pure sulfur at an S/L-9 impact site was a great surprise. The S_2 molecule had never before been detected on Jupiter (Noll et al. 1995). In fact it had only once before been seen in any astronomical object, Comet IRAS–Araki–Alcock in 1983. Could the sulfur have come from the comet itself? Possibly. But there was another, much more intriguing possibility. Theoretical models of the structure of Jupiter's atmosphere predicted that clouds of different chemical makeup were concentrated into relatively thin layers at different depths in the troposphere. The highest of these is the layer of ammonia ice crystals, readily seen on Jupiter from the Earth. About 30 km deeper below the ammonia clouds, a layer of clouds made of solid flakes of ammonium hydrosulfide (NII_4SII) was predicted to exist, though it had never been seen directly from Earth. If fragment G penetrated to the depth of this layer and heated it significantly, a chain of likely chemical reactions could easily lead to the formation of both sulfur and ammonia molecules.

Further examination of the FOS G-spot spectrum in the 180–220 nm interval revealed a sequence of bands of another molecule of sulfur, carbon disulfide or CS_2. Later in the week Keith's team also made some G-spot observations with the higher resolution GHRS instrument around 230 nm. In that set of data they discovered evidence of hydrogen sulfide (H_2S), the gas that produces the awful smell of rotten eggs. Taken together, the strength of all the sulfur molecules in the spectrum of the G spot suggested that the element was present in great abundance—far more sulfur than the parent comet of S/L-9 would have contained.

The presence of sulfur explained a great deal about how Jupiter looks to us Earthlings. The belts and spots normally seen on Jupiter, even though small telescopes, are often colorful. The Great Red Spot displays an orange–red hue, for example. A long-standing idea is that sulfur compounds from deep in the troposphere are roiled up into the high atmosphere by storms. When these compounds are exposed to sunlight, they are chemically altered into complex, sulfur-based molecules that have reddish-brown colors. Fragment G gave the first direct evidence that sulfur exists deep down in Jupiter's atmosphere; there was a reservoir of sulfur molecules that well up and, chameleon like, change colors, making Jupiter a beautiful sight.

The theoretical models that had predicted the presence of sulfur beneath the top layer of Jupiter's atmosphere appeared to be accurate. This suggested that other aspects of those theoretical models might also be accurate. The most intriguing facet of those models was that they predicted the presence of water.

At about seventy km below the ammonia clouds, the theoretical models predicted a lot of water (H_2O) ice clouds concentrated in their own layer, probably tens of kilometers thick. Before the S/L-9 impacts, planetary scientists had predicted, or at least had hoped, that a profusion of water would be detected. The expectation was that the more massive fragments might penetrate all the way to the water cloud layer, causing water vapor to be thrown into the upper atmosphere, confirming the existence of that deep cloud layer. There was some hope that the Hubble ultraviolet spectra would reveal H_2O vapor, but none was seen. NASA's Kuiper Airborne

Observatory and other ground-based telescopes made more sensitive measurements at infrared wavelengths. A weak detection was reported, but no one saw water in great abundance resulting from the comet impact events. The conclusion was that the fragments had not penetrated deeply enough into Jupiter to reach the water ice clouds, and the little bit of water that was seen had probably been carried in by the comet itself.

At the noon press briefing at Goddard on Wednesday, July 20, Roger Yelle described the other portion of the G spot spectrum taken with FOS, the part that Keith Noll had omitted the day before. The spectrum, showing the clear pattern of bands of pure diatomic sulfur molecules, could now be explained. It was one of the triumphs of that week.

The week's observing campaign with the Hubble spectrographs on the G spot came to a close on Thursday, July 21 with a final set of FOS observations. It was remarkable how rapidly the spectrum had changed over the course of just a few days. In the last set of observations, the evidence observed earlier in the week of ammonia, molecular sulfur, carbon disulfide, and hydrogen sulfide had largely faded away. Remaining in the final spectrum were spectral lines emitted by atoms or ions of magnesium, silicon, and iron.

Throughout the week reporters had been asking whether, in the dark impact spots, we were seeing material from the comet, from Jovian air, or both. The magnesium, silicon, and iron seen in that final spectrum certainly came from comet material. Those elements had never before been seen on Jupiter. Otherwise, the rest of the dark material spread out over the impact sites was in all likelihood composed of solid particles formed from dust and organic gunk, originating in the cooked mixture of Jovian air and comet debris. Figuring out the exact composition proved to be a complex and challenging problem. One thing was certain, however. The color of the dark stuff was the same in all the spots and in all parts of the spots, including material in the circular waves propagating away from the impact points. Whatever it was, the same material seemed to be created by every impact, big or small.

On Thursday afternoon a scientific celebrity paid a visit to the Science Observing Team at the STScI. The SOT was extremely busy that afternoon, but dropped everything to spend a couple of hours visiting with Carl Sagan. At least two members of the SOT, Melissa McGrath and Keith Noll, originally chose careers in astronomy because of Sagan's influence. This was an opportunity they couldn't miss. Sagan had slipped into town without fanfare. Notwithstanding his public popularity, gained through the Cosmos TV series and numerous visits to The Tonight Show, he was still a serious scientist, and he had come to the hub of the S/L-9 observing campaign with a serious scientific purpose: to discuss tholins.

In 1952 at the University of Chicago, chemists Stanley Miller and Harold Urey conducted a famous experiment, meant to simulate conditions on the early Earth. In their lab, a mixture of simple molecules—water, methane, ammonia, and hydrogen—was exposed to continuous electric sparks, simulating lightning. After a week, a rich mixture of complex organic molecules had accumulated in their apparatus. These included over 20 amino acids that are essential chemical components of cellular life.

Years later at Cornell Sagan did similar experiments showing that, when simple molecules that could be found on both the early Earth and on present-day Jupiter, as well as throughout the cosmos—methane, ammonia, water, and hydrogen sulfide—were exposed to ultraviolet radiation such as would be coming from the Sun, it produced reddish-brown tar-like substances, made up of heavy, complex organic molecules. Sagan called these substances "tholins." He postulated that their existence in the atmosphere of Saturn's moon Titan, for example, could explain that satellite's reddish-brown color. He and his Cornell colleague, Edwin Salpeter, speculated that life might exist in Jupiter's atmosphere, given its chemical composition, rich in simple organic molecules, and given the regular occurrence of lightning there. Could tholins form in Jovian clouds? Now S/L-9 had presented an opportunity to test the idea.

Sagan had come to Baltimore to talk to the SOT about whether tholins could have been created as the intense heat of the fragment impacts cooked comet dust and Jovian air. The dark reddish-brown material distributed around all the impact sites was similar in color to the "goo" that was formed in laboratory flasks during Sagan's own experiments. One or two of the SOT members gently tried to dissuade him, saying that at such low pressures and high temperatures found high up in Jupiter's stratosphere, heavy organic chains or rings might not survive for very long. But Sagan's idea that complex organic molecules, even amino acids, might have been produced in the highly energetic environment of a fragment impact probably would not be so readily dismissed today.

As the conversation drew to a close, Sagan made a statement to the team that stuck with me long afterward. He said, "I want to express my thanks to you all for pursuing this work so diligently and for being so effective in communicating the results to the public... It gives me a lot of confidence in the future to observe this group of smart, energetic young scientists, looking for all the world like freshly scrubbed members of the church choir." It was profoundly moving that this man, who had been so inspirational to members of the Hubble team early in their careers, could in turn be inspired by the work they had done.

He left town the next day, completely below the media radar. Two-and-a-half years later Carl Sagan died, far too young at age 62, a victim of myelodysplasia.

The week was drawing to a close. The last fragment, W, fell into Jupiter at 4:06 am EDT on Friday, July 22. At the concluding NASA media briefings the astronomers tried to tie up some loose ends. The impacts had raised many questions, only a few of which could be answered.

Hal Weaver had used WFPC2 to observe two fragments only ten hours prior to their demise. Although Jupiter's strong gravity had stretched out the dust surrounding the fragments along the direction of motion, there remained a dominant central body within each that held itself together. That was consistent with the power displayed by most of the impacts. Loose piles of rubble would not have produced the big fireworks shows we observed.

Andy Ingersoll described how the dark circular rings around the centers of five of the impact sites could be seen to expand outward at about 450 m s^{-1} (1000 miles per hour) in a time series of Hubble images. It was tempting to describe these as sound waves, compression waves moving outward through the Jovian air away from the initial fireball. Unfortunately, they were moving too slowly and were too narrow to be sound waves under the conditions thought to prevail in Jupiter's stratosphere. Although there was much speculation, the true nature of the waves was never conclusively explained.

One other surprising discovery came to light weeks after the "quick look" discussions in the NASA briefings. In all, Heidi's team was able to observe the plumes of four impacts. All the plumes went higher than had been originally thought and all of them reached approximately the same height, about 3000 km (about 1900 miles), whether large fragments or small ones produced them. The reason seemed simple. None of the fragments penetrated very deeply into the Jovian atmosphere. But a small fragment, with lower mass and less energy, penetrated less deeply and had less overlying Jovian air to blow skyward. A larger fragment, with larger mass and greater energy, penetrated more deeply and was able to propel a greater load upward. The difference in penetration depth tended to level the playing field for large and small fragments, so that each was able to eject its plume with about the same velocity, 10–13 km s^{-1}, to the same maximum height.

Comet Shoemaker/Levy 9 conducted a spectacular but poorly controlled experiment on Jupiter. We could observe it and marvel at it. But we could not always find a single, clear explanation for everything we saw. And we couldn't go back and repeat the experiment while controlling the variables as would be possible in a laboratory on Earth. Often it was a situation of, "it could be this, or it could be that." We learned a lot, but many of the things we observed remain mysteries.

Scientific results from the S/L-9 campaign were published together in an issue of the journal Science in 1995.[1] David Levy's first-hand account of the campaign is given in his enjoyable book, "Impact Jupiter: The Crash of Comet Shoemaker–Levy 9."[2]

What was the significance for us on Earth of the events that took place on Jupiter that week in July of 1994? Certainly Hubble and many other observatories on Earth and in space acquired a mother lode of scientific data. Scientists continue to analyze and interpret those data to the present day. But, valuable as they are, I don't believe the scientific results represent the most important outcome of the S/L-9 events.

During that week ordinary citizens began to realize that we live in a solar system that is still dynamic, still changing even after the 4.6 billion years since its formation. Collisions between solar system bodies still happen. There are thousands of asteroids whose orbits bring them close to Earth from time to time. On average, the time between really big asteroid and comet impacts on Earth, impacts large enough to wipe out a substantial fraction of life on our planet, is of order tens to hundreds of millions of years. On the other hand, the probability of being hit by a moderate size chunk of space rock, say 100 m in diameter, one that could decimate a city, sometime in the next few centuries is large enough to worry about. The sizes of the parent comet and the 21 fragments of Shoemaker–Levy 9 were consistently overestimated by the scientists studying it prior to and during the impact week. What is

---

[1] Science, 3 March 1995, Vol. 267, Issue 5202, pp. 1277–1323.
[2] 1995 (Cambridge, MA: Basic Books, A Member of the Perseus Books Group).

truly surprising is the fact that, even though the fragments were relatively small (0.5–1.0 km or so), they still did enormous "damage" on Jupiter.

On 1994 July 21, while the comet fragments were still raining down on Jupiter, the United States Congress added language to the NASA budget authorization bill requesting the Agency to develop a plan to search for and identify 90% of all objects greater than one kilometer in size whose orbits cross Earth's orbit around the Sun. NASA was asked to submit a report describing the plan back to Congress by 1995 February 1. Two years earlier, a committee of scientists led by David Morrison of NASA's Ames Research Center had recommended that a proactive search be undertaken to find and catalog near-Earth objects that could pose a potential collision hazard. The Morrison report was amplified by the actual catastrophic events of that week on Jupiter, residing so visibly on the front pages of newspapers across the country and on the evening national newscasts. The S/L-9 impacts undoubtedly stimulated Congress to take action.

In response to this and later Congressional mandates, a number of programs were established at American observatories to conduct sky surveys in search of near-Earth objects that might potentially cause us harm. Congress also directed NASA and the U.S. Department of Defense to investigate how we could respond if a comet or asteroid were discovered to be on a collision course with Earth.

On 2013 February 15 a brilliant fireball lit up the sky over the southern Ural region of Russia, blazing brighter than the Sun. It came in at a shallow angle at high velocity, 19 km s^{-1} or about 42,000 miles per hour. It exploded as an airburst at a height of about 30 miles above the ground over the city of Chelyabinsk. It generated a hot cloud of gas and dust as well as a powerful shockwave. The shockwave caused damage in six cities. It broke most windows and caused buildings to collapse in Chelyabinsk. Over 1000 people were injured, primarily from flying shards of glass. There was widespread panic.

The object responsible was probably a small asteroid about 20 m (66 feet) in diameter and weighing about 14,000 tons—nowhere near as large as the smallest S/L-9 fragment. It had not been detected prior to entering the Earth's atmosphere because it came from a direction nearly lined up with the Sun. It couldn't be seen in the glare of sunlight, even if someone had been looking for it. Later that same day, a completely unrelated asteroid, named 367943 Duende, came within 27,700 km (17,200 miles) of the Earth. It was estimated to be about 30 m (98 feet) in diameter. Such small objects clearly could do significant damage on a local or regional scale, if we were so unfortunate as to be hit by one in a populated area.

Comet S/L-9 sensitized the people of Earth to a small, but non-negligible threat from the cosmos. The Chelyabinsk event reinforced the concern. There is no cause for alarm, but we should remain vigilant and prepared.

On July 23 the Science Observation Team's work was done. There was nothing more to do. Some of them hadn't been away from the STScI for many days. So, they all decided to drive to Fells Point in Baltimore to have a good dinner together. At Fells Point there was a guy who had a modest 8 inch telescope set up on a street corner. He had a hat there. A person could throw a dollar into the hat and look at Jupiter. The whole team stood in line. He didn't know who they were. When they got up to the telescope, one-by-one, and looked, they could see the black spots. David Levy had been wrong about that.

Heidi Hammel said to her colleagues, "I have spent the whole week looking at fantastic Hubble images. But now that I see those big black spots with my own eye through the telescope, I am utterly blown away. The Hubble images came out of a computer. I can personally see this image of Jupiter with my own eye. The photons are coming directly from Jupiter to me. I'm having emotions about this that I haven't had all week."

The guy was handing out little yellow post-it notes. Heidi took one and later taped it into her notebook. The note had a cartoon drawing of Jupiter on it with a comet going "Splat!" It said, "1994 July 16–22, I survived the collision of Comet P/Shoemaker–Levy 9 with Jupiter! Fells Point, Baltimore, MD."

References

Noll, K. S., et al. 1995, Sci, 267, 1307

Weissman, P. 1994, Natur, 370, 94

AAS | IOP Astronomy

Life With Hubble
An insider's view of the world's most famous telescope
David S Leckrone

Chapter 9

From Darkness to Light: Black Holes and Quasars

The cover of the 1966 March 11 edition of Time magazine was a portrait of astronomer Maarten Schmidt. I was an astronomy graduate student at UCLA at the time. I was impressed that someone in my field of study could make it so visibly into the popular press.

The discovery that landed Schmidt on the cover of Time was a very big deal. In the 1950s astronomers at Cambridge University in England had used an array of radio antennas to map the sky at various radio frequencies. They published catalogs giving the position and brightness of each celestial source of radio waves that they detected. Some of these radio sources were very strong, but optical astronomers could not immediately associate them with objects that could be seen at visible wavelengths. The nearest objects to those positions just looked like stars.

In 1962 astronomers in Australia observed with a large radio telescope as the Moon occulted (passed in front of) one of the strongest of these radio sources, 3C-273, object number 273 in the "Third Cambridge Catalogue of Radio Sources." That gave them a very precise position for the source on the sky. It fell right on top of a medium-brightness "star" in the constellation Virgo. When Maarten Schmidt took a spectrum of this "star" with the 200 inch Hale reflecting telescope on Palomar Mountain, he saw a pattern of light spread out into its component colors unlike that of any star. It had broad, bright hydrogen emission lines that had been displaced to longer wavelengths, i.e., had been red-shifted, by about 16%. If the red-shift was, like that of galaxies, due to the expansion of the universe, then 3C-273 had to be at a distance of about 2 billion light years. It appeared extraordinarily bright, even though it was very far away. Its intrinsic brightness was about a thousand times greater than the total brightness of the two hundred billion stars in our Milky Way Galaxy, or about four trillion times the brightness of the Sun. It was in fact the most luminous, the most energetic single object astronomers had yet seen in the cosmos.

And yet it had the unresolved, point-of-light appearance of a star. Schmidt called it a "quasi-stellar radio source" or quasar for short.

In the decades that followed, astronomers identified and cataloged over 7000 objects like this. Some didn't emit radio waves and were dubbed simply "quasi-stellar objects" or QSOs. All of them were at enormous distances from the Earth, billions of light years away. 3C-273 turned out to be one of the closest to us. They were the most energetic objects in the visible universe—also the most puzzling. What were they? What produced their prodigious energy output?

Quasars were not the only powerful and mysterious energy sources to be found by astronomers in the 20th Century. A large number of relatively nearby galaxies, both spirals and ellipticals, contained nuclei that were super-energetic. They contained gas in motion at high velocities and emitted tremendous power in the form of radiation over the entire electromagnetic spectrum. Astronomers called these active galactic nuclei, or AGNs. Their discovery pre-dated the discovery of quasars. But astronomers began to wonder if the two could be related in some way. It was hypothesized that quasars might be much brighter, much more powerful and far more distant versions of AGNs, the intense glare of their central cores completely swamping out the faint light from the surrounding galaxy—AGNs on steroids.

In the 1970s and 1980s, two decades before the Hubble Space Telescope came on the scene, two important questions challenged observational astronomers working on quasars and AGNs: were the quasars merely distant AGNs, the ultra-bright nuclei of normal galaxies, undergoing exceptionally energetic outbursts; and what was the energy source powering these violent phenomena? Speculation commonly held that the quasars must indeed be faint normal galaxies, the brilliant centers of which were undergoing some kind of titanic upheaval. It was also commonly speculated that the most likely central engine powering these phenomena in both quasars and AGNs were supermassive black holes which were accreting, obliterating, compressing and heating up the surrounding material falling into them—gas, stars, smaller galaxies, other black holes.

Astronomers labored mightily at ground-based observatories to find observational proof of these two hypotheses, but to no avail. They carefully measured how the number of stars packed into a certain volume and how the velocity of motion of those stars increased as they looked closer and closer in toward the centers of radio-emitting giant elliptical galaxies. The measurements suggested the presence of a very large central mass in some galaxies. But the mountain-top telescopes just didn't have the angular resolution to see close enough to the galaxy centers to rule out other explanations—like dense concentrations of normal stars. They couldn't definitively prove the existence of central black holes of monstrous proportions. Similarly, the ground-based telescopes didn't have either the resolution or dynamic range to detect whether there were faint underlying galaxies being swamped by the intense brightness of the quasars far off across the universe. For the Hubble Space Telescope, in the first couple of years after SM1, finding definitive proof for what had previously been hypothetical was straightforward.

M87 is the 87th entry in Charles Messier's 1781 catalog of "fuzzy" patches of light in the sky, objects that might confuse astronomers looking for comets. It is an enormous elliptical galaxy; a spheroid of trillions of stars about 120 thousand light years across. It's the second brightest galaxy in the Virgo Cluster of galaxies some 54 million light years away, and one of the most massive galaxies in the local universe. From Earth it can readily be seen (as a fuzzy patch of light) through a small amateur telescope. Through the Hubble Space Telescope it is a magnificent sight, with a brilliant concentration of visible light at its center (Figure 9.1). A narrow blue jet extends 4900 lt-yr outward from its center. At its core resides an AGN, a powerful source of radio waves, X-rays, gamma-rays, and other intense electromagnetic radiation. M87 was a natural choice as the first target for scientists using the newly repaired Hubble to search for the engine that powers an AGN.

Richard Harms was the Principal Investigator on the Faint Object Spectrograph (FOS) and leader of the team at the University of California at San Diego that NASA had selected to develop the instrument and conduct its early scientific investigations. Holland Ford of the Space Telescope Science Institute (STScI) and Johns Hopkins University was the Deputy to Richard on the FOS science team, as well as being the Instrument Scientist designated by NASA to lead the COSTAR development effort.

Figure 9.1. Visible-light image of the giant elliptical galaxy M87 taken with Hubble's Advanced Camera for Surveys. The blue jet to the right of center emanates from an Active Galactic Nucleus (AGN). The many star-like points of light surrounding the galaxy are in fact globular star clusters, each containing hundreds of thousands of stars. (Credit: NASA, ESA, Hubble Heritage Team, STScI 2008-30.)

Now the two astronomers had a golden opportunity to put their labors of many years to the test. They wanted to point Hubble toward the very core of M87 to see what was really there. This observation was included in the original proposal for the FOS that Harms' team submitted to NASA in 1978. It continued to be a high priority objective for Harms and Ford as the launch date for Hubble approached in 1990.

To explore the nucleus of M87 with the FOS they needed a road map. They needed to use the new Wide Field and Planetary Camera (WFPC) to snap high-resolution images through a filter that covered a narrow band of wavelengths isolating light emitted by hydrogen and ionized nitrogen. The FOS had a very small entrance aperture and a one-dimensional detector (512 light sensitive elements lined up in a row). It could not take two-dimensional images. To figure out where to point the telescope to acquire light from an interesting bit of structure within a galaxy, they needed an actual high-resolution picture of the galaxy. But there was a problem of professional protocol standing in their way. The WFPC instrument team had highest priority to use their own instrument on targets they had previously designated and M87 was one of those targets.

In 1990 February, just two months before launch, the Hubble instrument teams and other observatory scientists with guaranteed observing time met at Princeton to compare notes on the observations they planned to make. The teams wanted to coordinate with each other to avoid "stepping on each other's toes" when it came to target selections and research objectives.

Holland Ford had driven from Baltimore to Princeton hoping to persuade the members of the WFPC team to allow the FOS team to take images of the nucleus of M87 very early on after the launch and deployment of Hubble. He explained the need as being essential to enable a search with the FOS for evidence of a supermassive black hole.

This request riled some member of the WFPC team. It would be an unseemly breach of accepted protocol. M87 was a prize target for them and they weren't happy that it might be "stolen away" by competitors from another instrument team. After the discussion had been going on for a while, Sandy Faber, a co-investigator on the WFPC team, spoke up. It seemed to her, she said, that what Holland was asking for was to be able to use Hubble as an observatory. And that's how observatories operate, with multiple instruments available to be used as tools to solve scientific problems. In that spirit she supported Holland's request.

It was dark when Holland and his STScI colleague Chris Blades headed from Princeton back to Baltimore. The weather was miserable, dangerous. Ice and snow covered the highway. It was difficult to see. Driving in such a white-knuckle situation, it was hard for Holland to think about what they had accomplished that day. In fact it had been a transitional moment for the astronomical community.

It was seemingly a simple observation—the recognition by Sandy Faber that Hubble was in fact an observatory, rather than a collection of individual instruments "owned" by individual instrument teams. The precedents went the other way. Major space astronomy missions prior to Hubble were indeed simply platforms on which individual science teams, competitively selected by NASA, could mount the instruments they had designed and built. Each team used its own instrument to carry out its own research. There were stipulations by NASA that the teams should archive their data in a way that would make them useful to the broader scientific community, but this was not always taken very seriously. The Orbiting Astronomical Observatories (OAOs) and the Orbiting Solar Observatories (OSOs) followed this model. The International Ultraviolet Explorer (IUE) was designated as an "observatory" and was in fact operated by NASA as a facility available to all astronomers. However, the IUE was a single-instrument (spectroscopic) platform, without the versatility of a traditional mountain-top observatory.

From the beginning of the program in the 1970s we NASA types referred to the Space Telescope as an "observatory." We used that terminology once or twice in the first "Announcement of Opportunity for Space Telescope" in 1977. But the significance of that designation didn't really sink in viscerally until that day in 1990 when Sandy Faber politely pointed out to her team mates that in an observatory they should be willing to share usage of their instrument with others. Thus, Harms and Ford got what they needed—the forbearance of another science team to use that team's instrument to observe one of that team's high priority targets early in the mission.

Then spherical aberration intervened. Neither the WFPC nor the FOS could provide the data they needed to solve the puzzle of the existence of a massive black hole in the core of M87 until a way was found to correct the telescope's bad vision. Like so many other primary science objectives for Hubble, this one had to be postponed until after a servicing mission could be flown.

In 1994 February, two months after the conclusion of SM1, the team finally acquired the long-delayed, beautiful, high-resolution WFPC2 images of the region around the nucleus of M87. The following May they put the FOS + COSTAR to work on M87 as they had originally planned long before.

What they observed had never been seen before except in the imaginations of astronomers—a small, thin circular disc of gas about 475 lt-yr in diameter, glowing brightly in the red light emitted primarily by hydrogen atoms (Figure 9.2; Ford et al. 1994). High-speed electrons spiraling outward around magnetic lines of force produced the long blue jet, which was easily seen in both ground-based and *Hubble* images of the galaxy. The jet clearly originated at the center of the disk. The disk was tilted with respect to the line of sight, so it looked elliptical—roughly football shaped.

The key question was, "Is the disk rotating, and if so, how fast?" The easiest way to tell would be to measure the Doppler shift in the wavelength of light emitted at selected points on the disk. Like a policeman can tell how fast your car is moving toward him or away from him by the shift of radar waves to higher or lower frequencies, Harms and Ford could look for the Doppler shifts in the glowing spectral emission lines at various points on the disk to measure its velocity of rotation.

The two astronomers used the FOS + COSTAR to take spectra of five carefully chosen points in the disk that they had mapped out in the WFPC2 image—one right at the center of the disk, two on either side and as close to the center as they could

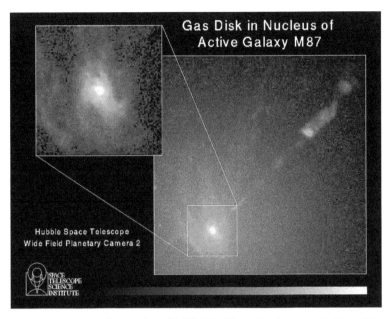

Figure 9.2. Image of the nucleus of M87 taken with WFPC2. The nucleus is surrounded by a thin rotating disk of hot gas, tilted at an angle to our line of sight. (Credit: NASA, H. Ford, R. Harms, the FOS Team, STScI 1994-23.)

point the telescope. Two other points farther off-center were also measured (Harms et al. 1994). Because of Hubble's terrific resolution, they were able to take measurements at points less than 60 lt-yr from the center of M87, much closer than had ever been done before. Sixty light years may seem like a large distance, but on the immense size scale of M87 (or of our own Milky Way Galaxy for that matter) they were practically rubbing elbows with the core of the galaxy. For comparison, the five brightest stars we can see in the evening sky from Earth range in distance between 4 and 98 lt-yr away. So, a distance of 60 lt-yr would practically be in the Sun's back yard.

Their spectroscopic data consisted of bright lines emitted by hydrogen atoms and ions (atoms with some electrons stripped off) of nitrogen, oxygen, and sulfur, isolated against a faint background of starlight. The emission lines were easily seen and their individual wavelengths were easy to measure (Figure 9.3).

The first important discovery was that the measured velocity of the disk at different distances from its center closely followed the same laws that Johannes Kepler had developed four centuries earlier to describe the motion of the planets around the Sun—the material in the disk was in Keplerian motion around whatever source of gravity lurked at its core. As we use the orbital motion of the planets to measure the mass of the Sun, Harms and Ford used the measured rotation of M87's disk to calculate the mass of that central object.

The result was breathtaking, though not really unexpected. At a distance of 60 lt-yr out from the center of M87, gas on one side was moving *toward* us at a velocity of

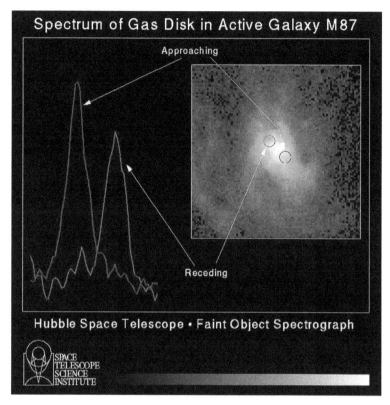

Figure 9.3. Faint Object Spectrograph observations of the rotating disk of hot gas surrounding the nucleus of M87. A bright spectral emission line produce by oxygen ions (oxygen atoms that have lost two electrons) is shown as measured at two positions on either side of the nucleus. The line is Doppler shifted to the red at the 11 o'clock position and to the blue at the 5 o'clock position. (Credit: NASA, R. Harms, H. Ford, the FOS Team, STScI 1994-23.)

about 500 km s^{-1} (1.1 million miles per hour) while gas on the other side was moving *away* from us at that same velocity. Correcting for the tilt angle of the disc, these measurements translated into 750 km s^{-1} actual rotational velocity.

It was simple to calculate from Kepler's laws of planetary motion and Newton's law of gravity that the central mass required to hold that gas in its orbit was about 2.4 billion times the mass of our own Sun (in other words $2.4 \times 10^9 \times 2 \times 10^{30}$ kg = 4.8×10^{39} kg). And that enormous mass had to be crammed into a very small volume. Hubble WFPC and FOC images taken even before SM1 had revealed that the bright, highly concentrated knot at M87's center had to be less than about 6 lt-yr across—perhaps much less. The only candidate known to science, based on general relativity theory, was a supermassive black hole. As Holland Ford said, "If it's not a black hole, I don't know what it is." Richard Harms chimed in, "A massive black hole is actually the conservative explanation for what we see in M87. If it's not a black hole, it must be something even harder to understand with our present theories of astrophysics."

How big (or small) could the black hole be? In the simplest model for a black hole—spherical, non-rotating, and not electrically charged—the event horizon, the distance at which the escape velocity due to the gravitational pull of the object exactly equals the velocity of light is called the Schwarzschild radius. For a black hole 2.4 billion times the mass of the Sun, that distance is about 50 Astronomical Units (au), that is 50 times the distance from the Earth to the Sun. Our solar system out to the outer edge of the Kuiper Belt is roughly 50 au in radius. So, the gargantuan black hole at the center of M87, all 2.4 billion solar masses of it, is crammed into a volume about the size of our own solar system.

Of course we can't directly observe M87's supermassive black hole. It is, after all, *black*. Light, electromagnetic radiation can't escape from it. But it is an immense source of energy. The material in the disk Ford and Harms discovered eventually spirals inward at higher and higher velocities, into that tightly concentrated central core and eventually across the event horizon into the black hole itself. As it does so, it is compressed and heated to very high temperatures. It emits prodigious amounts of radiation—visible light, X-rays, gamma-rays. This is the source of the radiation we observe in the active galactic nucleus of M87.

But is the situation the same in all the other AGNs found throughout the cosmos, are they all powered by monster black holes? In 1995 Holland Ford's doctoral student at Johns Hopkins, Laura Ferrarese, together with Ford and Walter Jaffe in the Netherlands, used WFPC2 and FOS observations of the elliptical active radio galaxy NGC 4261, similar to those obtained for M87, to establish that a disk of dust and gas surrounding the center of that galaxy was rotating in Keplerian motion around a black hole. That black hole had a mass 490 million times the mass of our Sun—less than M87's black hole mass, but still enormous (Ferrarese et al. 1996). That made two examples.

At about the same time that this work with Hubble was going on, radio astronomers used the Very Long Baseline Array (VLBA), a network of 10 large radio telescopes spread out over the United States from Hawaii, to New Hampshire to the US Virgin Islands, to pinpoint with incredible precision another massive black hole within the active galactic nucleus of NGC 4258 (Miyoshi et al. 1995). Instead of the visible light observations used by the Hubble astronomers, they looked for lines emitted as masers by water molecules orbiting in clouds around the nucleus of the galaxy. The term "maser" stands for microwave amplification by stimulated emission of radiation. Like a laser amplifies the intensity of visible light, a maser amplifies microwave radiation. And it so happens that water molecules can be stimulated to emit intense beams of microwaves at a wavelength of 1.35 cm. In the case of NGC 4258, the stimulation comes from radiation, produced by the active galactic nucleus, impinging on the clouds of water.

Radio astronomer Makoto Miyoshi and his colleagues found that the water cloud masers moved in perfect Keplerian orbits around the nucleus of NGC 4258. The central object keeping them in orbit had a mass of 39 million times the mass of the Sun, contained within a volume no larger than half a light year in radius. It had to be another supermassive black hole, independently discovered this time with an entirely different kind of observational tool than the camera and spectrograph used on Hubble.

In 1997 February we launched Servicing Mission 2 (SM2) to Hubble. The payload included a new, more sophisticated and versatile spectrograph that would replace both the FOS and the GHRS. This was the Space Telescope Imaging Spectrograph (STIS) designed and built under the direction of Principal Investigator Bruce Woodgate at Goddard Space Flight Center and his team. He advertised STIS as a "black hole hunter."

Because it was an imaging instrument with two-dimensional light detectors, STIS could take spectra of many points on the sky at the same time. In particular if astronomers lined up its long entrance slit on the center of a galaxy, perpendicular to the rotation axis of the galaxy, they could very efficiently measure how fast the central core of the galaxy was rotating. To one side the core was rotating toward us, and its light would be shifted toward bluer wavelengths; to the other side the core was rotating away from us, shifting its light toward the red. If a massive black hole were present, the rotational velocity of the gas and stars surrounding it in the nucleus would increase dramatically over a very small distance inward toward the center. The faster the material was rotating, the larger the blue and redshifts would be, as seen for example in spectral lines emitted by hydrogen atoms and ions of nitrogen and sulfur. At the very center of the galaxy, the position of the line would abruptly shift—a kind of astrophysical whiplash—creating a characteristic S-shaped curve (Figure 9.4). On the other hand, if no massive black hole were present, the spectral line would be tilted a bit, depending on the rotation of the galaxy as a whole, but it would remain essentially straight. STIS could find supermassive black holes at the center of galaxies 40 times more efficiently then FOS could—thus its designation as a black hole hunter.

Over the years a number of astronomers used STIS to carry out, in effect, a demographic survey of black holes at the centers of galaxies. As the numbers of massive black holes discovered continued to rise into the dozens, it became apparent that virtually every galaxy nucleus contained one. Their masses ranged from a few million up to a few billion times the mass of the Sun.

UCLA astronomer Andrea Ghez and her team carefully measured the motions of individual stars very close to the center of our own Galaxy, and showed that even the Milky Way has a central black hole of about 4 million solar masses (Ghez et al. 2005). It is crammed into a volume no larger than about 6.25 lt-hr in radius. It would easily fit within the orbit of Neptune. But our own massive black hole is a quiescent object. Apparently it has no material falling into it that could compress, heat up, and emit radiation as it approached the event horizon.

The most astonishing discovery to be gleaned from the ongoing census of massive central galactic black holes, however, is that their masses are tightly correlated with the masses of the large bulge structures of the galaxies whose centers the black holes occupy (Figure 9.5). The bulges can either be ellipsoidal structures containing tens of millions or more stars surrounding the centers of spiral galaxies, like the Milky Way, or they can be entire elliptical galaxies with up to several trillions of stars. These

Figure 9.4. Using Hubble's Space Telescope Imaging Spectrograph to find a massive black hole at the center of a galaxy. In the left frame STIS' long, narrow entrance slit (blue rectangle) is aligned across the nucleus of the elliptical galaxy M84 to measure the rotational velocity of a disk of hot gas and stars. In the right frame a spectral emission line of once-ionized nitrogen is measured at each spatial point along the slit. The spectrum is spread out (dispersed) over STIS' 2-D CCD light detector in the horizontal direction. The sudden zigzag very close to the nucleus results from large Doppler shifts and a high rotational velocity. The black hole in this galaxy nucleus is about 1.5 billion solar masses. (Credit: NASA, G. Bower, R. Green, B. Woodgate, the STIS Team, STScI 1997-12.)

bulge structures are huge in scale, well beyond the gravitational tug of the central black hole.

How strange! The biggest bulges contain the most massive black holes. Smaller and smaller bulges contain progressively smaller black holes. But each bulge is far too big to be "aware" of the existence of the black hole at its center, and the black hole has no "awareness" of the size and scale of the enormous assembly of stars around it. Yet, the two are somehow coupled together.

Continuing her work on supermassive black holes after receiving her PhD from Hopkins, Laura Ferrarese, and her colleague David Merritt primarily used STIS observations to demonstrate this correlation (Ferrarese & Merritt 2000). Somehow the origin of a galaxy (or at least of its bulge component) and the origin of the monster black hole at its center are tied together. Astronomers have a few ideas about possible mechanisms that would couple the two, but as of today no one knows with certainty. Clearly, however, Mother Nature is giving us an important clue about how galaxies, and black holes, form and evolve.

We now know from Hubble and VLBA radio observations that active galactic nuclei are powered by massive black holes with voracious appetites. They feed on whatever material—gas, stars, asteroids, comets, or other galaxies—happens to

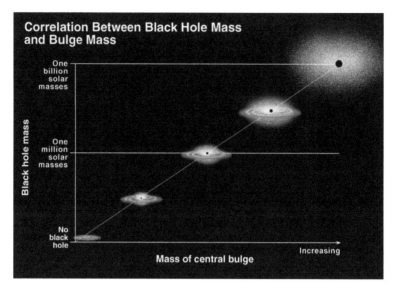

Figure 9.5. Artist's representation of correlation between the mass of a black hole in the nucleus of a galaxy and the mass of the central bulge of stars centered on the nucleus. (Credit: NASA, K. Cordes, S. Brown, STScI 2000-22.)

come close by and gets caught up in their immense gravitational pull. If no material is available for this kind of cosmic feast, the nuclei remain quiescent until the next "food" source comes along. If an enormous quantity of material becomes available, the black holes can gorge on it, giving off tremendous quantities of electromagnetic energy.

What about the quasars? Could they simply be galaxies in the distant, early universe too faint to be seen from the ground in contrast to the brilliant accretion disks around their central black holes? Presumably the accretion disks would generate enormous amounts of radiation as the black holes around which they rotate consume them. The process would continue for as long as the accretion disk was replenished with new material, say from a colliding galaxy. The Hubble Space Telescope provided the answer to the question.

Some of the earliest quasar observations from Hubble with WFPC2 showed them in the midst of collisions with other galaxies. This caused some speculation that there might be "naked" quasars—quasars that were originally not associated with a galaxy, but which acquired a host galaxy via a collision. However, as the sample size of quasars observed with Hubble grew larger, it became clear that this was not the case. John Bahcall and his colleagues produced a survey of 20 of the brightest nearby quasars, objects at redshifts less than $z = 0.30$, corresponding to lookback times of 3.4 billion years or less (Figure 9.6; Bahcall et al. 1997). Mike Disney from the University of Wales and the Faint Object Camera Science Team did similar work. They found that quasars existed in a variety of environments. They all were embedded in host galaxies that were always centered on the quasars to within the measurement uncertainties. In some cases the quasars were found at the centers of perfectly normal looking galaxies—both spiral and elliptical. Some quasars looked

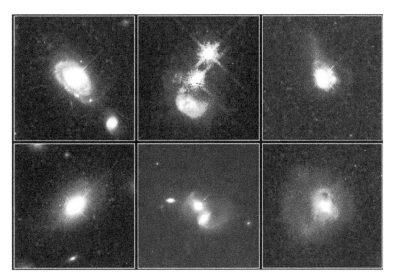

Figure 9.6. Examples of host galaxies of quasars observed with WFPC2. The two in the left-hand column are in the nuclei of normal looking galaxies—a spiral galaxy at the top and an elliptical galaxy at the bottom. The other four are in disrupted systems showing evidence of galaxy interactions or collisions. (Credit: NASA, ESA, J. Bahcall, M. Disney, STScI 1996-35.)

disrupted by the gravitational tug of nearby galaxies. In other cases they were observed in galaxies that were in the process of colliding with other galaxies. In 40% of the sample, with Hubble's exquisite resolution, Bahcall's team detected very close companion galaxies. Gravitational interactions and collisions with companion galaxies appeared to play a role in triggering the intense output of light from these distant objects.

The early universe was smaller than it is today. Galaxy collisions were far more common. A long-ago collision between two galaxies would partially dismember both, freeing up material—gas, dust, and stars—to fall onto a massive central black hole, resulting in the appearance in telescopes on Earth of the most luminous and energetic objects in the universe. That's what Maarten Schmidt was observing in that image and spectrum of 3C-273 that famously appeared on the cover of Time magazine on 1966 March 11.

References

Bahcall, J., et al. 1997, ApJ, 479, 642
Ferrarese, L., Ford, H. C., & Jaffe, W. 1996, ApJ, 470, 444
Ferrarese, L., & Merritt, D. 2000, ApJ, 539, L9
Ford, H. C., et al. 1994, ApJ, 435, L27
Ghez, A. M., et al. 2005, ApJ, 620, 744
Harms, R. J., et al. 1994, ApJ, 435, L35
Miyoshi, M., et al. 1995, Natur, 373, 127

AAS | IOP Astronomy

Life With Hubble
An insider's view of the world's most famous telescope
David S Leckrone

Chapter 10

How High Is the Sky? Measuring the Universe

The primary job of the Space Telescope Science Institute (STScI) was to define the scientific research program of the observatory. In the early 1980s its first permanent director, Riccardo Giacconi, began to ponder what the most productive use of the telescope would be to assure that the most important science would get done. Many critical research programs required the accumulation of a lot of observational data over a long period of time: observations of faint objects required long exposure times, some programs required observation of numerous cosmic targets or a succession of observations of the same target spread out over days, weeks or months. But the observatory was limited to roughly 3000 hours of on-target observing time available each year. And a wide range of scientific programs did not require such large allocations of that precious time. What if the Hubble spacecraft had a mission-ending failure early on? Would the most important science that only Hubble could address be left undone?

In 1983 Riccardo appointed a Space Telescope Advisory Committee (STAC), chaired by Jeremiah Ostriker from Princeton, to consider the options. At the time a lame joke went around about the names of STScI committees—we had a STIC, a STAC, and a STUC. In the end the STAC recommended that extensive amounts of observing time, early in the mission, be devoted to three especially important Key Projects: determine the distance scale and rate of expansion of the universe to 10% accuracy; map the distribution of the gas surrounding and between nearby galaxies, the intergalactic medium; randomly survey the sky to medium depth, to fainter limits than could be achieved from the ground just to see what's there.

Of these three Key Projects, the first was historically the most important for the Hubble mission. The basis for recalibrating the distances to far-away galaxies was the ability to observe classical Cepheid variable stars in those galaxies. The telescope's 2.4 m aperture size was set by the requirement to see individual Cepheid variables in galaxies at least as far away as the Virgo cluster of galaxies.

Cepheids are yellow giant stars that pulsate, growing larger and smaller, brighter and fainter in a regular, well-understood cycle. The more massive and more luminous Cepheids pulsate more slowly, taking longer to complete a cycle, than the less massive, less luminous ones. There are numerous examples in our own Milky Way Galaxy, including their namesake, the star delta Cephei.

There is a well-understood tight relation between the rate of pulsation and the intrinsic light output of the classical Cepheid variable stars. This period–luminosity relation is called Leavitt's law in honor of Henrietta Leavitt.

In 1892 Henrietta Leavitt graduated from an institution that later became Radcliff College. A course she took in her senior year motivated her interest in astronomy. She volunteered to work as an assistant at Harvard Observatory in 1895, becoming a permanent member of the staff in 1902. Working under astronomer Edward Pickering, her job as a "human computer" was to determine the brightness of the stars imaged in a collection photographic plates taken of various regions of the sky. She did extraordinary work in establishing a standardized system of stellar magnitudes (brightness on a logarithmic scale) and applying that system to stars in large swaths of the sky.

Later Leavitt was assigned to measure the brightness of variable stars in hundreds of photographic plates taken of nearby satellite galaxies of the Milky Way, the Large and Small Magellanic Clouds (LMC and SMC). She measured nearly 1800 variable stars in these two dwarf galaxies. Since all the stars in each galaxy were approximately at the same distance from Earth, their apparent relative brightness corresponded to their relative intrinsic light output. If star A looked twice as bright as star B that was because it was intrinsically twice as bright, since both stars were at about the same distance from Earth. Leavitt discovered that some of them, the Cepheid variables, displayed a tight, linear relationship between the logarithm of their brightness and the logarithm of their period—the time over which they completed one cycle of brightness variations. Her discovery, Leavitt's law, was published in a paper submitted under her boss' name, Edward Pickering, in 1912 (Leavitt & Pickering 1912). The first sentence in Pickering's paper read, "The following statement regarding the periods of 25 variable stars in the Small Magellanic Cloud has been prepared by Miss Leavitt."

By measuring their pulsation periods, astronomers could directly measure the intrinsic brightness of Cepheid variables, according to Leavitt's law. Comparing the intrinsic brightness of a Cepheid to its apparent brightness at Earth (which varies as one over the square of the distance from the source) gives a direct measurement of its distance. One complication with this technique is that there are actually two different types of Cepheid variable stars, as recognized by astronomer Walter Baade in the 1940s. One type is systematically fainter than the other and each follows a different period–luminosity relation. The two types were mixed together in the set of data Henrietta Leavitt used to establish Leavitt's law. So, early on, distances to other galaxies derived from the original version of Leavitt's law were significantly underestimated. To use Cepheids to measure distances accurately, an astronomer needed to understand which type he or she was measuring. After Baade's discovery, the distances to other galaxies had to be re-determined.

In the autumn of 1923 Edwin Hubble was doing research on the statistics of Novae in the great spiral nebula in Andromeda, M31. Novae were stars that rapidly brightened (often appearing "out of nowhere," hence the name meaning "new star"), and then slowly faded over weeks or months. Novae were known to occur frequently in M31. Hubble was looking for changes in the brightness of stars seen in photographic negatives taken on glass plates at the 100 inch Hooker telescope—the largest telescope in the world at that time—at Mt. Wilson Observatory north of Los Angeles. In comparing a pair of these plates he stumbled upon two rather bright variable stars that he took to be novae and one fainter variable star about which he was less certain. Checking back through a larger set of plates that covered a longer period of time he found the latter to follow the light curve of a classical Cepheid variable similar to those found by Henrietta Leavitt in the Magellanic Clouds.

In 1923 the makeup of the universe was still being vigorously debated. In particular, astronomers differed on whether the great spiral nebulae like M31 were part of a single large all encompassing stellar system, the Milky Way, or rather were individual "island universes," each separated by great distances from the others. By 1929 Hubble had measured the light curves of 40 Cepheids in M31. Applying Leavitt's law, he conclusively proved that M31 and a number of other "great nebulae," were in fact separate, individual conglomerations of stars, gas, and dust—"island universes" far away from our own Milky Way system.

Cepheids could be discerned in only a handful of nearby nebulae (or "galaxies" as they were later called). To push out to greater distances, Hubble used the Cepheid-based distance measurements for these nebulae to estimate the intrinsic light output of other, brighter standard objects seen within them. For example, if he assumed that there is an upper limit set by physics to the maximum brightness that stars can achieve, then the brightest stars observed in any of the nebulae might have about the same intrinsic luminosity as those observed in any other nebula. So, measurements of the apparent brightness of the brightest stars observed in any two nebulae should reflect the relative distance of those nebulae from Earth. If the brightest stars were four times fainter in nebula B than in nebula A, then B must be about twice as far from us as A.

If a nebula was too far away to allow him to resolve its brightest individual stars, then Hubble used the apparent total brightness of the nebula as a whole to estimate its distance, assuming that on average all the nebulae of a given shape and size probably have about the same total light output from their stars and glowing clouds of gas. The Cepheid light curves provided more accurate distances to these Island Universes than the brightest stars, and the brightest stars yielded more accurate values than the total integrated light from faint individual nebulae. By the early 1930s Hubble had estimated distances to about 100 isolated nebulae and to about a dozen clusters of nebulae.

At the same time several other astronomers were hard at work at various observatories trying to measure how fast these nebulae were moving through space. They used spectrographs that dispersed or spread the light from the nebulae out into their component colors. Typically, they could see in these spectra that certain narrow bands of color were missing, having been absorbed by atoms of hydrogen, iron, and calcium. The astronomers took advantage of the Doppler principle that light observed from a source that was moving away from the observer would be shifted to redder wavelengths and light from a source moving toward the observer would be seen shifted toward the blue. The amount by which the wavelength was shifted from its position at rest was directly proportional to the velocity of the object.

Between 1912 and 1925 astronomer Vesto Slipher, working at Lowell Observatory in Arizona, spectroscopically measured velocities for 45 of the brightest great nebulae. With only a few exceptions all the light from each nebula was shifted to the red. They all appeared to be moving away from Earth with the fainter and presumably the more distant ones moving away faster than the brighter ones. Hubble combined his measured distances with Slipher's measured velocities to come up with the earliest version of his velocity–distance relation—the farther away the nebula was the faster it was receding from us. Velocity of recession was directly proportional to distance.

However, the major step in firmly establishing "Hubble's law" was enabled by the meticulous measurements of the spectra of very faint nebulae at the 100 inch Hooker telescope by Milton Humason. He was a remarkable person. While Edwin Hubble had received an elite education at the University of Chicago and as a Rhodes Scholar at Oxford in England, Humason was a high school dropout. As a youth, he loved exploring the mountains near Los Angeles. As Mt. Wilson Observatory was being constructed, Humason got a job driving mule trains up the mountain, carrying supplies to the construction site. In 1917 he was hired on as a janitor at the new observatory. Attracted to the work he saw going on there, he became a volunteer night assistant, supporting the work of the professional astronomers. His talent caught the attention of George Ellery Hale, the founder of the observatory, who amazingly hired this poorly educated man to the full-time staff in 1919. That was a good move!

Humason developed observational techniques to obtain photographs of the spectra of very faint objects. He used spectrographs designed to be extra sensitive and a method of acquiring a target, e.g., a nebula (galaxy), which was too faint to be seen by eye looking through the telescope, into the entrance slit of his spectrograph. A long exposure time would then yield a usable spectrum of that faint target.

Humason provided Hubble with velocity measurements for virtually all of the nebulae (galaxies) for which the latter had estimated distances—velocities ranging from a few hundred kilometers per second up to a value just shy of 20,000 km s^{-1}. The velocity–distance relation, Hubble's law (or perhaps more appropriately, in my opinion, called the Hubble–Humason law) could then be extended 18 times farther out in distance than in its earliest incarnation. At these greater distances it remained true that velocity was directly proportional to distance.

In a paper entitled, "A Relation Between Distance and Radial Velocity Among Extra-Galactic Nebulae", Hubble (1929) cautiously inferred that the velocity–distance relation might indicate that our universe is expanding. By 1936 Hubble, in his famous book "The Realm of the Nebulae,[1]" spoke more confidently about the

---

[1] 1983, Sillman Milestones in Science (New Haven: Yale University Press).

expanding universe. He wrote, "It may be stated with some confidence that red-shifts are velocity-shifts or else they represent some hitherto unrecognized principle in physics. The interpretation as velocity-shifts is generally adopted by theoretical investigators, and the velocity–distance relation is considered as the observational basis for theories of an expanding universe. Such theories are widely current." Nevertheless, Hubble was still careful to state in his book that it remained to be proven that the velocities are due to motions of the nebulae through space. He referred to them for convenience as "apparent velocities."

Hubble's conclusion conformed to a theoretical prediction by the Belgian priest and cosmologist Georges Lemaître, based on his solution of equations from Einstein's General Relativity. Lemaître published his conclusions in 1927 in French in a relatively obscure Belgian journal. As a result his work was not widely known to other scientists. It turns out that Lemaître had actually scooped Edwin Hubble. Using Hubble's published galaxy distances and Vesto Slipher's measured velocities of those galaxies, Lemaître had demonstrated what later became Hubble's law in his 1927 paper. In 1931 Lemaître was invited by the Royal Astronomical Society of England to translate his paper into English for publication in the Society's widely read journal. In his translation, Lemaître omitted the part of his paper that utilized the observations obtained by Slipher and Hubble. Apparently, he thought they were no longer relevant, since Hubble had already published his more thorough study in 1929. In 2018 78% of the membership of the International Astronomical Union voted to recommend that the name of Hubble's law be changed to the Hubble–Lemaître law.

When I was in school, I learned to think of the expansion of the universe in terms of a loaf of raisin bread rising and expanding as it is being baked in an oven. From the point of view of each raisin, every other raisin in the loaf is moving away or receding from it, and the farther away a raisin is, the faster will it appear to be moving away from the "home" raisin. This is characteristic of a uniformly expanding volume (e.g., one doubling in size every hour). It doesn't mean that the "home" raisin (Earth, the Milky Way Galaxy) is at the center of the loaf (the universe), or is in any way special. Observers on any other raisin (sufficiently far away from the boundary provided by the bread pan) will have the same perspective that the loaf is expanding away from it. The raisins are not physically moving through the loaf of bread, but rather are being carried along by the loaf's general expansion in volume. Similarly, the large redshifts and large apparent velocities of galaxies measured by Slipher, Humason and later generations of astronomers do not reflect physical motion of these island universes through space, but rather the expansion and stretching of space itself.

Hubble derived the constant of proportionality (later called the Hubble constant), the slope of the straight line relating distances of galaxies to their apparent velocities that best fit the observational data. He estimated that for every million parsecs (megaparsec or Mpc, equivalent to 3.26 million light years) of increased distance, the velocity of recession increased by about 550 km s^{-1}. He estimated an uncertainty in this constant of 10%. We now know that this value is much too large.

In Hubble's time the intrinsic light output of Cepheid variables was not accurately known, though their brightness relative to each other could be measured and related to their relative period of variability with reasonable accuracy. Nailing down their absolute brightness was a critical step in the century-long effort by astronomers to determine the distance scale of the universe. There was also the issue of separating out the two different types of Cepheids, of which Hubble had no knowledge.

Hubble's distance measurements were also hampered by the limited resolution of his images. Even a relatively large mountaintop telescope like the 100 inch Hooker couldn't overcome the turbulence of the Earth's atmosphere that smeared out the light from the nebulae (galaxies) he was photographing. He mistook the bright, tiny images of clusters of hundreds of bright stars in the more distant nebulae as being single bright stars. This introduced large errors in his bright-star method of estimating nebular distances. Observing individual stars in distant galaxies would have to await the factors of 10 or 20 improvement in the clarity of galaxy images provided by the Hubble Space Telescope some 70 yr later.

The Hubble constant, H_0 (the subscript zero signifying the value of the constant of proportionality between distance and velocity, H, *at the present time*), has a special significance in our understanding of the universe. By inverting it, calculating $1/H_0$, we can estimate how long this expansion has been going on—a rough measure of the age of the universe (called the Hubble time). Let's assume the universe has been expanding at the same constant rate over its entire history. A large H_0 suggests that the universe is expanding rapidly, and hasn't taken long to reach its present size—the universe is young. On the other hand, if H_0 is small, the universe is expanding more slowly and has required a longer time to grow to its present scale—the universe is old.

The Hubble time is a "rough" estimate because it is likely that the universe hasn't expanded at a constant rate. That assumption is an over-simplification. To calculate a more realistic age one has to combine the measured Hubble time with a theoretical cosmological model—a model that describes all the salient properties of space and time. How much gravitational matter exists in the universe? Is its mass sufficient to slow down the expansion? Or bring it to a complete halt? Or reverse it so that the universe ultimately collapses back onto itself. Is spacetime curved or flat? If it's flat, two parallel beams of light would remain precisely parallel no matter how far out they traveled. If curved, the parallel beams would slowly bend toward each other or spread apart, depending on the nature of the curvature. Are there other forces in addition to gravity that affect the expansion rate? These factors, embedded in Einstein's General Theory of Relativity, make a difference in the age one computes for the universe. It's all of these factors, not H_0 alone, that describe the structure and history of the universe. Finding the cosmological model that most accurately describes the observed properties of the universe is a "holy grail" for astronomers.

When Einstein first announced his general theory of relativity in 1915, he made the assumption that the universe is static—it neither expands nor contracts. But how could that be, if every bit of matter in the universe was gravitationally tugging on every other bit of matter? Gravity should cause the universe to collapse on itself. So, he reasoned that there must be some other force that would exert a positive pressure offsetting the inward tug of gravity, pushing outward to keep the cosmos perfectly static. Einstein arbitrarily introduced into his equations a "Cosmological Constant" that would provide this countervailing force against ordinary gravity. Later, he learned of Edwin Hubble's observations that demonstrated that the universe was not static—that it was expanding. He then famously referred to his cosmological constant as "... my biggest blunder," and removed it from his equations. However, cosmologists continued to retain the constant, designated by a capital Greek letter lambda (Λ), in the equations for the sake of completeness and to express the possibility that there could be other forces that influence the expansion rate of the universe in addition to gravity. They simply set its value to zero.

If we take the reciprocal of Edwin Hubble's estimated value of H_0, 550 km s^{-1} Mpc^{-1}, we calculate a Hubble time for the universe of about 1.8 billion years. In the 1920s and 1930s geologists measured the ages of some of the oldest rocks found on the Earth's surface, using the known rate of radioactive decay of uranium into lead and measuring the abundances of those two elements in the rocks. They determined that those particular rocks were formed at least 3 billion years ago. The universe couldn't be younger than the Earth itself! (Today we know the age of the Earth is about 4.6 billion years.) This reality check showed that, though Hubble had used the best techniques available at the time, his estimate of H_0 was considerably off the mark.

By the 1970s the accuracy of the Hubble constant had been considerably improved mostly due to a more accurate calibration of the intrinsic brightness of Cepheid variables, the separation of the two varieties of Cepheids, and the establishment of additional distance indicators. Published determinations of H_0 ranged from about 40 to about 100 km s^{-1} Mpc^{-1} with scores of studies finding values scattered in between. Astronomers were divided into two major warring factions. One, led by Sidney van den Bergh and Gerard de Vaucoleurs, clung tenaciously to a value of 100 km s^{-1} Mpc^{-1}, and a younger universe. The other, led by Allen Sandage and Gustav Tammann, believed just as fervently in 50 km s^{-1} Mpc^{-1} or less and an older universe.

The best hope for breaking this impasse was sharp, high-resolution camera images from the Hubble Space Telescope. A major scientific objective stated in a foundational Hubble program document issued in 1977 was the "precise determination of distances to galaxies out to expansion velocities of about 10,000 km s^{-1} and calibration of distance criteria applicable at cosmologically significant distances." The telescope was sized to have enough light gathering power and resolution to be able to discern individual Cepheid variable stars at least as far out as the Virgo Cluster of galaxies.

At that time the distance to the Virgo Cluster was highly uncertain. Published estimates ranged from about 50 million light years (15 Mpc) to about 80 million light years (25 Mpc). So, it would be a major accomplishment for Hubble, if it could be used to locate Cepheid variables in some of the galaxies in the Virgo Cluster. A modern version of Leavitt's law could then be applied to derive an accurate distance to the cluster. The Virgo galaxies would be rich in other, brighter distance indicators, such as supernovae, whose light output could be accurately calibrated using the Cepheid distance. These additional brightness standards, or "standard candles," would allow astronomers to extend the distance scale much farther out across the universe beyond the Virgo Cluster. Jerry Oistriker's Space Telescope Advisory Committee generalized this into a Hubble Key Project to measure H_0 to an accuracy of 10%—a major step forward from the then existing uncertainty of a factor of two.

In the summer of 1985, a group of leading experts on the distance scale of the universe met at the Aspen Center for Physics in Colorado, ostensibly to hold a conference on the current state of their field. In reality their main purpose was to see if they could organize themselves into a single, unified team to propose a plan to the STScI to carry out the H_0 Key Project. As described by Harvard astronomer John Huchra, "In the end there really was no way to get the old timers to work with the young Turks." So, it was the young Turks, thirteen of them, who coalesced into the team that submitted the proposal to the STScI that was officially selected to use the Hubble Space Telescope to recalibrate the distance scale of the universe and to measure the present rate of expansion, H_0, to 10% accuracy or better.

The Principal Investigator leading the team was Marc Aaronson, a highly regarded and much honored young astronomer at the University of Arizona. However, on the evening of 1987 April 30, tragedy befell not only his team, but also the entire astronomical community. Marc was preparing to observe that evening on the 4 m Mayall telescope at Kitt Peak National Observatory outside Tucson, Arizona. He wanted to check on the weather outside and so made his way around the dome to a heavy door that opened out onto a catwalk. The dome was rotating at the time, but electricity to the motors driving it was automatically switched off when that door was opened. Nevertheless, the dome's inertia kept it moving for a few seconds more. A ladder used for climbing up onto the dome protruded downward a few feet from the bottom of the structure. The timing of things was simply horrific. Marc opened the door and started to step out. The dome continued to rotate. The end of the ladder struck the door slamming it into Marc's body. He was crushed. He died instantly.

Word of this calamity spread rapidly throughout the astronomical community. There was widespread shock and sorrow. Sadness engulfed Marc's Hubble Key Project team. Emotional recovery took a few weeks. But the team realized that they had to press on, to honor Marc Aaronson, if for no other reason. Leadership responsibility for the program divided naturally among three members. Wendy Freedman took over responsibility for the H_0 science program. Jeremy Mould and Rob Kennicutt would share the management duties.

The Challenger disaster in January 1986, had led to a halt to the Shuttle program and a delay of Hubble's launch indefinitely. The loss of the Key Project team's popular leader had placed an even darker cloud over this critical research initiative. But the hiatus of several years allowed the team to regroup and to hone its procedures and strengthen its computing resources.

One important problem was how to most efficiently collect accurate light curves for the Cepheid variables in distant galaxies while not needlessly eating up precious Hubble telescope time. The cycle of light variations in the Cepheids ranged from a few days for the intrinsically fainter examples to a couple of months for the brighter ones. To measure the full range of brightness variations for the low luminosity, rapidly varying Cepheids in a galaxy, the telescope would need to take exposures every few hours for two or three days. But the team couldn't afford to take camera images every few hours for the 70 or 80 days they needed to measure the light curves of the high luminosity, longer period variables. The rest of the community would be waiting impatiently for its turn on the telescope. One scientific program, no matter how important, couldn't monopolize Hubble's calendar.

To resolve this conundrum, Wendy Freedman and her colleagues invented a formula for how to space their observations of a galaxy over 10–12 weeks to give them optimally timed brightness measurements and light curves for a wide range of possible Cepheid periods. They would not waste telescope time. Without the extra time afforded by the delay in Hubble's launch date, they might not have been ready for such an efficient implementation of their Key Project. Even the darkest clouds have a silver lining.

Spherical aberration also delayed the start of the Key Project. Detecting and accurately measuring the brightness variations of Cepheids in the galaxies of the Virgo Cluster required the exquisite image quality that the Hubble telescope was supposed to provide. After all, the telescope had been designed explicitly for that purpose. But now there were skeptics. Some astronomers had recently scaled the brightness and variability of Cepheids in our own Milky Way galaxy out to the distance of the Virgo Cluster and had predicted that they wouldn't be detectable at that great distance within the complex, crowded, and confused fields of thousands of other stars in which they were embedded. Uncertainty about the viability of the H_0 Key Project lingered over Freedman and her team. It could only be resolved after the telescope's optical performance had been corrected. They had to live with that worrisome unanswered question for the more than three years it took to mount the Hubble repair mission in late 1993.

Wendy Freedman was a native of Toronto, Canada. She attended the University of Toronto, receiving her PhD in astronomy in 1984. Her doctoral dissertation was on the subject of star formation and star populations in other galaxies outside the Milky Way. Her observational studies of nearby galaxies with the newly opened Canada–France–Hawaii telescope on Mauna Kea on the big island of Hawaii made her well equipped for the H_0 Key Project when the opportunity came along. In 1984 Wendy became a post-doctoral fellow at the Carnegie Observatories in Pasadena California. Three years later she was appointed a permanent faculty member at Carnegie, the first such position for a woman on its staff.

Down the hall from Wendy's office in the Carnegie Observatories' headquarters building on Santa Barbara Street in Pasadena was the office where Alan Sandage resided. Sandage was one of the most famous, respected, and influential astronomers in the world—a lion of the profession. His doctoral advisor at Caltech had been Walter Baade. Sandage received his PhD in 1952. He worked as an observing assistant to Edwin Hubble at the Palomar Observatory from 1949 until Hubble's death in 1953, after which he carried on Hubble's research. He became recognized as a leading authority in galactic and extragalactic astronomy, studying globular star clusters, galaxies and the wider universe in general. Sandage had been working for decades on improving the accuracy of the distance scale of the universe, its age, and the rate of its expansion. He had chosen not to join Marc Aaronson's team of younger astronomers selected to execute the H_0 Key Project with the Hubble Space Telescope.

At a conference in Milan, Italy in 1992 September—a little over a year prior to the Hubble repair mission—Alan Sandage presented his latest work on the Hubble constant and made a closely argued case as to why his numbers had to be correct. Based on all his research, Sandage settled on a value $H_0 = 44 \pm 12$ km s^{-1} Mpc^{-1}. If the universe was expanding that slowly, it must have taken a very long time to reach its current size. Indeed, if that had been the steady rate of expansion through all time, it would have taken about 22 billion years—the Hubble time.

Sandage had recently completed work to estimate independently the age of the oldest objects astronomers knew about, the globular star clusters—spherical structures made up of hundreds of thousands of old stars, believed to have formed earlier than any other objects in the Milky Way Galaxy. He determined, using observations of the colors and brightness of stars in globular clusters and the latest theoretical models of stellar evolution—models that calculated how long it would take for the population of stars in globular clusters to burn up their nuclear fuel—that they were "only" 14 billion years old. If he allowed, say, a billion years for the earliest stages of the universe before the globular clusters came into existence, then the true age of the universe must be about 15 billion years.

The ratio of this "True Age" of the universe to the Hubble time, 15/22, was almost exactly two thirds. That ratio was precisely what the then-current Standard Model of Cosmology required.

The Standard Model postulated that the universe had started with an incredibly short period of Inflation followed by the uniform expansion of the kind reflected in the Hubble law. The geometry of the universe was flat. The density of matter in the universe must be very high, high enough to assure that ultimately the expansion would gradually come to an end—spacetime was "closed." Moreover, there were no unusual repulsive forces at work in the universe; Einstein's "cosmological constant" was zero.

Sandage concluded that his measurements of the rate of expansion of the universe and the ages of the globular clusters effectively validated the Standard Model. Under the Standard Model the low rate of expansion implied by an H_0 value of 44, coupled with such a "young age" for the universe found from the globular clusters, required that the original expansion rate after the Big Bang had been much higher and that the expansion had decelerated rapidly—the universe had slammed on the brakes as it began to expand. The universe must be filled with an enormous amount of mass all of which would tug so strongly on itself through the force of gravity that the expansion would be forced to slow down. There must be enough mass in fact to cause the expansion to stop eventually.

As far as Alan Sandage was concerned, all was well. He could tie his best observational work, together with a well-understood model of the cosmology of the universe that flowed directly from Einstein's general relativity, into a neat and tidy package. The guys on the other side of the argument, those coming up with much larger values for H_0, were simply mistaken.

Sandage noted that his results were in direct contradiction to the results to be presented in the next paper on the program at that conference by astronomer R. B. Tully—"the other side of the debate." Sandage had used three separate techniques to measure the distances of galaxies. Tully used three entirely different methods. He concluded that the best value for H_0 was 90 ± 10 km s^{-1} Mpc^{-1}.

Such was the state of affairs in humanity's quest to understand the "big picture" of the universe—how big it was, how old it was, how rapidly it was expanding, what its structure was, what its ultimate fate would be—at the time Hubble's blurry vision was repaired in 1993. Competent, experienced, highly regarded scientists disagreed on the basics by a factor of two! Thus, the stage was set for a clearer look at the situation through the eyes of the Hubble Space Telescope.

In the autumn of 1993 Ray Villard, the lead public outreach specialist at STScI, approached Wendy Freedman about recommending a galaxy that could be observed with Hubble's new and improved wide field camera (WFPC2) shortly after SM1 had been completed. The idea was to demonstrate the improvement in the telescope's clarity of vision in a way that would be obvious to even a casual spectator. Wendy suggested M100, a gorgeous face-on spiral galaxy that, as it so happened, resided in the Virgo Cluster. It was the before-and-after picture of M100 that Senator Mikulski held up to the press in 1994 January to pound home in an unmistakable way that "the trouble with Hubble is over!"

The new M100 image also served an immediate scientific purpose. Thousands of individual stars could be resolved in the images (Figure 10.1). Wendy and her colleagues used a computer program to carefully measure the colors and brightness of all those stars. They easily covered a range of values that included the known color and brightness of Cepheid variables. From that first small set of images it seemed certain that Cepheids would be easy to detect and their light curves could be readily measured in the Virgo Cluster. But the proof was in actually doing it.

During the two months beginning in April of 1994 the Key Project team obtained a series of 12 one hour exposures with WFPC2 of a field in M100 where they expected to find Cepheids (Figure 10.2). The exposures were carefully spaced in time following the formula Wendy had derived earlier to keep the program efficient. They designed the observations to capture light curves for Cepheids with periods ranging from 10 to 60 days.

Wendy was sitting in her office on Santa Barbara Street when the first set of data came in. She ran a program that automatically searched for brightness variations in individual stars in the images, searching through about 40,000 stars in one field. Another program was designed to plot graphs of the variations with time. Out popped several absolutely beautiful light curves so characteristic of Cepheid variables—a saw tooth pattern where the star rapidly brightens and then slowly dims. She was euphoric! Here in a few graphs was the proof she, the astronomical community and the Hubble program needed. There were more examples—about 20 Cepheids in all. The naysayers who had earlier argued that it couldn't be done had been proven dead wrong. And most importantly, a critical scientific performance mandate for the Hubble Space Telescope set in the 1970s had been achieved.

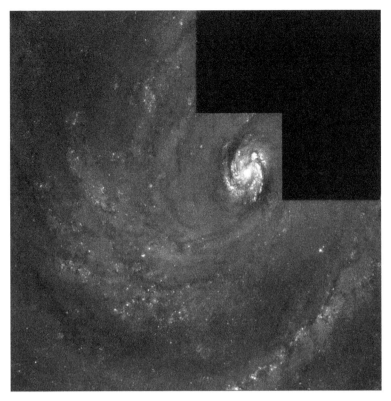

Figure 10.1. The grand spiral galaxy M100 in the Virgo cluster, imaged with WFPC2 shortly after the first servicing mission. Many individual stars are resolved, showing it would be feasible to find and measure Cepheids. (Credit: NASA, W. Freedman, the WFPC2 Team, STScI 1994-02.)

Wendy could tell immediately from these first observations that M100—and likely the assemblage of several thousand galaxies that made up the Virgo Cluster as a whole—was closer than previously believed. M100 was about 56 lt-yr (17.1 Mpc) away. To Wendy, nearer than expected distances immediately suggested that the value she derived for H_0 was going to be on the high side.

About six times farther away than the Virgo Cluster was another rich grouping of galaxies—the Coma Cluster. It was far enough away that the recessional velocities of its member galaxies due to the cosmic expansion (the Hubble flow) dominated the much smaller velocity contributions due to the galaxies' random motions in space (their individual "peculiar velocities"). The measured galaxy velocities due to the expansion of the universe were of course much lower in the much closer Virgo Cluster and the peculiar motions of its member galaxies made a relatively larger contribution to the total, complicating the business of deriving an accurate value for H_0. The farther out across the universe one could measure distances and velocities, the better.

From both clusters Wendy derived a preliminary value of $H_0 = 80 \pm 17$ km s^{-1} Mpc^{-1}. That range of uncertainty, reflecting the many possible sources of error that might be in play, was huge. It admitted a probable value of the Hubble constant somewhere between 63 and 97. But even that large error bar didn't encompass the value of 44 in

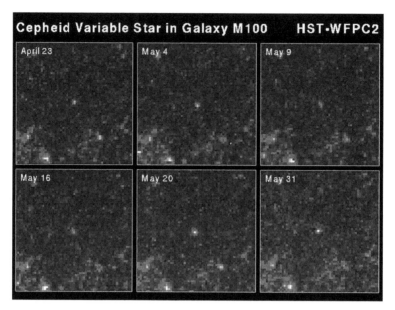

Figure 10.2. A Cepheid variable star in M100 varying in brightness over a portion of its cycle. The period of the Cepheid, 51.3 days in this case, can be used to derive its intrinsic luminosity by means of an updated version of Leavitt's law. (Credit: NASA, W. Freedman, the H_0 Key Project Team, STScI 1994-49.)

which Alan Sandage had such great confidence. And there was an even more awkward complication. A Hubble constant of 80, when plugged into the Standard Model of cosmology, the model that required a high-density of matter in the universe, resulted in an age for the universe of about 8 billion years! Sandage had estimated an age of 15 billion years, based on the oldest stars in the galaxy having a calculated age of 14 billion years. It appeared from the work of the Key Project team that the universe as a whole was much younger than the oldest stars within it. This was nuts! How could that be?

It helped somewhat to depart from the Standard Model and assume a cosmology with a much lower concentration of matter. But the most extreme low-density model could only increase the measured age of the cosmos to 12 billion years, still apparently too young.

The paradox of these early results was immediately evident to Wendy and her team. They struggled to find an explanation. The quality of their observations with Hubble was superb. The Cepheid light curves were beautifully defined. They could find no obvious errors in their analysis. In early publications and lectures about their results they speculated that perhaps the age estimates for the globular clusters, based on the most modern models of the physics of stellar evolution, were in error. Or perhaps there was a problem with the current understanding of cosmology and the models that were constructed from general relativity to describe it. Maybe there was a Cosmological Constant after all. Both possibilities were considered radical and engendered a lot of skepticism in the community.

As Wendy described it to me in our recorded interviews, Alan Sandage became incensed by this turn of events. He had been working on this problem for decades at

Mt. Wilson and Palomar observatories. His zeal about the correctness of his conclusions was nearly religious. Wendy was a new junior staff member at his institution, a young woman relatively fresh out of school. As she saw it, Sandage believed her to be impertinent in challenging his beliefs. As time went on, Wendy felt that he was becoming very nasty and very personal toward her, that she was under attack. He wrote to her what she considered to be nasty notes. When they both were invited to give talks about their work at scientific conferences, Sandage would call the conference organizers and try to persuade them to drop her from the program. Internally he wrote to the Director of Carnegie insisting that she not be allocated telescope time at the institution's observatory at Las Campanas in Chile. When the Director discussed it with her, she offered that "maybe the explanation is that there is a non-zero Cosmological Constant," to which he replied, "get serious, you're not talking like a serious scientist." Later his successor in the Director's position was asked by a reporter to comment about the conflict between two scientists in the same group. He responded, "Wendy has better data and better analysis methods, but Sandage is clearly right!"

It was the holiday season near the end of 1994. The offices on Santa Barbara Street were nearly empty. People were home celebrating with their families. Wendy sat alone working in the Carnegie library. In walked Sandage. When he saw her there, he became furious. He had just returned from a trip. On the flight back home, he sat next to another senior astronomer. That person showed Sandage a recent article written by Wendy and her team giving some new results from the Key Project. Apparently, Sandage took umbrage at the possibility that other senior professionals in the astronomical community were taking her work seriously, and by implication challenging his own. When he encountered her, alone, in the library he laced into her.

Sandage began berating Wendy, growing more and more angry as he spoke. "You've done *this* wrong, and *this* wrong, and *this*," trying to pick apart the details of her work reported in the article. Wendy did her best to remain calm. But the thought passed fleetingly through her mind that Sandage was considerably larger and probably much stronger than she. As his rage increased, might it get physical? No one else was there to witness what was happening? That thought quickly subsided. And then he got to a point in his diatribe that gave her a natural entry. He shouted, "You don't even know the metal content of these Cepheids! You don't know if their period–luminosity relations are sensitive to that." Wendy responded, "You're absolutely right. We don't know that for certain. It's something we need to study further." Sandage turned from rage to exasperation. "What is it with you?" he shouted. "I can't even get a rise out of you!"

That's when Wendy knew what he most wanted, a "rise" out of her. He wanted her to fight back, as if that would somehow validate his rage. But she remained doggedly calm. She would not be intimidated. What had attracted her originally to science was that it was an objective process. Science was not someone's opinion. Excellent quality data and rational thought would ultimately win the day. But science was also a human endeavor, and as she was witnessing, personalities and human passions can enter the picture.

Wendy and the Key Project team pressed on undeterred. They had to discover even the subtlest systematic errors, everything that could possibly be wrong with the analysis and interpretation of their data, and beat them down.

Over the next several years they observed Cepheids in about 20 spiral galaxies. The team used them to calibrate several other secondary distance indicators that could be observed in even more distant galaxies whose Cepheids were too faint to be of use. Most notable among these was a particular type of exploding star, Type Ia supernovae. These were easy to identify and remarkably, their light curves could be calibrated so as to allow an accurate measurement of their intrinsic peak luminosity. They were perfect standard candles. They would play a pivotal role in the astonishing, completely unexpected changes to come in cosmology just a few years later.

As to the "age paradox," the initial discovery that the universe seemed to be younger than its oldest stars, several factors began to close that gap. Discovering and eliminating systematic errors, and combining results from five different distance indicators, allowed the Key Project team to reduce their measured value of the Hubble constant to 72 ± 8 km s^{-1} Mpc^{-1}. Using the cosmological models of the day, the Team derived 9–13 billion years as the range of possible ages for the universe, depending on how dense they assumed the matter contained in the universe to be. But that was still puzzling, still surprisingly young.

The theorists who worked on models of the evolution of stars in globular clusters improved their computer codes, making them more detailed and physically realistic. This caused the ages estimated for these clusters, when their stars were compared to the predicted brightness and colors from stellar evolution models (the technique used by Sandage), to drop somewhat. However, the biggest correction to the cluster ages came about because of a new space mission—the European Space Agency's Hipparcos satellite, launched in 1989. Its mission was to measure the positions, motions, and distances of millions of stars distributed around us in the Milky Way Galaxy. These precise astrometry measurements allowed a redetermination of the distances to several globular clusters without the need to use computer models of stellar evolution. The results indicated that the previous distance estimates, such as those made by Alan Sandage, were in error. In general, the clusters were farther away than originally believed, farther by as much as 25%. Being farther away meant that the stars in the clusters were intrinsically brighter than originally thought. More luminous stars meant more massive stars, younger stars. The globular clusters were younger than Sandage had calculated.

Taken together, the improved computer models of stars plus the precise new measurements of stellar distances from Hipparcos allowed astronomers in the mid-1990s to reduce the estimated ages of globular clusters to 12–13 billion years. So, the universe itself must be in the neighborhood of 13–14 billion years old.

This still left the Hubble Key Project Team in a quandary. Their calculated age of the universe, based on their new and improved distance scale and the measured expansion rate of the universe, was still too young!

Sometimes in science when experiments or theoretical calculations produce puzzling results, it's because you've overlooked some source(s) of error in the way the experiment was set up and performed, or in the assumptions that went into interpreting the measurements, or perhaps in the physical realism of the computer codes that produced theoretical predictions of what might be observed in nature. Sandage and his collaborators were convinced that Wendy Freedman's measurements of H_0 were somehow flawed by such sources of systematic error. There was such a satisfying internal consistency to his own measurements and interpretation of the data that he was totally convinced he had to be correct—everything seemed to fit so perfectly together.

In fact, his work was susceptible to errors some of which were beyond his control. He used theoretical models of the evolution of stars in globular clusters that were the best available at the time. Later those models were updated and improved, causing Sandage's age estimate for the universe to be too old. His distance measurements for globular clusters were superseded by the results from Hipparcos, resulting in a further reduction in their ages from the values he originally derived. Some of his measurements of star brightness, both for Cepheids and supernovae in spiral galaxies, came from photographic plates. Converting the density of the image of a star in the emulsion of a photographic negative into accurate measurements of the star's actual brightness is fraught with possible error. In the end, his results couldn't match the quality of the Hubble data, and the internal consistency of his results eventually came unraveled.

Fortunately, Wendy did not fold under the intense emotional pressure placed on her by Sandage and his colleagues. She held true to her belief that accurate observational data and the logic of the scientific method would ultimately reveal the truth. Paraphrasing that Carnegie Director, Wendy did have the better data and the better analysis methods and SHE and her team were proven right.

So, what did explain the paradoxical difference between the two independently derived ages of the universe—one calculated from the expansion rate of the universe and one determined from the best estimates of the ages of the oldest objects in the universe? How could the universe be a billion or more years younger than its oldest stars? Neither Wendy Freedman nor Alan Sandage could have known in 1994 how this tension would be resolved. Sometimes when a result looks strange, there is a good reason for it; some important physics is missing!

Mother Nature had been keeping a secret from Freedman and Sandage. In the late 1990s cosmology was turned on its head by the mind-blowing discovery that the expansion of the universe was actually speeding up. The Standard Model was wrong. The universe was not decelerating as the Standard Model required, but was accelerating due to some mysterious force that worked against normal gravity. This could best be described by Einstein's "biggest blunder"—the cosmological constant—or something similar to it. The models of the universe that set the cosmological constant to zero were incorrect. It had to be added back into the equations of general relativity in order to describe an accelerating universe. Wendy's original intuition about this proved prescient.

There is today a consensus view among most cosmologists that the universe as we observe it is best described by what they call the ΛCDM model. That is a model that has a flat geometry and is dominated by the gravity of cold dark matter (CDM) and a cosmological constant (Λ), the repulsive force that is now called "dark energy."

Neither cold dark matter nor dark energy is understood. The physics remains mysterious and elusive. But in the absence of understanding, astronomers have at least given them names and determined that together they make up about 95% of the mass + energy content of the universe. The remaining 5% is the ordinary hydrogen, helium, and heavier elements from which we are all made.

Plugging their value of $H_0 = 72$ km s^{-1} Mpc^{-1} into a ΛCDM model, with cold dark matter plus ordinary matter making up about 30% of the content of the universe and dark energy comprising the remaining 70%, Wendy and her colleagues calculated that the universe is about 13 billion years old with an uncertainty of 10% (Freedman 2001). With later refinements and with the addition of new data from other space missions, particularly missions dedicated to studying the cosmic microwave background (WMAP, Planck), astronomers have settled on 13.8 billion years as the age of the universe since the Big Bang. The globular clusters are now estimated to be a little over 12 billion years old. At last this all makes sense!

Now, perhaps, when a small child asks one of her parents, "Mommy, how high is the sky?" Mommy can be prepared with an answer that has a solid basis in science. But first the question would need to be transposed into, "Mommy, how big is the universe?" When astronomers observe the cosmic microwave background, or the very first clumps of proto-galaxy material, they are of course seeing the universe as it was 13.4–13.8 billion years ago. The distance, 13.8 billion light years, corresponds to the distance light can travel over the entire age of the universe up to now. It is the radius of the sphere that contains everything we can see **now**, the radius of the visible universe. But in the intervening 13.8 billion years, as those primal photons of light have traveled to us, the universe hasn't been standing still; it has continued to expand. So, how big is the universe right **now**? Or rather, how far away from us **now** is an object that was 13.8 billion light years away when it emitted photons of light that we see today? About 47 billion light years, if the universe has a low density of matter as the ΛCDM cosmology predicts, according to UCLA astronomer Ned Wright's Cosmology Tutorial.[2]

This is a beautiful story. Wendy Freedman ran headlong into new physics—a fundamental property of nature that was unknown at the time she and her Key Project team began their work. They are the lucky ones. Most scientists never have the experience in their lifetimes of serendipitously coming upon something completely new and unexpected that helps all the pieces of the puzzle they are working on fall into place. But that's what makes science fun.

References

Freedman, W. L. 2001, ApJ, 553, 47
Hubble, E. 1929, PNAS, 15, 168
Leavitt, H. S., & Pickering, E. C. 1912, Harvard College Observatory Circular, 173, March 3

---

[2] http://www.astro.ucla.edu/~wright/CosmoCalc.html.

Chapter 11

You Can Almost See Forever: The Hubble Deep Fields

In October of 1992 Riccardo Giacconi announced to the staff of the Space Telescope Science Institute (STScI) that he was resigning as Director, effective at the end of the year, to take up the position of Director General of the European Southern Observatories (ESO). Pete Stockman took over as Acting Director. It was then up to AURA to find a suitable replacement for Riccardo. This time the Hubble Project at Goddard was invited to weigh in on the choice. When AURA told NASA of its first choice, Bob Williams, then Director of the Cerro Tololo Inter-American Observatory in Chile, I checked with some colleagues who knew him. They gave him high marks for his good scientific judgment and diplomatic style. He was well liked. There could not have been a starker contrast in personality and leadership style than that between Bob and his predecessor. Bob was soft spoken, respectful, courteous, and empathetic. He took up his new position in Baltimore in September of 1993, just three months before the first servicing mission. One of his first initiatives was to begin to foster improved working relations between the Institute and the Hubble Project.

Bob Williams may have been mild mannered. But he was not timid. He was not averse to taking bold scientific risks, if he thought there could be high rewards. In the early months of his new job he had to weigh the risks presented by the upcoming encounter of Jupiter by the fragments of Comet Shoemaker/Levy 9 in July of 1994. Many astronomers, including some on the science team established to implement the observing campaign on the fragment impacts, were publically skeptical that anything of interest would be seen. But the community had lobbied hard for hundreds of precious Hubble orbits to be devoted to collecting observations as each fragment crashed into Jupiter. How badly would the recently repaired Hubble look to the public, if it devoted so much observing time to the comet campaign and then drew a big blank when the images came in? Not wanting to make such a momentous decision in isolation, Bob established an advisory committee of astronomers, chaired by Bob Millis of Lowell Observatory, to recommend the nature and extent of Hubble's S/L-9 observing program. As I recounted in Chapter 8, the results were spectacular, both scientifically and to the public. Had Bob acquiesced to the skeptics and declined to risk over 100 hr of Hubble observing time, the Hubble Program would have been humiliated by its lack of proactive planning to observe impacts that were, in fact, visible in small amateur telescopes from the surface of the Earth. Bob Williams 1, naysayers 0.

As early as 1984 NASA, the STScI and the Hubble Science Working Group (SWG) began to wrestle with issues of who should get how much observing time on the telescope once it was launched and checked out. Everyone's self-interest (and paranoia to some extent) went on display. Everyone became a lawyer. Turf had to be protected. The instrument teams and other members of the SWG were guaranteed observing time to implement the first scientific investigations with the instruments created for the observatory whose development they had overseen. That group was designated "Guaranteed Time Observers" or GTOs. But the question was, how much observing time did they have coming? The GTOs argued for more time, the STScI on behalf of the large community of general observers (GOs) argued for less. The situation was confused by the ambiguity of how observing time was to be accounted for, e.g., by number of on-target hours or by number of total spacecraft orbits used. The problem was further complicated by the difficulty of estimating how efficient the spacecraft would be in providing usable observing hours. It was ultimately up to the NASA Program Office at HQ, which had selected the instrument teams and other members of the SWG in the first place, to iron this out. In the end compromises were beaten out, documented, and revised as events like the Challenger disaster and the discovery of spherical aberration dictated. Writing the US Constitution couldn't have been much more difficult.

Out of this legalistic process emerged an interesting concept, that of observatory Director's Discretionary Time (DDT). Up to 10% of the available HST observing time could be reserved as DDT. The STScI Director could allocate observing time outside the normal peer review process for such purposes as: observations of transient phenomena where there was not time to submit a normal, peer-reviewed Target of Opportunity proposal; high-risk observations with a low probability of success but of considerable scientific importance if successful; and exploratory observations using Hubble's scientific instruments in an innovative fashion. Bob Williams allocated DDT for the Comet S/L-9 observing campaign. He would later exercise his discretion to allocate blocks of time to other observing programs of the "high-risk, high-reward" variety, which would prove foundational to future directions of science and our fundamental understanding of the universe.

Williams' research throughout his career had focused on relatively nearby things in our own or neighboring galaxies—gaseous nebulae, the clouds of gas and dust out of which new stars form, and novae, explosively erupting white dwarf stars. Extragalactic astronomy, the deep universe, was not his field. But when he took responsibility for implementing Hubble's overall science program as Director of the STScI in 1993, he suddenly was dealing with an observatory that had been sold to the public and Congress as a "Time Machine," capable of observing objects far out across the universe as they appeared eons ago. This of course results from the fact that light travels at a finite speed, 186,000 miles (300,000 km) per second. It takes a very long time to travel across the vast distances of the universe to get to us.

After he had been Director for about a year, Williams started wondering why Hubble was not being exploited as it could or should be for deep extragalactic work. Wendy Freedman's team working on the cosmic distance scale Key Project was mostly focused on searching nearby galaxies for Cepheid variable stars. Another Key Project, the Medium Deep Survey, carried out while Hubble still suffered from spherical aberration, was done in "parallel mode." That is, it captured images of random fields in the sky in parallel as another instrument pointed at a primary target. These "catch as catch can" images only probed to a redshift of approximately $z = 0.4$, corresponding to a lookback time of about 4.4 billion years. Other Hubble observing programs had also probed about that far back but no farther.

When astronomers talk about looking out across the universe at distant galaxies, they refer to a quantity called "redshift" symbolized by the letter "z." When Hubble and Humason were making the distance and apparent velocity measurements that led to the discovery of Hubble's law (the more distant a galaxy is, the faster it appears to be moving away from us) they were referring to the shift in observed absorption or emission lines in the spectrum of each galaxy toward redder wavelengths. The quantity "z" is the measure of how large that shift is as a fraction of the original un-shifted wavelength emitted by the galaxy. If $z = 0.1$, the wavelength of every line in the spectrum is shifted by 10% toward the red. For example, the wavelength of a prominent line of atomic hydrogen (the most abundant element in the universe) measured in the laboratory to be 656.3 nm, would appear at a redder wavelength, 721.9 nm, for a galaxy at $z = 0.1$. That light left the galaxy about 1.3 billion years ago.

Two factors contribute to such a redshift: (1) The intrinsic velocity of the galaxy as it moves through space relative to other galaxies (called its "peculiar velocity"), producing a Doppler shift of the light waves, and (2) The stretching of the light waves to longer wavelengths due to the expansion of space and time between the galaxy and us. For galaxies very close to our Milky Way the intrinsic motions of the galaxies through space dominate the value of z. In fact some galaxies are moving toward the Milky Way, their spectra are blue shifted, and they have negative z values. The Andromeda galaxy M31, 2.5 million light years away, is moving toward the Milky Way at a velocity of 110 km s^{-1} ($z = -0.0004$). M31 and the Milky Way are expected to collide and merge together in a few billion years. But farther out the expansion of space and time takes over and becomes the dominant contributor to the value of z. Out there, well into the "Hubble flow," z is always positive and the light observed from those distant galaxies is always red-shifted. The higher the z-value, the farther back in time we're seeing.

Given the "time machine" aspect of Hubble, one wonders why there had not been more emphasis on going deep to observe very distant galaxies far back in time. One obvious reason is that such observations require very long exposure times and use up a lot of the observing time available to astronomers with the telescope. The Time Allocation Committees that met yearly to prioritize observing proposals undoubtedly wanted to allow as many programs as possible—small amounts of observing time allocated to many programs, rather than burning up a lot of time on relatively few programs. It seemed more democratic that way.

Another possible reason is the effect one published scientific paper may have had on the Hubble observer community, a paper published shortly before Hubble was launched (Bahcall et al. 1990). The authors were John Bahcall, Raja Guhathakurta, and Don Schneider. Raja and Don were post-doctoral fellows working with John Bahcall at the Institute for Advanced Study in Princeton. Their paper was entitled, "What the Longest Exposures from the Hubble Space Telescope Will Reveal."

John, Raja, and Don posited the following question. Suppose very distant galaxies are similar to the spiral, elliptical, and irregular galaxies we see close by in our own part of the universe; how would they appear at those great distances? Their working hypothesis, to be tested by Hubble observations, was that "everything in the HST universe has previously been revealed by ground-based observations." Using this "parochial principle," they derived predictions for the numbers, colors, and types of faint galaxies that would appear in the HST data. Based on ground-based data, they assumed how many galaxies of each type they would expect per unit area of sky and how large in angular size those galaxies would be. Their third assumption was that at great distances any fine details (for example, spiral arms) would be lost from view so that galaxy images would smooth out into homogeneous ellipses. Their calculations indicated that at greater and greater distances such galaxies would simply fade from view into the background. Astronomers would be unable to detect extremely distant galaxies.

In the same paper, the authors stated, "It will be especially exciting if HST observations reveal objects or structures not predicted in this article." In other words their paper should have been viewed as a null baseline. If the distant universe were similar to the nearby universe, Hubble would see little of interest. But if new types of galaxies or other objects were seen in deep Hubble exposures, it would be a discovery of something very unusual and of great interest.

Unfortunately, some astronomers lost sight of this and interpreted the paper as a prediction that Hubble would not see much that is new and interesting insofar as high-z extragalactic research is concerned. I recall John Bahcall bringing a preprint of the *Science* paper to a meeting of the Hubble Science Working Group some months before launch, and representing it as showing that extragalactic research in the deep universe probably is going to be a waste of time. "That's no fun," I thought at the time.

When Bob Williams came upon the scene, he had not really been exposed to that argument. As a "naïve" outsider, he was keen to push Hubble as far as it could be pushed, to great distances across space and back in time, to see what's out there. The case for proceeding with deep image exposures was clinched for him by an observation made by Mark Dickinson, a young post-doc at the STScI. An observing proposal submitted by Dickinson to the annual peer review process in 1994, after spherical aberration had been corrected, bore fruit. His proposal to spend 32 orbits of Hubble observing time doing a deep (18 hr) exposure with WFPC2 in a single filter on a single field in the sky was accepted. His target field was a rich cluster of galaxies associated with a strong source of radio waves, a peculiar galaxy named 3C-324. Ground-based observations of this field had previously provided a redshift measurement for the radio galaxy $z = 1.2$. The corresponding lookback time was 8.5 billion years. Thirty-two orbits of telescope time was far and away the biggest observing program accepted for Hubble up to that time. It would produce the deepest image taken by the telescope to date. The Time Allocation Committee chose to take a chance that this observation would pan out as something important.

When Mark Dickinson showed Bob Williams the Hubble image of the 3C-324 cluster, Williams was blown away. It was a spectacular image, showing thousands of galaxies, many of which looked like "train wrecks" (Figure 11.1). They were small, fragmented, disrupted, and asymmetric—nothing like we see in the local universe. As far as Williams was concerned, that image clinched the case; more extragalactic research, probing to higher redshifts, needed to be done with Hubble. The next issue was how to do it.

Though Williams was not an extragalactic specialist in his research, he was lucky to be surrounded at STScI by a cadre of young, ambitious and very bright astronomers who were. In addition to Mark Dickinson, there was Harry Ferguson, who held a prestigious Hubble Fellowship, Associate Astronomer Mark Postman, Post-doctoral Fellow Andy Fruchter, and Mauro Giavalisco, a Post-Doc on assignment from the European Space Agency. This was Williams' team, which regularly met every morning at "science coffee" to thrash out ideas for a new Hubble initiative for high-z research.

The first duty of the Space Telescope Science Institute was to support the international astronomical community in carrying out the most important scientific research with Hubble. As he had done for the Comet S/L-9 observing campaign, Williams called together a small panel of some of the world's most accomplished

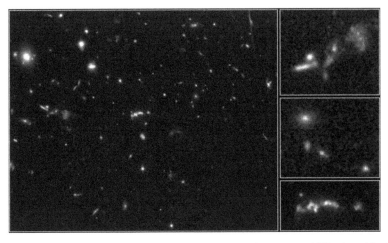

Figure 11.1. Rich cluster of galaxies surrounding the radio galaxy 3C-324 at a redshift of $z = 1.2$, observed by Mark Dickinson with WFPC2 in 1994. Many of the galaxies were small and had distorted structures as shown in the column of images on the right. This image persuaded STScI Director Bob Williams that Hubble needed to probe to even higher redshifts. (Credit: NASA, M. Dickinson, STScI 1994-52.)

extragalactic astronomers to represent the community in advising him on how best to do a high-z program. The assumption they were asked to consider was that Williams would contribute a large portion of his Director's Discretionary Time (100–200 Hubble orbits) in 1 year for a program of deep extragalactic observations. What should such a program look like?

Williams convened the panel at STScI early in 1995 for a 1 day brainstorming session addressing that question. The panel met with Bob alone in the morning and then with the larger STScI team during the afternoon. In the morning session each of the eight panel members had 15 min to present their ideas for a high-z program. They discussed a lot of options: a single deep exposure on one target location in the sky or shorter exposures on multiple targets; and one filter pass band to be used to go as deep as possible or three (which would allow a color image to be created). But the single most divisive question the panel considered was whether the target field should contain a known galaxy or cluster of galaxies, analogous to Marc Dickinson's 3C-324 observation, or should it be a totally ***blank*** field? Observing a field with known objects would be the conservative course. At least one could be certain that ***something*** would be seen. Observing a blank field would be a total crapshoot. If Bahcall and his colleagues were correct in their predictions in that 1990 *Science* paper, there was a real possibility that a totally blank field would end up looking just like that—totally blank. One hundred or more precious observing orbits with Hubble might go straight down the drain.

The panel's discussion was not heated, but there was clearly tension in the room. Trying to come up with a consensus proved frustrating. In the end the panel split 50–50 between those who wanted to focus on a field containing a known target and those recommending a totally blank field. At the end of the day, Bob sent them home with, "Thanks, I'll think about it." Some of the expert panelists concluded that the meeting had been a failure. They had been unable to reach a consensus on an approach that would satisfy the Director's desire to start using Hubble to explore the deep universe.

Over the next several weeks at the morning science coffees, Williams and his STScI team wrestled with how to proceed. Eventually they reached a pragmatic solution. They, better than anyone, knew what they wanted to do and with all their experience programming Hubble, they knew how best to do it. They would go for a very deep exposure on a single blank field. To mitigate criticism that this was a "data grab" by STScI staff members for their own personal scientific use, they decided that the entire set of data would be fully processed and released immediately on the Institute archives for access by the entire community. Williams would form an advisory team of astronomers to represent the community in guiding the planning of the program. If the observations were successful, they would be a foundational contribution to the study of how the universe had evolved. If they were not successful, Bob Williams would shoulder the full responsibility.

The five young astronomers working with Williams did so at some professional risk. None of them had advanced far enough in their careers to gain academic tenure—job security that would allow them to pursue cutting edge research in the future. If this Hubble Deep Field (HDF) observation did not yield new and important science, especially if it were a complete dud with nothing of any consequence being seen at all, that could be especially damaging to their reputations. But there was no dissent among them. They were excited by the prospect of using the world's most powerful telescope to explore that which was completely unknown. They wanted to press on.

Over the ensuing months Williams and his HDF team concentrated on planning the details of the observations. They settled on devoting 150 orbits of observing time—about 10 days' worth—as that would yield an optimal signal quality relative to the background detector noise. They would observe the field with four filters, bands transparent to blue, visible (yellow–orange), and deep red light, plus a near-ultraviolet filter that passes light not visible to the naked eye. The blue, visible, and red images could be combined to make a full color picture of the HDF suitable for release to the public. The near-ultraviolet images would hopefully provide information about the rate at which stars formed at various times in the distant past.

Perhaps the trickiest issue was the choice of where in the sky they should observe. The field of view of the WFPC2 was about 2.6 min of arc across—approximately the diameter of a baseball seen at 100 yards. The team wanted to choose a field that looked completely blank in ground-based surveys like the Palomar photographic sky survey. The field should be well away from the plane of our own spiral Milky Way, so as to minimize the effects of foreground interstellar dust that scatters light and skews its color toward redder wavelengths. The field should be as dark as possible, away from the light pollution of Milky Way stars. No known source of radio waves should be near the field so as not to interfere with radio astronomers who might want to do their own deep exploration there. To locate and lock onto a field, there had to be at least two nearby guide stars accessible at all times with Hubble's Fine Guidance Sensors. Finally the team wanted to choose a field that stayed visible for as long as possible throughout an orbit. That would give the most efficient use of observing time—the maximum total signal in a set number of orbits.

There are sweet spots in the sky where astronomers love to observe with Hubble, caps around the north and south poles of the spacecraft's orbit about 36° across where stars and galaxies are visible continuously—where they do not rise above or set below the Earth's horizon. These are called the continuous viewing zones or CVZs. Locating the HDF field within or near a continuous viewing zone would enable long exposure times during every orbit. Of course such a good thing can't last forever. Hubble's orbit precesses or wobbles around the Earth's rotation axis once every 56 days. So, once every 56 days the north and south continuous viewing zones sweep out bands around the sky perpendicular to the Earth's axis. A target in a continuous viewing zone won't stay there any longer than 5–6 days. Targets just outside a CVZ, while not viewable for all 94 min of a Hubble orbit, are still available for viewing for a large fraction of Hubble's orbital period. The HDF team chose an ideal field in the northern CVZ in the constellation Ursa Major just a short distance above the bowl of the Big Dipper. They planned a 10 day observing campaign.

Sometime in the early spring of 1995 I got a call from Bob Williams. He wanted to tell me about his plan for the HDF. In particular he wanted to know if the Hubble

Project would have any problems with it. My first reaction was to get excited. I thought pushing back frontiers was exactly what our observatory should be doing, and I told him so. I also assured him that he would have the support of the Project more generally. By that I meant that the Hubble operations team at Goddard would give its best effort to help make these observations successful. Observing the same target field continuously over a long period of time has its own challenges for spacecraft operations. For example, maintaining proper temperature control while the spacecraft remains in a single orientation in the heat of the Sun could be a challenge. The ops team would need to do some advance planning and preparation.

The directors of major ground-based observatories usually retain blocks of discretionary observing time that they normally use, for example, to enable observations of targets of opportunity—comets, supernovae, etc—or to support high-risk observing programs that a risk-averse time allocation committee might be reluctant to recommend, or to pick up an interesting observing proposal that just missed the cut. Director's Discretionary Time on Hubble is used for similar purposes. What made the Hubble Deep Field program unusual, and potentially controversial, was the fact that it had been initiated as a major science undertaking by the Director himself, and was to be implemented in-house by astronomers on his staff. The scientific community, which jealously guards the paradigm of competitively selected, peer-reviewed research proposals, might well collectively lift an eyebrow and look askance at this departure from the usual norms. Bob Williams was fully aware of the social delicacy of what he was doing and took every opportunity to be fully transparent in informing the community. This was a program to be fostered by the STScI Director, using his own resources, on behalf of the entire community, with data to be readily accessible to anyone from the get-go, and with the advice and oversight of a representative panel of outside experts in the field.

One such occasion to be transparent came along in early June of 1995 when Williams briefed the Space Telescope Institute Council (STIC) on his plans at one of their periodic meetings at the STScI. The STIC was an important community-based advisory committee instituted by the AURA Board of Directors to keep close tabs on the performance of the STScI. AURA was the consortium of universities that was under contract to NASA to build and run the organization that was the STScI. The STIC regularly sought inputs from the Goddard Project and NASA Headquarters, for example, to make sure we were satisfied customers.

As Bob described it in our recorded interviews, after the June briefing the members of the STIC seemed enthusiastic about the scientific potential of the HDF plan Williams had set forth. This one set of new data might well foster a major step forward in humanity's understanding of the universe. There was, however, a lone voice of dissent, one that could not be ignored. Lyman Spitzer, revered by all as the father of the Hubble Space Telescope, spoke up in his usual polite and understated manner to ask, "Bob, are you really sure you want to do this?" He went on to note that there was a risk that the observations would fail to detect very high-z galaxies and that could lead to a bad reaction from a public that had been so skeptical about the telescope in recent years. Bob replied that the STScI astronomers felt very strongly that it was important to try.

The negative implication of Spitzer's question was not lost on Goetz Oertel, the President of AURA, who regularly sat in on STIC meetings. Later Oertel said to Williams, "Bob, I basically support what you want to do with the HDF, but we have to take Lyman's concerns seriously." Williams replied, "I strongly believe in the HDF, and if it doesn't work out, I'll fall on my sword and resign as Director."

A day or two after the STIC meeting, Bob Williams took a phone call from John Bahcall. John said he needed to take the train down to STScI soon to attend to some business regarding planning of one of his observing programs. He wondered if he could meet with Bob for an hour or so while he was there. John and Lyman Spitzer both resided in Princeton, John at the Institute for Advanced Study and Lyman at the Princeton University Observatory. They were brothers-in-arms, having fought many battles over the years to win support for the Hubble Space Telescope within the astronomical community, NASA, and Congress. Both clearly felt the protective instincts of parents, given Hubble's troubled childhood. It seems likely that Lyman had briefed John about the plan for an HDF after he returned from the STIC meeting. And that's what Bahcall wanted to discuss with the Director.

On that June afternoon John Bahcall sat in Bob William's office, very serious in demeanor. He spoke to Bob cordially, calmly, without emotion. He said he had heard about William's plan for a very long observation with Hubble on a blank field in the sky with the objective of detecting galaxies at high redshifts. He thought there was a real likelihood that such a deep observation would not detect any galaxies at high z, based on the work he and his colleagues had published in *Science* in 1990. NASA had just finished servicing Hubble at great expense to correct spherical aberration. Now the STScI Director was about to take a big risk. If you spend so much time and only find a blank field, it's going to look very bad. It will damage Hubble's public reputation, maybe fatally so. It may set Hubble back, perhaps permanently. This is not the time to be doing this—maybe in a few years, but not this soon with Hubble's tarnished reputation still on peoples' minds.

Williams came to the meeting well-armed by his young STScI collaborators with factual arguments supporting the case that distant galaxies, perhaps entirely new kinds would be seen. The major argument was, of course, the strange distorted clumps of stars Marc Dickinson had found in his WFPC2 observation of 3C-324 at a redshift of 1.2. That observation in itself made a compelling case to try going deeper.

A characteristic of John Bahcall's thought process, well known to his associates at Princeton, was that, if he were presented a set of facts and drew a conclusion from those facts, then he believed anyone else presented with the same set of facts would draw the same conclusion. That was the nature of rational thought. John remained adamant in his strong opposition to the HDF, while Bob Williams remained equally strongly convinced that the HDF was the right thing to do for science. The two remained at loggerheads, even after Bahcall visited Williams a second time a few weeks later and tried again to dissuade him. And so the decision was final and the deed would be done. Bob Williams would take the hit to his career, if it turned out he was wrong and Bahcall had been right.

In planning the HDF, the STScI team was aware of course that the WFPC2 deep images would be of little use if they did not have redshift measurements for the galaxies they might discover. They would have no idea how deeply across space and how far back in time they were looking. Obtaining accurate redshifts required spectroscopic observations. Spectrographs mounted on telescopes break the light from a galaxy apart into its component colors and spread it out across the surface of some kind of light-sensitive electronic detector. Atoms and molecules emitting or absorbing light in the outer layers of the stars that make up a galaxy have well known signatures. Dark lines where some of the light has been absorbed or bright lines where extra light has been emitted are spread apart from each other in wavelength by well-defined amounts, creating patterns similar to those produced by the same elements in a chemistry lab on Earth. From those patterns the atoms or molecules involved can be identified. Once they've found the patterns, and identified the chemical elements producing them, astronomers can measure at what wavelengths the lines are seen in the galaxy spectrum and compare those to the wavelengths of the same lines of the same elements measured in the lab. For distant galaxies, the patterns are always shifted to longer, redder wavelengths. The light seen in the visible part of the spectrum by Earth-based astronomers, started out as ultraviolet light when it left its home galaxy at a redshift of $z = 1$ or 2 or more. So, astronomers have to be familiar with the patterns of spectral lines produced by common elements, like hydrogen, helium, carbon, etc, at ultraviolet wavelengths in order to be able to discern them in the visible spectra of distant galaxies and to use them to measure the redshifts of those galaxies. The magnitude of the redshifts and the associated apparent velocities of recession tells them, through Hubble's law, how far away the galaxies were when they emitted the light and how long the light has taken to reach Earth (the lookback time).

But how does one begin to get redshift measurements for galaxies that haven't been seen yet in the blank Hubble Deep Field? At that time the largest optical telescope in the world, the first of two 10 m telescopes of the Keck Observatory on top of the dormant Mauna Kea volcano in Hawaii, had just come on line. The University of California, Caltech and the University of Hawaii jointly operated the observatory. Bob Williams contacted colleagues out west who had access to Keck. He proposed that the STScI provide a short, two-orbit observation of the HDF, a pre-HDF, looking for the brightest galaxies in the field. Astronomers at Keck might then follow up with spectra and measured redshifts for those galaxies. Since a spectrograph on a telescope spreads light out, that is the light from a galaxy is no longer concentrated into a small image, this kind of observation is usually done on relatively bright objects where the total amount of light to be spread out is plentiful. The question of how to get redshift measurements for fainter galaxies in the HDF would have to wait.

Over the next months the astronomers at Keck began to collect spectra for dozens of galaxies found in the two-orbit Hubble image. Each night after they finished observing, they loaded their spectroscopic data onto the Keck public website to give anyone who wanted to use them access to the observations. Thus, they were following the new paradigm of the HDF program to make all the data public immediately.

For ten consecutive days in 1995 December 18–28, Hubble pointed with few interruptions at the tiny field in Ursa Major, collecting 342 separate images ranging from 15 to 40 min in exposure time through four different color filters with WFPC2.

For 150 spacecraft orbits around the Earth, Hubble stared at one spot on the sky. The HDF team stacked all of these images together to create a single, spectacular full color image. They counted about 3000 individual objects in the image, only 10 or so of which were foreground stars. Every other object in the field was a distant galaxy. Some looked familiar, like the spiral and elliptical galaxies we see in the nearby universe. But many didn't look familiar at all. They were tortured clumps of stars and gas, much smaller than the Milky Way—"train wrecks" similar to what Mark Dickinson had seen in his pioneering image of 3C-324. One could scarcely call them galaxies—more like protogalactic clumps.

In the end the Keck astronomers acquired good redshift measurements for approximately 140 galaxies seen in the pre-HDF and HDF images. The most distant galaxy measured with Keck had a redshift $z = 4.22$, corresponding to a lookback time of over 12 billion years. That was surprisingly deep for a ground-based telescope. But the Keck was, after all, a new and very large telescope. Any worry about whether the HDF observations would turn up a blank field in the high-redshift universe was clearly off the mark.

The 342 HDF images, all stacked together into a single composite deep exposure, produced a gorgeous multi-color image (Figure 11.2). What was at that time humanity's farthest view out into the cosmos would, in my opinion, have made a compelling news story. Both Ray Villard, the News Director at STScI, and I suggested to Ed Weiler's outreach team that NASA should immediately issue a press release about it. Much to my surprise, given his past predilections for beautiful Hubble images on the front pages of the New York Times and Washington Post, Ed vetoed that idea. "It's just a pretty picture with no science to go with it," he said. In principle he was right. We had a rule that NASA news releases would only follow when a paper about the work had been produced and peer reviewed for publication in a scholarly journal. In this case the HDF team at the STScI had vowed not to do science with the data, at least not right away, but rather to release it immediately for the entire community to work on. So, by definition there would be no immediate scientific conclusions drawn from the HDF (though some were obvious from simple inspection, like the fact that there were clearly many objects out there that were unlike anything seen in the nearby universe). I likened Ed's decision to a conclusion that Hillary and Tenzing's summiting of Mt. Everest was not newsworthy because, after all, it was just a mountain. But his direction was of little consequence. The STScI team produced its own press release about the HDF in time for the upcoming meeting of the American Astronomical Society in 1996 January in San Antonio. Bob Williams discussed the HDF in a scientific paper delivered at that meeting. One day later, on 1996 January 16, an article by Kathy Sawyer, entitled "NASA Takes Portrait of Universe; Hubble Team Releases Most Detailed Image Yet of Deep Space," accompanied by the image of the HDF, appeared on page A-3 of the Washington Post. An excellent first-hand account of the birth of the HDF by Bob Williams himself can be found in his book "Hubble Deep Field and the Distant Universe.[1]"

---

[1] 2018 (Bristol: IOP Publishing) (http://iopscience.iop.org/book/978-0-7503-1756-6).

Figure 11.2. The original northern hemisphere Hubble Deep Field taken through four filters with WFPC2 in 1995. (Credit: NASA, R. Williams, the Hubble Deep Field Team, STScI 1996-01.)

How far away were the faintest objects in the HDF image? How deeply back in time had the HDF penetrated? Most of the research subsequently done with the HDF observations relied on the Keck spectroscopy that had reached beyond $z = 4$. The universe was only 1–2 billion years old when some of the faintest galaxies emitted the light that we see in the HDF image. Most of the objects seen in the field were too faint to be amenable to traditional spectroscopy. However, astronomers used the HDF observations to invent a new, very powerful technique for measuring reasonably accurate redshifts from separate images taken through multiple (four or more) color filters. I call it the "now you see it, now you don't" technique.

Each color filter used for the HDF was transparent to a particular wavelength band of light 100–200 nm wide. The filter absorbed or reflected all other colors outside that band. The near ultraviolet, blue, visible, and deep red filters had their maximum transparency centered on wavelengths around 300, 450, 606, and 814 nm, respectively. The farther away a galaxy was, the higher its velocity of recession and the greater was the shift of its light output toward the red, toward longer wavelengths. When they arrived at Earth, photons that were originally emitted with, say, far-ultraviolet wavelengths might well find themselves shifted into the pass band of the near ultraviolet, or blue, or visible, or red filter, depending on how far away the galaxy was and thus, on how large its redshift was.

Now here is the really clever astrophysics part of this scheme. Hydrogen, the lightest chemical element, makes up approximately 90% of the ordinary matter in the universe. (We're not including dark matter here, the composition of which remains unknown.) The outer layers of the stars in distant galaxies, the layers that emit the photons of light ultimately seen by HST, are composed largely of hydrogen atoms—one electron bound to one proton. Moreover, the interstellar gas distributed among the stars in a distant galaxy is also composed mostly of hydrogen. Photons of light with wavelengths in the interval 91–120 nm have a very high probability of being absorbed by the hydrogen gas, producing strong dark lines in the spectrum of a galaxy. At wavelengths less than 91 nm, the photons have enough energy literally to rip hydrogen atoms apart, producing free electrons and protons. In the process the energy of the photons is absorbed—the photons disappear.

Galaxies near and far drop off precipitously in brightness—effectively they go dark—at wavelengths shorter than about 120 nm. This sudden transition from light to dark, produced by absorption of light by hydrogen atoms, is a key marker in the spectrum of any galaxy. It's called the Lyman break. An astronomer observing the brightness of light emitted by a nearby galaxy such as M31 in Andromeda through a set of color filters would have to extend her/his coverage deep into the far-ultraviolet to find the wavelength where the galaxy goes dark. But if that galaxy were more distant, say at redshift $z = 3$, that transition from light to dark would occur in the wavelength band covered by the WFPC2 blue (B) filter used for the HDF observations. In the longer wavelength visible (V) and deep red (I) filters the galaxy would be seen with its full brightness. In the shorter wavelength near-ultraviolet (U) filter it would not be seen—it would be dark. In the blue filter the galaxy would be seen but fainter, since the light-to-dark transition occurs within the blue filter bandpass (Figure 11.3; Dickinson 1998).

A galaxy with redshift $z = 4$ would have its drop-off to darkness within the bandpass of the visible filter. It would be seen at full brightness in the deep red filter, fainter in the visible filter, possibly barely detected in the blue filter and unseen in the near-ultraviolet filter. Now you see it, now you don't. By observing a very distant galaxy with a set of color filters and looking for the filter images in which it gets faint or disappears entirely, astronomers could estimate the redshift of the galaxy.

These measurements were called "photometric redshifts." The original HDF images were ideal for proving that this technique worked with reasonable accuracy, as the photometric redshifts could be compared to the more precise spectroscopic redshifts measured with the Keck telescope. With time astronomers refined the photometric technique by using a larger set of filters, spanning a wider range of wavelengths and carefully selecting which filters should be used. Eventually it was demonstrated that photometric redshifts accurate to better than 5%–10% were achievable—plenty good enough for most extragalactic research.

The original Hubble Deep Field set precedents that were followed by future STScI Directors. In 1998 Bob Williams himself sponsored an HDF-South, replicating the 150 orbits of WFPC2 imaging of the northern HDF in a field in the constellation Tucana (named after the South American Toucan bird) in the southern sky. The HDF-North and HDF-South images looked quite similar, supporting the

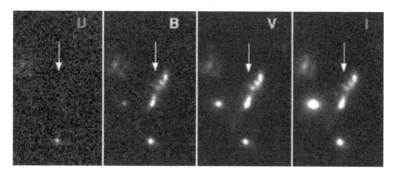

Figure 11.3. Example of using the Lyman break to measure the approximate redshift of a distant galaxy. In the U-filter image nothing of the galaxy is seen. In the V and I images it is seen brightly. In the B-filter image the galaxy is visible but appears fainter. From this information an astronomer would deduce that the "photometric redshift" of the galaxy is about $z = 3$. (Credit: M. Dickinson, reproduced with permission © Cambridge University Press.)

Cosmological Principle that on large scales the Universe is homogeneous and isotropic. It looks the same in all directions. As before, all the data were acquired by the STScI using Director's Discretionary Time and were released to the public very shortly after they were acquired.

There were numerous follow-up observations of the two HDFs by other observatories and by other Hubble instruments, including NICMOS (the Near Infrared Camera and Multi-Object Spectrometer) and STIS (the Space Telescope Imaging Spectrograph), both installed in the second Hubble servicing mission in 1997. But the Advanced Camera for Surveys (ACS) installed in 2002, and the Wide Field Camera 3 (WFC3) installed in 2009 took the biggest steps. Both of these cameras were much superior to WFPC2 in sensitivity, resolution and field of view. So, they could go deeper across space and time. For these two cameras the deep field observations were taken in a different location from the original HDF, in the southern constellation Fornax (the furnace). Whereas the original HDF probed to $z \sim 4$, the Hubble Ultra Deep Field (HUDF), imaged by the ACS in about one million seconds over 400 Hubble orbits in 1993–1994, contained images of galaxies as deep as $z \sim 7$, looking back about 13 billion years to a time about 800 million years after the Big Bang.

In 2009–2010 the Infrared Channel on the brand new WFC3 produced yet another deep view of the same southern field in Fornax that had been the object of the HUDF in 2002. This was called the HUDF 2009–2010 (Bouwens et al. 2011). Combining these observations with images taken with ACS revealed objects with photometric redshifts in the range $z = 8-9$ (Figure 11.4). These early protogalaxies emitted the light we now see when the universe was only about 600 million years old.

The *piece de resistance* of deep universe exploration, however, was the Hubble Extreme Deep Field or HXDF (Illingworth et al. 2013). Australian astronomer Garth Illingworth and his colleagues at the University of California at Santa Cruz stacked deep field images taken with the infrared channel of the new WFC3, together with the original HUDF and images of the same field from multiple other

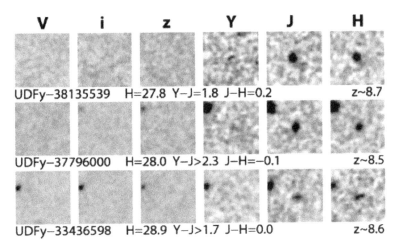

Figure 11.4. Three extremely faint protogalaxies observed in the Hubble Ultra Deep Field 2009–2010. The first three thumbnail images on the left came from ACS observations. The remaining thumbnail images on the right were taken with the WFC3 IR Channel. From left to right the six images were taken through a visible, two deep red, and three near-infrared filters. The photometric redshifts are in the range $z \sim 8.5$–8.7, corresponding to lookback times of about 13.2 billion years. (Credit: R. Bouwens, G. Illingworth: reproduced with permission © Springer Nature: Bouwens et al. 2011.)

Hubble observing programs spanning the years 2002–2013. This amounted to about 3000 separate images with about two million seconds of accumulated exposure time, taken with Hubble's two premier cameras. The HXDF is to date the deepest look back across space and time ever achieved by human beings (Figure 11.5). Supplementing the Hubble data with observations from the infrared Spitzer Space Telescope to extend the photometric redshift measurements farther into the infrared than Hubble alone could reach, Illingworth and his colleagues were able to detect some of the earliest protogalaxies—clumps of stars and gas pulled together by the gravity of blobs of cold dark matter at $z \sim 10$–11, when the infant universe was only 400–500 million years old.

What have we learned from almost two decades of Hubble deep field observations? The Universe is rich in galaxies. Extrapolating the total count of galaxies in the HUDF, the HUDF 2009–2010, and the HXDF over the area of the whole sky, we can infer that there are perhaps as many as 300–500 billion galaxies in the observable universe.

The first clumps of stars and gas were very small compared to modern-day galaxies, some as small as the globular clusters we see in the night sky or the bright star-forming regions we see in the Milky Way and nearby galaxies. The Universe was a much smaller place 13+ billion years ago, and it was common for these small proto-galaxy clumps to collide and merge with each other. As time went on small clumps merged to form bigger clumps. This process of collision and merger continued over the eons until galaxies grew to the sizes we see nearby today. Meanwhile, the Universe continued to expand and collisions became less common.

Figure 11.5. The Hubble Extreme Deep Field (HXDF), combining images taken by ACS and WFC3 over the period 2002–2013 of the central portion of the same southern field in the constellation Fornax that was originally observed for the Hubble Ultra Deep Field in 2002. In total 3000 separate images were added together amounting to a total exposure time of two million seconds. This is the deepest image of the universe ever acquired in visible and near-infrared light. (Credit: NASA, G. Illingworth, et al., the HUDF09 Team, STScI 2012-37.)

The formation of galaxies in the first few hundred million years after the big bang apparently got off to a slow start, but the process was accelerating rapidly by the time the universe was 600+ million years old. The first stars were probably very massive and very hot, composed almost entirely of hydrogen and helium. They burned through their nuclear fuel rapidly and ultimately ejected much of their nuclear processed material, now enriched in heavier elements like carbon and oxygen, back into the interstellar medium from which the next generation of stars would coalesce. The several generations of Hubble Deep Field observations showed that the rate at which new stars were formed within the galaxies reached its peak at $z \sim 2$, about 10 billion years ago. After that the supply of raw material for new star building in the interstellar medium began to be depleted and stars formed less frequently.

The next chapters in our account of the history of the Universe will be written by the successor to Hubble, the James Webb Space Telescope (JWST), which was designed expressly to do this kind of early-universe science. But the Hubble Space Telescope, with its succession of ever more modern cameras and electronic light sensors, probed far more deeply than any of us might have expected in 1990 when it was first deployed. The pessimistic views of some astronomers that Hubble would see little or nothing beyond $z \sim 2$ or 3 were dead wrong. We could almost see forever, almost to the beginning of time.

References

Bahcall, J., Guhathakurta, P., & Schneider, D. 1990, Sci, 248, 178
Bouwens, R., et al. 2011, Natur, 469, 504
Dickinson, M. 1998, in The Hubble Deep Field, ed. M. Livio, et al. (Cambridge: Cambridge University Press), 219
Illingworth, G., et al. 2013, ApJS, 209, 6

AAS | IOP Astronomy

Life With Hubble
An insider's view of the world's most famous telescope
David S Leckrone

Chapter 12

A Bad Sunburn: The Second Servicing Mission

The worldwide community of optical astronomers who observe with the Hubble Space Telescope numbers about 8000. This population is divided into several tribes. The two largest tribes are the imaging astronomers, who use cameras to take pictures through color filters to do their work, and the spectroscopists, who use spectrographs to tease out the fine details of the light sent our way by cosmic sources. These two groups are not mutually exclusive; many astronomers are skilled at using both kinds of instruments. But typically they specialize.

Both groups were being well served by the new scientific instruments to be installed in Hubble in the second servicing mission, SM2, in February of 1997. NASA had selected both the Space Telescope Imaging Spectrograph (STIS) and the Near Infrared Camera and Multi-Object Spectrometer (NICMOS) for development into flight instruments in 1985. The plan was that they would fly on the first servicing mission, perhaps four or five years after Hubble was first deployed. However, fate intervened. The crisis of spherical aberration led to a higher priority for SM1 to get the image quality of the observatory up to snuff with the corrective optics of the Wide Field and Planetary Camera 2 (WFPC2) and Corrective Optics Space Telescope Axial Replacement (COSTAR). Work on NICMOS and STIS was slowed down, their funding stretched out to help finance the contingency call-up mission that SM1 had become.

SM1 in 1993 turned out to be a huge success. Now it was time to focus on NASA's original intent for normal servicing of the observatory. SM2 would be the first mission whose primary aim was to expand Hubble's scientific capabilities, to update its technology and to carry out needed maintenance and repairs. Construction and testing of the two new instruments and all the other hardware required for the mission was accelerated dramatically (Figure 12.1). New tools were created to enable the work of the astronauts who would be doing the spacewalks. They began training in the enormous swimming pool, the Neutral Buoyancy Lab (NBL) at JSC, using full-scale mockups of Hubble and the instruments and other

Figure 12.1. The completed STIS and NICMOS instruments at Ball Aerospace in 1996. (Credit: NASA, Ball Aerospace.)

modules they would be installing during four planned extra-vehicular activities (EVAs) in orbit. They also spent time at Goddard, training with the Hubble mechanical and electrical simulators originally created for SM1.

In addition to these physical preparations, SM2 and every other Hubble servicing mission required extensive preparation of mission strategies. Many decisions had to be made regarding what items should be included in the payload manifest, what their relative priorities were, what the most efficient way was to fit them into the EVA timeline, what procedures would be followed in case contingencies arose, and finally what subset of the work to be done in orbit was considered "good enough" to allow NASA to declare the mission minimally successful or fully successful. We might plan some lower priority tasks to be done if time permitted. These would not be considered essential for the mission to be deemed successful. They were for "extra credit," so to speak.

Underpinning the planning process were two overriding considerations. First, the probability of catastrophe, loss of mission and crew, for any shuttle flight was about one in seventy, even with the most careful planning and execution. That was a fact of life for human spaceflight. The astronauts on Hubble servicing missions understood that risk and accepted it because they recognized the value to humanity of what they were doing. Nevertheless, their safety always came first. If a mission had to be scrubbed before launch or terminated early in orbit, so be it. We could fly another day. That's why we prepared contingency plans for abbreviated missions. Beyond concerns for astronaut safety, other contingencies also could lead to a shortened mission: astronaut illness, orbital debris, fuel shortage, critical systems failures, etc. What if we could only have one EVA, or two, instead of the planned four or five? Our payload priority lists guided the planning for how we could best use the available time in a curtailed mission.

Second, shuttle flights and servicing missions were expensive. The typical cost to NASA Space Science of a full-up servicing mission was about $450 million, not counting the costs of flying the shuttle, which were borne by NASA's Human Spaceflight Program. The amortized cost of a shuttle mission (the total cost of the Shuttle Program over all years divided by the total number of shuttle missions) was also about $450 million. The mission-specific cost to conduct an EVA-intensive shuttle flight such as SM2 was about $150 million. That included the costs of planning, training, shuttle preparations, non-reusable components (the External Tank, etc), fuel, and mission operations. So, the stakes for these Hubble missions were very high, both in terms of the financial investment and in terms of their importance to the future of science. Every one of the roughly 2000 min available for spacewalks on each mission was precious and must not be wasted. Efficient choreography of the EVAs was of equal importance to the priority of the various tasks the spacewalkers had to accomplish. Sometimes, tasks of lower priority would get performed before those of higher priority because that would allow the EVA to flow most efficiently, simplify the work of the crew, and hopefully allow every task on the mission to be successfully completed.

For over two years prior to SM2 all of this technical analysis, discussion, debate, and decision-making occupied the Hubble Project at Goddard, the Space Shuttle organization at Johnson, and the team at Kennedy responsible for launch preparations. The Hubble Program office at NASA Headquarters monitored the process as it went along. Briefings and reviews were held regularly for the Directors of the three responsible NASA centers and for NASA Headquarters. Several independent review teams periodically monitored our plans and our work. For example, at NASA's request the Space Telescope Institute Council (STIC) organized an independent review in 1995 of the Project's scientific readiness for SM2, chaired by Malcolm Longair of Cambridge University.

The results of this laborious process were documented in a bookcase full of three ring binders containing Flight Rules, detailed Mission Timelines, Test Plans, Management Plans, Contingency Plans and Procedures, etc. It was a mountain of paper reflecting a highly disciplined approach to making sure the mission would succeed. We had a plan for anything and everything we could think of that might go wrong. The protocol was that we would follow these plans during the mission, but with the understanding that they might be modified in their details to reflect what was actually going on in real time. In general, improvisation in orbit could lead to mistakes and wasted hours of precious EVA time. That was to be avoided except under highly unusual circumstances. It turned out in fact that improvisation would become necessary during SM2.

The relative priority of the scientific instruments and of the various pieces of engineering hardware in the payload manifest was the subject of strongly held opinions on the part of the astronomers, managers and engineers, and the topic of extensive discussion and debate. There was always a logical tension between spacecraft maintenance work, without which Hubble might stop working properly between servicing missions, and the top science goals, which were the main reason for Hubble's continued operation in the first place. It was a complicated subject. Just because the insertion of a new scientific instrument into Hubble was deemed the most important task on the mission, it didn't follow that it would be the first task done. Of great concern also (beyond the top priority to protect the astronauts from harm) was the very limited time available for the EVAs—nominally six hours, but with some latitude depending on how fast the astronauts were consuming their oxygen and other expendables.

In what order should the work be done so that no EVA time would be wasted, so that every task would be as simple and efficient as possible for the spacewalking crew, and so that the chances of getting all the work done in the four or five EVAs available would be maximized? It required a considerable amount of time to rotate the telescope from one work site to another and to open and close doors. Doors on the spacecraft could not be left open overnight. So, the logic of the sequence of work to be done throughout the entire mission had to be carefully thought through to avoid unnecessary loss of EVA time to overhead. Here again mission priorities served as guidelines but did not dictate the precise ordering of the work.

A good example of the planning process at work was that for SM1 in 1993. Getting the new WFPC2 and COSTAR instruments mounted into Hubble with their ability to correct the telescope's spherical aberration should logically have been the most important tasks of the mission. After all, that was why this call-up mission was being done in the first place. But they were placed in the construction of the EVA sequence as the number 3 and 4 tasks behind the spacecraft maintenance tasks of replacing the old solar arrays and failed gyros.

As SM1 was actually implemented the first EVA included two Gyro Packages, two Electronics Control Units for the Gyro Packages, and some replacement Fuses. EVA 1 also included about 50 min worth of advance preparation for the replacement of the solar arrays the following day. The need for time-consuming preparations for solar array change out is what necessitated the delay of that task to EVA 2. The WFPC and COSTAR were inserted on EVA 3 and EVA 4, respectively. Packing the EVA timeline in this efficient way allowed the crew to complete the full list of 14 tasks in five spacewalks. The SM1 mission was a resounding success.

In the preparations and training for SM2, a more nuanced example of the tension between payload priorities and timeline efficiency emerged, unfortunately, in the form of a small family dispute. It happened at our first Joint Integrated Simulation (JIS) the week of 1996 October 28.

Before each servicing mission, the entire team would come together five or six times at JSC and GSFC for a full dress rehearsal of the mission. The astronauts took up their positions in a highly realistic Shuttle simulator across the JSC campus. The Shuttle management team sat at their console positions in Mission Control. The Hubble operations team remained in residence in the Space Telescope Operations Control Center (STOCC) at Goddard. And the Headquarters and Goddard Program and Project management teams resided in their normal spots in the Customer Support Room at JSC.

The JISs were purposely designed to be intense and challenging. The hope was that, after six or seven difficult JISs, the actual mission would seem relatively easy, perhaps even "boring." Each JIS simulated one or two flight days, including the planned EVAs. We would be following the normal timeline for a particular flight day when, out of the blue, the JIS Director, working behind the scenes, would throw an imagined contingency or failure into the mix that would require all parties to respond. We had to practice working together as a cohesive team to solve these problems.

For this mission we had designated NICMOS as the number one priority and STIS a close number two. That was because NICMOS would significantly expand the range of wavelengths of light that Hubble could see into the near infrared, out a bit beyond 2000 nm, more than doubling the observatory's previous coverage. That would open up entirely new vistas for Hubble imaging science. The STIS was vastly more sensitive and efficient than the two original spectrographs, enabling spectroscopists to collect a much wider wavelength span in a given amount of time in ultraviolet and visible light. With its two-dimensional light detectors STIS could also observe extended objects, like the nuclei of galaxies, obtaining spectra simultaneously for every point seen in its field of view. But we already had decent, though slow, one-dimensional spectrographs working on Hubble. NICMOS offered something completely new. That gave it a slight edge in priority.

The layout of the timelines for EVAs was the job of the Shuttle EVA team at JSC. They measured the amount of time the astronauts needed to perform each servicing task as the crew practiced in the big NBL swimming pool. Over many trials they determined a best estimate and assigned that to be the likely amount of time required in orbit (taking account that tasks tended to require more time in orbit than in the NBL). They then tried out several alternative timelines, based on mission priorities, the logical flow of the mission (i.e., completion of Task B was a prerequisite for Task A), and the recommendations of the astronauts who had to perform the work in the payload bay. The graphics illustrating the recommended timeline (and alternatives that might be considered) were distributed to the entire Hubble team many weeks in advance of the first JIS. Also included were timelines for various contingency scenarios. What if we could do only one EVA, or two, or three? How would those EVAs be structured?

When I first saw the proposed SM2 EVA layout and timeline, I was ecstatic. Both of the new scientific instruments, the two highest priority items in the payload, would be installed on Hubble during EVA 1. No matter what contingencies might arise thereafter, Hubble would have its new, advanced scientific capabilities. Great for science! It never occurred to me to question the order of insertion **within** the planned EVA 1. STIS came before NICMOS. So what? The chances of some undefined problem that would require the mission to be terminated after STIS insertion was completed, but before NICMOS insertion was done, were so infinitesimally small that it wasn't worth giving them any thought. If one instrument were successfully mounted into Hubble, both would be.

Monday morning, 1996 October 28, the first day of JIS 1, began with the JIS Director throwing us a problem. Due to some issue with the shuttle, the mission would have to be terminated sooner than planned. We would have time for only two EVAs.

We checked the contingency plan and saw that STIS and NICMOS installation would still be done on EVA 1. EVA 2 would be extended by 1 hr from 6 to 7 hr. That would allow the new Fine Guidance Sensor, its new Optical Control Electronics Kit, and the new Solid State Data Recorder to be installed. There would also be time to replace one of the old Engineering and Science Tape Recorders. This combination of updates to Hubble would provide a major scientific and engineering upgrade. It would also be sufficient to allow NASA to declare full mission success.

I walked over to the table in the CSR occupied by the Headquarters Program Management Team to discuss what was happening with Ed Weiler. I assured him that we would still get the two new instruments in even in this shortened mission. "Okay, then," he said, "But I want NICMOS to go in first." "Ed," I replied, "That isn't the plan. We have to follow the plan. STIS is supposed to go in first, then NICMOS." This turned into a heated discussion. Ed's perspective was that of Headquarters, that the mission priorities they had blessed should be the primary factor used in executing the mission. It was also clear that he suspected that I, as a spectroscopist, was biasing the decision. What Weiler did not understand was the extensive planning and thought process that had gone into our preparations for the mission.

About this time Mike Hauser, my former boss at Goddard, intervened. Mike was now the Deputy Director of the Space Telescope Science Institute (STScI). He was participating in the JIS representing Bob Williams, the Institute Director. Mike was highly respected in the community for his scientific leadership and for a quiet, thoughtful dignity that made him a good diplomat. Having overheard our conversation, he suggested that we table the issue for the time being and work on it later, after the JIS was over.

The following week the Space Telescope Users' Committee (STUC) held a previously scheduled meeting at Goddard. Weiler broached the subject of the EVA 1 insertion sequence with them. After some discussion, Program Manager John Campbell and I agreed to bring it up with the JSC people, to ask what the implications of switching the order of installation of STIS and NICMOS would be.

On Tuesday, November 12, Campbell, Rud Moe (the Goddard Servicing Mission Manager on console during SM2), and I had a teleconference with Jeff Bantle and Mark Lee. Jeff was the lead Flight Director for the mission and Mark Lee was the lead EVA astronaut. We laid out the issue for them and said some of the scientists would prefer to see NICMOS inserted first, out of concern that something might go wrong that would prevent it from being put into HST after STIS had already been installed.

Mark Lee then made what to us was a compelling case for leaving the ordering as it was—STIS first, followed by NICMOS. NICMOS was designed to expel the nitrogen gas that sublimated off the block of solid nitrogen ice used to keep its detectors and some of its optics very cold. This expelled gas had to be vented to space through a hose that was to be attached to an opening at the bottom of the Hubble spacecraft. That orifice was immediately in front of STIS. Mark said that the astronauts could not attach the hose before STIS was installed. Otherwise, it would be hard to see what they were doing, the operation would be awkward, and it would place both STIS and NICMOS at risk of damage. Even if NICMOS were latched in and electrically connected, its installation couldn't be finished until STIS was in place. Putting NICMOS in first, taking steps to secure the vent line so that it wouldn't dangle loose, then returning to complete the NICMOS installation after STIS had been latched in could take up to an extra hour of EVA time that would be better used on other tasks.

Mark Lee's argument was an excellent example of important factors that go into planning Hubble servicing beyond the priority order of the hardware that makes up the payload. After our teleconference I wrote a memo to all the members of the STUC, and to Weiler, explaining the astronaut's strong preference for leaving the plan for EVA 1 unchanged. The subject never came up again.

Ed Weiler's impulse to insert NICMOS first during JIS 1 was understandable. To an outsider it would seem like common sense. Your mission was being severely curtailed. You want to be sure the highest priority scientific task gets done no matter what. The problem, however, was that Ed came into the JIS unfamiliar with the rationale behind the servicing plan. This might have been rectified with a brief tutorial to bring him up to speed.

I think the experience of JIS 1 and the follow-on discussions were a learning opportunity for all of us. It gave all the astronomers involved a better understanding of the risks and potential unforeseen consequences of changing a well thought out plan in real time without a full understanding of all the factors that had gone into developing the plan. Success of SM2 and all the servicing missions required the discipline to *follow the Plan.*

It was Valentine's day, 1997 February 14—Flight Day 4 of STS-82, carrying the crew and payload for SM2 on board shuttle Discovery. Launch three days before had been awe-inspiring. As with the launch of the first Hubble servicing mission on STS-61 in 1993, the early morning sky had been dark—until Main Engine and SRB ignition. Then you could have read a newspaper by Shuttle light.

I was sitting at my station in the customer support room (CSR) in Building 30 at the Johnson Space Center about 10 pm Central Standard Time (local time in Houston) anxiously waiting for EVA astronauts Mark Lee and Steve Smith to emerge from Discovery's airlock to begin their first spacewalk. Their task was to remove Hubble's two original spectrographs, the Faint Object Spectrograph (FOS) and the Goddard High Resolution Spectrograph (GHRS), and to replace them with the new, high-tech spectrograph, STIS, and Hubble's first instrument intended for near-infrared observations, NICMOS. This was a big day for science.

Leading our Goddard Hubble Management Team in the CSR was Program Manager John Campbell. Frank Cepollina (Cepi) was also sitting nearby. We all had our headsets on, monitoring the communications between the crew in orbit and the shuttle team on the ground. In addition, Campbell, whose earlier experience on Hubble had been mainly in flight operations, was also paying close attention to the communications backchannels carrying conversations among various engineers and managers in the STOCC back at Goddard. Although HST was now firmly attached to the Flight Support System (FSS) structure in Discovery's payload bay and was hooked up to shuttle electrical power and communications, internally it was still an active spacecraft that needed to be monitored and controlled by the operations team in the STOCC. John wanted to keep close track of the precious spacecraft as well as the ongoing shuttle activities. Cepi, who had full responsibility for the Hubble hardware in the payload bay, and whose team had trained the EVA astronauts and built their tools, monitored the communications of his team of engineers residing in a control-center-like room one floor above us.

Several years before, the Project made the decision that the ESA solar arrays should be kept fully deployed during this mission. The solar arrays were huge (40 × 8.2 feet) flexible blankets that could be unrolled and rolled up rather like window blinds. They were extended and kept taut by cleverly designed booms that were compactly stored as flat steel sheets within a cassette, but that took on a stiff cylindrical form when unrolled. During SM1 one of those bi-stem booms split while the array blanket was being rolled up. That caused the array to stick in place. The roll-up process couldn't be completed, and thus the array could not be returned to the ground as planned. Later in that mission astronaut Kathy Thornton discarded the broken solar array wing overboard.

The Project did not want this kind of problem to come up again during SM2. But the wings were fragile structures. The Project's lead mechanical engineer, John Decker, and his team had spent over two years performing an exhaustive program of analyses and tests to verify that the fully extended solar arrays could handle the various forces they would experience throughout the SM2 mission without breaking.

The EVA astronauts were trained to avoid any contact with the wings while they were working on Hubble. At certain times during this mission the wings were to be oriented perpendicular to the telescope and the telescope would be rotated to give the EVA crew easy access to particular equipment bays. Hubble had to be tilted downward a bit before it could be rotated, to prevent the solar array panels from bumping into Discovery's tail.

It was planned that Hubble would be given a re-boost to a slightly higher orbit using Discovery's small reaction control thrusters. This would be a test run for a larger re-boost in the next mission, nominally in 1999. That's when the Sun would be near a maximum in its activity during the 11 yr solar cycle, causing the Earth's upper atmosphere to heat up and expand. The added atmospheric drag would speed up the rate at which Hubble's orbit would decay, causing it to lose altitude. The plan was to boost it high above these denser (though still ephemeral) atmospheric layers. Could the fully open solar array wings handle the loads that would result from the shuttle firing its thrusters?

Decker's team concluded that the solar arrays would be safe throughout the mission, if we planned to be very careful with them. With Discovery's payload bay fully occupied on STS-82 there was no room available for spare solar arrays. Hubble could not operate without the electrical power that two healthy arrays would provide. And so, the decision was taken to keep the bi-stems fully extended and the arrays fully deployed throughout the mission.

As we were preparing for the first EVA to begin, Hubble was standing erect on the Flight Support System near the back of the payload bay with its two beautiful golden solar wings unfurled. A new wrinkle to the STS-82 mission was that Discovery had been modified to prepare it for later missions to construct the International Space Station (ISS). In particular this was the first flight of the new

Figure 12.2. An astronaut at work in the payload bay of shuttle Discovery during STS-82, SM2 in 1997. At the bottom is the new external airlock, being flown for the first time on this mission. The Canadian Remote Manipulator System arm can be seen at top center. All of the spacewalks on this mission were performed with the two solar arrays fully deployed. They were oriented this way to reduce the chances an astronaut would bump into them while working in the various compartments in the aft shroud and equipment section below the arrays. (Credit: NASA.)

external airlock. Rather than being enclosed in the orbiter cabin as before, the airlock now protruded out into the payload bay. The idea was to allow the orbiter to dock its airlock directly onto a corresponding structure on the ISS so that the crew could pass freely between the orbiter and the space station. Now, for the first time, EVA astronauts would pass through the new airlock and egress into the payload bay to begin their day's work on Hubble (Figure 12.2).

Lead EVA astronaut, Mark Lee, and his partner, Steve Smith, fully suited up, were in the airlock following the procedure to depressurize—to expel the air from the airlock out into space. The only concern the crew noted was that the rate of depressurization was a little faster than expected. Otherwise all seemed normal.

Then, suddenly, we lost much of the telemetry from the telescope, as seen both at Goddard and by all of us at JSC. Hubble had shut down. We didn't know why. "What the hell happened?," John Campbell exclaimed. The Shuttle Flight Director in the Mission Control Room got on the loop and asked the Hubble team, "Did you guys do anything?" A chaotic din of conversations began within the HST team, both face-to-face and on the communications links. Campbell got on the "all HST" channel that the entire team would be monitoring and said, "Calm down everybody. Let's take some time and figure out what happened."

About a half hour went by, but it seemed longer than that. Then the astronauts said they had seen one of the solar arrays move, but they didn't know why. When they first noticed the movement, the solar array closest to the crew cabin was spinning. It passed through 45° from horizontal, then to 90° (parallel to HST's long axis), then over to 30° from horizontal on the other side and finally rebounded slowly back to about 45°, ending up a full 135° from where it was supposed to be. The crew estimated that the movement took about 15 s all together.

The STOCC had enough basic telemetry from the spacecraft to verify that the two solar array wings were now mismatched; they were no longer aligned as they should be, parallel to each other and perpendicular to the long axis of the telescope. One wing was correctly oriented, but the other was not. When the spacecraft's DF224 computer detected that the two array wings were misaligned, it automatically put Hubble into a safe mode. That's why, suddenly, we were not receiving normal telemetry.

Because the Hubble solar arrays were off-kilter, the JSC shuttle guys became concerned. After all, if our telescope's arrays were mismatched, we must have done something wrong, issued an errant command or something. That was their working assumption.

It was Cepi's team that first came up with the correct answer. Engineers on console in the CSR, Rud Moe and Mike Weiss, pointed out that we were dealing with a new airlock, and the event happened while it was being depressurized. The plume of air flowing out from the airlock might have hit that array wing, forcing it into an un-commanded rotation.

Lee and Smith waited in the airlock while the solar array anomaly was being assessed. The crew inside the shuttle cabin was discussing the situation further with Mission Control. They said, "We all agree here that the movement and even the dampening oscillations [when the rotation stopped] were more a movement of the canister [the central cylinder from which the arrays were originally deployed and to which they were still attached], rather than the solar array tips, indicating a rotation of the solar array itself rather than a bi-stem deflection. And it also seemed that the dampening coincided with the closing of the depress valve which would confirm your earlier hypothesis of that being the cause."

The situation was summarized later by Mission Commander Ken Bowersox in his STS-82 Crew Report: "Although the crew did not immediately associate the rapid slew [of the solar array wing] with airlock depressurization, the depress had been stopped because the crew was concerned about the rapidity of the airlock pressure decrease. It was subsequently determined by the ground that the depressurization had caused the un-commanded slew. The ground slewed the arrays back to 0° and the crew was given the go for airlock egress using modified depressurization procedures. An additional one and one-half hours were spent by the EVA crew in the airlock as a result of the solar array anomaly."

My first thought after hearing Moe and Weiss describe their hypothesis about the airlock was, "Poor John Decker! He spent the better part of the last two years worrying about how much mechanical stress these fragile arrays could take. And the very first thing that happened on the first EVA of the mission was that they got wacked by a stream of air from the shuttle." Sigh! But in the final analysis, it seemed that no real damage had been done and so EVA 1 proceeded as planned.

For EVA 1 on Flight Day 4, the telescope was rotated 90° so that the +V2 side (under one of the solar wings) was facing forward toward Discovery's cabin. Two aft shroud doors on that side had to be opened to gain access to the GHRS and the FOS that were to be replaced with STIS and NICMOS. EVA astronauts Lee and Smith completed the instrument change-outs without any significant complications. After the EVA was over the shuttle crew used a camera at the end of the long remote manipulator arm to do a photo survey of the +V2 side. Those images gave us the first hint that something didn't look quite right about the condition of the multi-layer insulation (MLI), the silver-colored, foil-like material covering the equipment bays that ringed the spacecraft.

Flight Day 5 brought EVA 2. The telescope was rotated so that the −V2 side (under the other solar wing) was facing forward. This gave direct access to the Fine Guidance Sensor (FGS) that was to be replaced. It was astronauts Greg Harbaugh and Joe Tanner's job to remove one Fine Guidance Sensor (FGS1) and to replace it with our only spare FGS. The master plan was ultimately to change out all three FGSs in a round-robin fashion. The removed FGS1 would be brought back to the ground, refurbished, and re-flown on the next servicing mission about three years hence to replace FGS2, which in turn would be refurbished for yet another servicing mission sometime in the future. When I first suggested the FGS round robin to John Campbell a few years before, he had doubted that it would be feasible, because the Perkin-Elmer engineers who had originally designed the FGSs were no longer on the scene. However, Perkin-Elmer and its successor company, Hughes Danbury Optical Systems, owed us compensation for the spherical aberration debacle. It was not difficult to bring them back on line to support Hubble servicing.

The three FGSs were at the end of a chain of five hardware and software components used to slew the spacecraft to a desired point on the celestial sphere, brake its motion, keep it pointing in that direction, lock on to the astronomical target, and hold it precisely and stably in the field of view of one of the scientific instruments. The chain consisted of star trackers that made large scale star maps of the sky to determine where Hubble was pointed; reaction wheels, massive fly wheels that used angular momentum to drive the spacecraft's coarse motion in pitch, roll and yaw; magnetometers to measure the orientation of the spacecraft relative to the Earth's magnetic field so that electromagnets on the sides of the spacecraft could interact with the field to "dump" momentum and slow the motion; a set of precision gyroscopes that determined the spacecraft's attitude when it moved from one pointing direction to another and which held the pointing direction while the FGSs did their work locking onto guide stars. The pointing accuracy and stability required by Hubble's cameras and spectrographs was roughly the equivalent of hitting a tee shot westward from Washington, DC and sinking a hole-in-one in Denver. It was like pointing a laser from the Washington Monument toward Manhattan and precisely hitting the center of a dime on the top of the Empire State Building. The FGSs were the instruments that provided that ultimate level of pointing control.

In routine operations, we typically used two of the FGSs (FGS1 and FGS2) for pointing control of the spacecraft, reserving the third one (FGS3) for use as a scientific instrument to make precise measurements of the positions and motions of stars in the sky. This is the science of precision astrometry. Hubble's FGSs were the very best tools for that kind of scientific research available prior to the launch of the European Space Agency's Gaia satellite in 2013.

Spherical aberration had produced two negative outcomes for the FGSs. Every pointing required locking on to two guide stars, one in each of two FGSs. With spherical aberration we were limited to using brighter guide stars than originally planned. This was a consequence of the fact that the alignment of internal optics within an FGS proved highly sensitive to the presence of spherical aberration in the telescope's light beam. It took only a tiny misalignment of FGS optics to badly smear out the aberrated image of a star, so that it was no longer bright enough to be useful for pointing control.

Second, the FGS design envisioned the final pointing on guide stars to be in a configuration called Fine Lock. To get to Fine Lock, the FGS passed through a configuration called Coarse Track. In routine operations with spherical aberration, we often chose to remain in Coarse Track and do our astronomical observations continuously in that state rather than attempting to proceed into Fine Lock. This gave reasonable pointing stability on fainter guide stars, stars that were too faint to be held in Fine Lock. The problem was Coarse Track was very hard in a mechanical sense on some of the mechanisms inside the FGSs. It caused their bearings to wear out rapidly. This excessive, unplanned use of the Coarse Track mode, which provided a way to work around the adverse impact of spherical aberration, now required the FGSs all to be brought back to the ground eventually to have their bearings replaced.

Included in the refurbishment of each FGS was a new adjustable mechanism to allow the internal optical components that had been so sensitive to spherical aberration to be very precisely aligned in orbit. That would allow us to detect fainter guide stars and routinely begin to use the Fine Lock mode again. In addition to installing the new FGS (called FGS1R), Harbaugh and Tanner installed a new component called the Optical Control Electronics Enhancement Kit (OCE-EK) that allowed control of the precise internal optical alignment of the FGS. In the future each refurbished FGS would be accompanied back to orbit by its own OCE-EK.

Finally, the EVA crew replaced a failed tape recorder, one of three Engineering and Science Tape Recorders (ESTRs), with a new one. These were old-fashioned spacecraft data recorders that literally used reel-to-reel magnetic tape. Later in the SM2 mission, the crew would also install a new Solid State Recorder (SSR) in place of one of the other old ESTRs. The SSR was new technology. It could record and transmit data at the same time, had no moving parts and could hold about 10 times more data than the old ESTRs. But because it was new, we didn't want to rely on it completely before it had been proven out in normal Hubble operations. That's why we also installed a newly manufactured, but old technology ESTR as well.

Now the subject of conversation between the crew and the ground team turned to the issue of the apparent problems with Hubble's skin. On Flight Day 3, following the rendezvous, after remote arm operator Steve Hawley had grappled Hubble and berthed it to the FSS, the crew used the camera on the end of the remote arm to record a photo survey of the −V3 side of the HST spacecraft. That's the side where the white radiator used to keep the WFPC2's CCD light sensors cooled sticks out a bit from the side of the spacecraft. In normal operations that side is always oriented away from the Sun to avoid unwanted heat falling on the radiator. Pilot Scott "Doc" Horowitz summarized what they saw, "I guess the verbal assessment here is that the telescope and the [solar] arrays look like they're in pretty darn good shape. Sox [Flight Commander Ken Bowersox] may want to talk to you later about the comparison between [STS] 61 [SM1] and now. But for the uneducated observer like me, it looks really in fine shape." From Mission Control the Capsule Communicator replied, "We copy. That sounds great." And Doc was right. That side of Hubble was like new.

But during EVA 2 the scene looked different. Greg Harbaugh described what he was seeing to the flight controllers on the ground. "Steve, while we have a second here, I could give you a verbal on this MLI on these bays. In a number of places it's cracked (Figure 12.3). It looks like it's just gotten old. It looks like it's cracked. I don't see any place where it's crumbling, but I would recommend I guess that we be

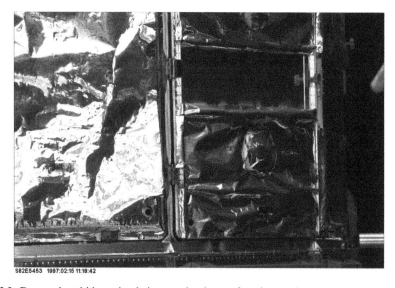

Figure 12.3. Damaged multi-layer insulation on the doors of equipment bays 5 and 6. (Credit: NASA, Wes Ousley.)

extra careful and not touch any of this MLI on these bays up here because I think there's good potential for pieces to come off."

A few minutes later Houston called, "Discovery/Houston with a question on the MLI. It's hard to tell from the pictures we're seeing here, but it looks like it might just be only the outer layer of the multi-layer insulation that is torn and pulling away. Can you give us some insight on that?" Joe Tanner replied, "I can definitely verify that. It's just a very thin layer that looks almost like aluminum foil. Underneath that is the quilted blanket and it's fully intact. I don't believe we have a problem protecting the [bay] door. It's just a question of contamination from this stuff that could well break off."

At the conclusion of EVA 2, the Goddard team asked the crew to rotate the telescope so that the +V3 side was facing forward and a complete photo survey of that side could be done. As the −V3 side had to be shielded from the Sun as much as possible, the +V3 side bore the brunt of the sunlight shining on the spacecraft.

A large gaggle of reporters had gathered at KSC and then at JSC to cover STS-82. After the edge-of-the-seat suspense of Hubble's first servicing mission in 1993 and the epic spectacle of the impacts of the fragments of Comet Shoemaker/Levy 9 on Jupiter in 1994, it appeared that Hubble had retained its status as potentially big news, and they were there en masse to cover it.

The JSC Public Affairs Office arranged for a media briefing to be held after the conclusion of each EVA shift. Panelists were selected to provide brief remarks and then to take reporters' questions. These included the JSC Lead Flight Director, Jeff Bantle, and representatives from Headquarters and Goddard. For the latter assignments we rotated. John Campbell (GSFC, Hubble Program Manager) and Ed Weiler (HQ, Hubble Program Scientist) took the briefings following EVA 1 and EVA 3. Ken Ledbetter (HQ, Hubble Program Executive) and I (GSFC, Hubble Senior Project Scientist) were scheduled to handle the post-EVA briefings for EVA 2 and EVA 4.

It was Ledbetter's and my misfortune to have drawn the short straw for the EVA 2 briefing, the first interaction with the media following the discovery of the cracks and tears in Hubble's MLI. The reporters had been following the air-to-ground communications between the astronauts and the ground team with rapt attention. So, it was no surprise that word had gotten around. In addition, JSC had invited a couple of the SM1 astronauts, Jeff Hoffman and Story Musgrave, to help the media follow what was going on during SM2, to serve as "color commentators." Story in particular was a media favorite. He was both brilliant and eccentric, and could be counted on for great sound bites. When Story realized that the EVA crews were seeing what looked like significant damage to Hubble's outer skin, he was reported to have said, "Hubble is falling apart. I doubt that it will make it to the end of the century [2000]," or words to that effect. Some of us speculated that he was angling for a new flight assignment to go up and repair the MLI.

Flight Director Bantle, Ken Ledbetter, and I sat down on the stage to begin the post-EVA 2 briefing. Our opening remarks were plain vanilla. I talked about the importance of the upgraded FGS to Hubble operations and science. Ken made some general comments about how well the mission was going. We said nothing about the grungy MLI.

Then came the moment for questions and answers. Many hands shot up. The questions were urgent in tone. What's wrong with Hubble? Is it coming apart? Will it last much longer? What are you going to do about it? The reporters wanted serious engineering answers and they wanted them now!

Ken and I both replied that we had just learned about the MLI issues, that our engineers were looking into it, and we didn't have any feedback from them about it as yet.

That was not good enough. The whole world was following the mission, often in real time on television. The reporters had to know what to tell their readers and viewers. This line of questioning got more intense. Ken and I quickly reached a point where all we could do was repeat ourselves.

Then I observed Ken go speechless, blankly staring ahead. He was an experienced, veteran manager, normally soft spoken but very articulate. He was on the panel to deal mainly with Hubble program questions. I was also hoping he could deal with high-level engineering issues, as John Campbell would do on the alternate panel with Ed Weiler. I was to field questions about Hubble science. Neither of us was prepared for this. The reporters continued—insistent, demanding.

After a very awkward and protracted pause, with Ken saying virtually nothing, I felt very uncomfortable and thought one of us should speak up. That's when I made a big mistake. I decided to try levity as a means of defusing the situation. The quick quip is my style, part of my personality. Maybe it would work here. So, I said with intentional irony something like, "Look. I'm just a 'simple' astronomer. Ken here is a 'simple' manager. Be kind. We're really not the ones who can answer these questions." Of course that fell flat with this audience, and I immediately regretted trying to feed humor to the lions. They wanted red meat.

Finally, one of the more senior reporters said that they really needed for us to give them **something** to tell the public that afternoon. Ken spoke up, replying with something as anodyne as before: "We first learned of the degraded MLI just a few hours ago. There has not yet been time to assess the situation. We have an engineering tiger team looking at it as we speak. But we expect that Hubble should be okay at least to 1999 when the next servicing mission is scheduled. We'll get back to you tomorrow with a full update." Ken's reassurance was just hopeful speculation since we had no technical damage assessment at that point. But there was nothing more we could say. That seemed to be enough to allow the reporters to back off.

After the briefing, when we got back to the CSR, Weiler was waiting for me. I couldn't tell how angry he was. He said in kind of a dazed voice, "I can't believe you said that. I really can't believe you said that." I replied, "Ed, your guy (Ledbetter) was dying up there. I thought I had to say something to try to bail him out." And Ed repeated, "I can't believe you said that."

After that Weiler instituted a process that we should have been following all along. After each EVA a group of us met to discuss what questions might come up at that day's briefing, and to plan how we would answer those questions. Also, if there were any technical issues likely to come up that the briefers were not well qualified to handle, we would include an engineer on the panel who could provide whatever relevant information was available.

Wes Ousley grew up in a space-faring household in Maryland. His father, Gil, graduated from the University of Maryland with a degree in Mechanical Engineering. Gil went to work for NASA at Goddard Space Flight Center when it first opened in 1958. He ultimately became NASA's first representative to Europe and the Chief of the International Projects office at Goddard, leading numerous space missions in collaboration with other nations. Following in his father's footsteps, Wes also graduated from the University of Maryland with a degree in Mechanical Engineering. He went to work at Goddard in 1977. He always wanted to build satellites. It was a family thing.

It didn't take long for Wes to latch onto Frank Cepollina's team at Goddard, creating engineering concepts for servicing satellites in space. From 1978 on he witnessed Cepi turn into a real innovator of "space stuff." As usual, Cepi was a decade or two ahead of everyone else. He inspired his team, telling them to "Ignore what people say. Ignore the naysayers. **It's** going to happen!" And in 1984 **it** did happen with the successful repair of the Solar Maximum Mission (SMM) spacecraft in the payload bay of Shuttle Challenger.

Wes was the lead thermal engineer for the Solar Maximum Repair Mission (SMRM). It was his job to see to it that none of the equipment—the SMM spacecraft when it was opened up, or the new hardware going into the spacecraft—got too hot or too cold in a space environment that was alternately a furnace and a freezer. Planning the SMRM had provided the impetus to Cepi and his team to transition from conceptualizing about how to do servicing to actually doing it. So, when planning for Hubble's first servicing mission began in 1990, Wes and the rest of Cepi's team were ready for it.

On SM1 Wes was the thermal engineer responsible for assuring that all the delicate and expensive hardware being carried in protective enclosures, mounted on carrier structures inside the payload bay, would experience a benign environment that was neither too hot nor too cold. The carriers and protective enclosures themselves also required protective insulation and heaters. And when the astronauts on an EVA removed an instrument or a new spacecraft component from its protective cocoon, Wes had to assure that the EVA procedures protected that hardware from temperature extremes.

On the third EVA of SM1 in 1993, Story Musgrave and Jeff Hoffman successfully replaced the original WFPC with the WFPC2, containing the optics needed to correct spherical aberration. That was a huge accomplishment for science. At the end of the EVA, the two astronauts were supposed to do a smaller, secondary task that would help improve Hubble's pointing control operations. They needed to mount two new magnetometers on top of the original units that had become annoyingly noisy during their three years in orbit. The remote arm lifted the astronauts upward to the top of the telescope where the magnetometers resided. The view of the Earth and the Shuttle up there was incredible!

After a few minutes of absorbing the awesome scenery, the two astronauts turned their attention to the task of mounting the two new magnetometers on top of the older pair. Hoffman noted that a piece of the insulation covering one of the old units had come loose. Inexplicably he tore it off. That immediately raised an issue of the insulation further disintegrating over time, producing a cloud of particle contamination around the telescope's aperture. Something had to be done about this before Hubble was deployed from the shuttle several days hence.

Cepi had returned to Maryland shortly after the launch of SM1 (STS-61) to attend to the funeral arrangements for his daughter who had just passed away. He was following the mission in the STOCC during the EVA shifts in the late night and early morning hours. When he heard about the degrading insulation around the magnetometers, he immediately called Wes Ousley at his station on console at JSC and said, "Wes, you have to find an expendable piece of [MLI] blanket that is already out there somewhere on the carriers and come up with a way to fabricate small protective covers from it that the crew can fit over the magnetometers. Write the detailed how-to-do-it instructions and pass those along to the JSC guys so that they can create the procedures in their own format and send them up to the crew. Hopefully, Musgrave and Hoffman can put them in place at the end of EVA 5. You have 24 hr to get this done."

Wes enlisted the help of Diane Schuster at Goddard, who had experience sewing the MLI blankets for all of the servicing payload hardware and support equipment. They identified a suitable piece of MLI on one of the instrument carriers. With input from Diane, Wes put together step-by-step instructions for retrieving the MLI and jury-rigging small covers that would fit over the magnetometers. He then put the instructions into the communications flow that would get it to the right JSC person, who would translate it into a form that could be uplinked to Endeavour.

The communications path that allowed the Goddard Hubble team to pass information on to the JSC Shuttle team was not exactly efficient. While the GSFC people and the JSC people worked together as a closely knit Hubble team prior to the mission, once the mission began, the GSFC team became *the customer*. The JSC team took charge of all communication with the crew in orbit. If it became necessary to relay technical information from the GSFC engineers to the astronauts, that information had to be sent through a rather lengthy path beginning with the Goddard Servicing Mission Manager on console in the Customer Support Room at JSC. He was the only Goddard person allowed to directly communicate with the JSC Shuttle team in Mission Control. So, Wes's instructions would start there and be relayed to JSC "Payloads" who would relay it to JSC "EVA" who would relay it to JSC "EVA Support" who would call on other colleagues as needed within the JSC team to translate those instructions into a procedure that would then be relayed by the JSC Capsule Communicator up to the crew in orbit.

If this chain of communications was not managed with tight discipline, it could easily devolve into the old game of "telephone" where the message coming out of the end of the chain of players was very different from what was initially put in at the beginning. During the early years of Hubble servicing this lack of a mode of authorized direct communications between the technical experts on the ground who really understood Hubble, and the astronauts in orbit who needed to deal with some kind of problem with Hubble (such as the badly damaged MLI encountered during

SM2) proved very frustrating to Hubble managers, not least John Campbell, the Hubble Program Manager on the scene at JSC during SM2.

Thus, as might have been expected, Wes Ousley's carefully wrought instructions had become a bit garbled by the time the crew on Endeavour received them. Pilot Ken Bowersox had a question about them. Wes, monitoring the conversation on air-to-ground thought to himself, "Okay, let's see how the JSC guys handle this." Suddenly he heard the voice of the Hubble Servicing Mission Manager, Rud Moe, in his ear. "Wes, what are they supposed to do here?" Wes thought, "So they worked the procedure, but they didn't understand it." While on the phone with Diane, Wes responded on the payload loop with several suggestions, some of which made their way up to the crew. In the end astronauts Bowersox and Claude Nicollier were able to craft small hoods that could be used to cover the magnetometers to prevent any further debris from escaping.

When the magnetometer covers were in place, one of the crew called down to congratulate one of the JSC guys, the one who had written the final uplinked procedure that contained Ousley's instructions. Wes thought nothing of this. Clearly the JSC guys like to get called out by name with a pat on the back from astronauts in orbit. This could be good for their careers. But outside that Shuttle world, no one would notice or care. However, Cepi *did* notice and was irked. The next day he called Wes again and asked, "Did you hear what they said? They complemented the JSC guy who just wrote out your procedure and sent it up."

What had really set Cepi off was his irritation that the JSC guys had taken what Wes wrote, mistranslated it, and then sent it up as a procedure to the crew. When something went wrong with it, they didn't know how to respond. This guy got a pat on the back for not doing a very good job. That was Cepi. He was a perfectionist and couldn't abide poor performance by anyone working on Hubble, even a guy from another center.

After the photo survey of the +V3 side following the second EVA of SM2 was completed, Wes Ousley downloaded the images. The crew had hinted at a problem with the top layer of the MLI during the EVA. But what Wes saw in the high-resolution images was a shock, completely unexpected. In one of the first images he thought he was seeing shadows. But soon he saw in other views that the MLI was badly torn, in some cases even sticking out from the surface of the spacecraft (Figure 12.4). Near the top of the Sun shield, the tube containing the telescope optics, the surface layer was badly ripped in two places. The largest tear was nearly two feet long with its edges curled at an angle outward (Figure 12.5). To Wes the second tear was more ominous. It was close to the very top of the Sun shield, parallel to the entrance aperture of the telescope. Its edge was curling directly upward in a way that suggested that the torn MLI might eventually roll over into the field of view of the telescope, if nothing were done to prevent it. That would have a negligible affect on Hubble's optical performance, but the greater risk was that of MLI particles flaking off and eventually contaminating the interior of the telescope.

The outermost layer of Hubble's MLI was a thin film of silver-coated Teflon, chosen because it reflects the Sun's rays and radiates heat away very efficiently. A pristine metalized Teflon layer would remain near room temperature even when the

Figure 12.4. The torn surface layer of MLI sticking out from the Bay 8 equipment section door. (Credit: NASA, Wes Ousley.)

Sun was directly shining on it in orbit. However, the next layer down in the 18 layer MLI was aluminized Kapton. With the Teflon surface layer torn away, the exposed Kapton below could reach a temperature of 250°C, allowing a huge heat input (50 W m^{-2}) into the spacecraft. As a thermal specialist, Wes knew that was a very bad thing.

The Earth's atmosphere is extremely thin at the altitude of Hubble's orbit, but it isn't a complete vacuum. Particles of atomic oxygen impact the spacecraft's surface and chemically react with it as it moves through the extremely thin air. Almost constant exposure to low-energy solar X-rays, solar ultraviolet radiation, charged particles—mostly protons and electrons from the Sun—and to atomic oxygen had caused the Teflon on the sunlit side to become brittle. The constant changes in temperature as Hubble orbited between daylight and darkness, over a range of 100°C, had caused the fragile MLI to expand and contract orbit after orbit, until it began to crack and tear. It was as though Hubble was badly sunburned and its skin was starting to peel.

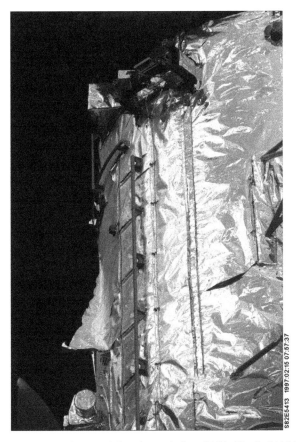

Figure 12.5. Torn MLI near the top of the telescope's Sun shield. (Credit: NASA, Wes Ousley.)

Wes realized that something had to be done to staunch the damage being inflicted to Hubble's skin. If the problem became significantly worse with time, electronics and other sensitive components inside the spacecraft might be damaged as they became progressively hotter. It might become necessary to constrain where the telescope could point in the sky to limit the incessant exposure of the +V3 side to the Sun. It was a concern to be taken seriously. Without mitigation of some kind, there might eventually be regions of the sky that astronomers could no longer observe regularly, potentially limiting Hubble's scientific capabilities.

Wes immediately went to Cepi and Campbell to report this urgent concern. Remembering the unfortunate episode with the magnetometer covers on SM1, Cepi exclaimed, "Okay Wes, this time you're going to take charge of this. You're going to be the guy who makes the presentation to the teams. And you're going to make sure that the stuff gets sent up to the crew correctly."

Wes's first move was to contact the master of Hubble thermal engineering at Goddard, Yukio Yoshikawa—better known as "Yuke." During the 1980s Yuke had been the lead thermal engineer for Hubble while it was still on the ground at Lockheed. His job was to make sure that, regardless of how hot or cold it was outside the spacecraft, the insides were kept constantly at a benign temperature, usually room temperature (20°C or thereabouts). He accomplished this with a combination of insulation, electric heaters, radiators and the laws of physics. Yuke knew instinctively which parts of Hubble were the most vulnerable to damage from excess heat. He also had access to a library of detailed original photos of Hubble's surface from before it was launched.

Wes asked Yuke to download the new +V3 photo survey files and compare them to how the spacecraft had looked more than a decade before. Together they found about 30 spots where the surface MLI layer had been damaged in orbit—far too many places to consider repairing during one EVA. So Wes asked Yuke to get opinions from the engineers at Goddard about which locations were the most critical.

It was Flight Day 6 and EVA 3 was underway. Astronauts Lee and Smith were working on replacing equipment in several of the equipment bays. This included a new Data Interface Unit that routed commands and data to various components on the spacecraft; one of the archaic tape recorders replaced with a new Solid State Recorder; and one of the four Reaction Wheel Assemblies. Their running commentary included observations about the condition of the MLI. "The whole surface of the telescope looks different on this side because of the constant exposure to the Sun. The handrails are a different color. The MLI has little spider web cracks in it." As the EVA went on, Flight Director Bantle asked Lee and Smith to stay focused on getting their pre-planned work done, assuring them that the team on the ground would schedule a time later for a meeting during which everyone could provide their inputs and the whole issue of the MLI condition could be fully discussed.

Wes rescheduled his on-duty time so that he was working 12 hr shifts. He found an empty conference room with tables where he could spread out the survey photos. While the engineers at Goddard were assessing the damage, Wes pulled together a small team with the relevant expertize at Johnson. The latter team's job was to figure out what to do about the highest priority damaged areas. What were the most threatened things? If it gets a lot of Sun on it, what's going to get closest to its thermal red limit, so that it's really going to get screwed up. Wes told the group, "Okay, tomorrow is the last EVA [number 4]. Whatever we can come up with overnight they'll implement tomorrow at the end of the EVA. Then they're coming home. We've got to figure out what the astronauts can do, using only what they have on board Discovery."

Based on Yuke Yoshikawa's and Wes's own preliminary assessment, they needed the EVA crew to repair the two large tears at the top of the Sun shield and the MLI covering three of the electronic equipment bays—Bays 10, 8, and 7. Bay 10 was in the greatest jeopardy. It contained a critical piece of equipment, the Scientific Instruments Command and Data Handling (SIC&DH) computer. A failure of the SIC&DH would make continued use of Hubble's scientific instruments impossible. Hubble would be dead as an astronomical observatory.

Both Bay 10 and Bay 8 had already started to heat up. If the MLI degradation continued, Yuke expected both bays to heat up to unacceptable levels over the next few years. This was what Wes reported in a jam-packed conference room to the all-hands meeting that convened shortly after EVA 3 had concluded. The outcome of the meeting was that EVA 4 would proceed the next day as originally planned, but with the MLI repair work added at the end. Wes's team was to determine what the astronauts needed to do and have those instructions faxed up to the crew by the JSC Payloads engineer—a much more direct communication path than the one that had impeded the magnetometer repair in SM1.

After the big meeting, Wes remembered a famous scene from the 1995 movie, "Apollo 13." After a catastrophic explosion in an oxygen tank, the three Apollo 13 astronauts, Lovell, Swigert, and Haise, had taken refuge in the Lunar Excursion Module (LEM), as the main Command Module (CM) cabin rapidly lost its life-support capabilities. But the LEM, which had to serve as their lifeboat on the four day journey back to Earth, also had limited provisions. After a while, its ability to remove excess carbon dioxide from the air they were breathing, by passing it through a canister of lithium hydroxide, was exhausted—a potentially fatal situation. The solution was to bring a replacement canister from the CM into the LEM. But the CM unit was a cube, while the LEM unit was a cylinder. In other words they needed to fit a square peg into a round hole. On the ground at JSC a small engineering team was assembled to solve this problem before the crew expired. In the scene Wes was recalling, the leader of that team rather forlornly looks at the other members and says, "Okay people, listen up. We have to find a way to make **this** [the square CM canister] fit into the opening for **this** [the round canister from the LEM], using nothing but **this** [a pile of assorted odds and ends that could be found in either the LEM or the CM, which he unceremoniously dumped on a table]."

Of course the situation with Hubble's tattered MLI was not a matter of life or death. But still, what was at stake was the potential loss of Hubble scientific capabilities in the upcoming years. And the approach to a solution was similar to that in the movie. Wes's team needed to bring into the conference room and place on a table all the odds and ends available to the crew in orbit that might be used to repair the worst areas of torn MLI.

On board was an "MLI Repair Kit," which contained an assortment of tape, wire, tie wraps and two 1' × 3' sheets of MLI. The crew had trained to use this kit to cover any holes in the Sun shield produced by meteoroid impacts that they might discover during the mission. Holes like that could allow sunlight to enter the telescope, producing an unwanted background of stray light. If there had been more such spare MLI on board, it might have been enough to do all the repairs they needed to do. Alas, that wasn't the case. However, the two sheets were sufficient to patch the two large tears at the top of the Sun shield.

The team searched for more material that could be used, particularly to protect Bays 10 and 8. Bay 7 wasn't in bad shape. The MLI covering its door was just starting to curl up, but hadn't torn as yet. They only needed to restrain it somehow, to push it back down in place. But the MLI on the other two bays was badly damaged and needed serious attention.

There was some good fortune in their search. ESA had included on board eight pieces of Teflon insulation, about 8" × 16" in size, which were designed to wrap around the metal bi-stems that held the solar arrays taut. The bi-stems on the replacement solar arrays installed on SM1 in 1993 were wrapped in MLI in an attempt to limit their rapid heating and cooling as Hubble passed into and out of darkness during its orbit. It was those sudden temperature fluctuations that had caused the bi-stems on the original arrays to "twang" and the spacecraft to jitter. So now, for SM2, ESA sent along some patches to be used in case the bi-stem MLI got damaged. They were already equipped with cords, clips, and Velcro allowing them to be mounted securely in place on the bi-stems.

Wes's team designed a rectangular patch for the Bay 8 door using a pair of the bi-stem Teflon pieces connected together end-to-end with Kapton tape. The clips and cords that were already attached (properly shortened) could be used to mount two of the jury-rigged patches to the Bay 8 door. For Bay 10 two of the bi-stem Teflon pieces were to be attached together side-by-side, creating a square. Two squares were to be mounted as patches on the Bay 10 door. In hand-drawn sketches team members carefully documented the design of each of the patches and how it was to be installed on its door.

The lead electrical engineer on the Hubble team pointed out that the crew had electrical wire available on board. His idea was to string the wire tautly in several places over the door of Bay 7. That was all that was needed to hold the MLI on that door firmly in place, preventing it from curling up any further.

Wes's team began writing instructions and giving them to the JSC Payload Officer to fax up to the crew. It seems there were many pages of instructions, so much so that Doc Horowitz later joked that they had almost used up the entire supply of fax paper on the shuttle. In the end only a few pages were really needed, the hand-drawn diagrams showing how to construct and install the Bay 8 and Bay 10 patches (Figure 12.6). The attachment of wires over Bay 7 and the installation of the patches at the top of the telescope were straightforward and didn't require detailed instructions.

Wes hurried his team along to design the procedures, get them documented and sent up to the crew, realizing that the crew would need many hours to assemble the Bay 8 and 10 patches and collect together everything else they would need to complete all the MLI repairs at the end of EVA 4. The instructions were in the astronauts' hands approximately seven hours after the end of EVA 3. That left the crew, specifically Horowitz and Lee, approximately six hours to get things ready on board Discovery before Harbaugh and Tanner had to suit up for EVA 4.

Meanwhile, the team on the ground was busy trying to replicate what needed to be done in orbit on the mockup of Hubble in the training facility at JSC. Astronauts Jerry Ross and Jeff Hoffman asked what they could do to help. They checked over everything Wes's team had planned and gave it a thumbs-up. Then with a stopwatch they methodically timed every step of the procedure to determine how much time the crew needed to complete each task, to establish that the plan was feasible. This was the NASA way—test before you fly and fly only what you have tested.

As the lead EVA astronaut on STS-82, Mark Lee was responsible for assuring that all the spacewalks went well and that the needs of the four EVA astronauts were understood. At some point during the hours leading up to EVA 4, the radio crackled in Mission Control. The Capsule Communicator recognized Lee's voice. "We've been going over the procedures the Goddard folks sent up. In a nutshell there is just

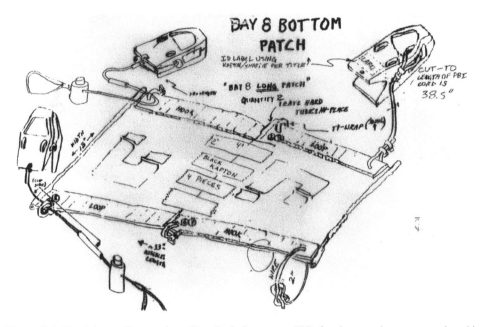

Figure 12.6. Hand-drawn diagram from Wes Ousley's team at JSC, faxed up to the astronauts in orbit, showing them how to construct a patch from available materials to cover torn MLI at the bottom of the Bay 8 door. (Credit: NASA, Wes Ousley.)

too much to do here to be tacked on to the end of EVA 4. So, here's what we're going to do," he said with authority in his voice. "We're going to add a fifth EVA to do the repair work on the MLI." There was silence for a little while. Then the Capsule Communicator, having gotten an affirmative nod from the Flight Director, responded, "Okay then. That's what we're going to do." Mark Lee was in charge of the EVAs. It was his command decision. What were they going to do? Tell him his crew couldn't do it?

And so, driven by the impetus of Mark Lee, EVA 5 was added to the mission. The cost was loss of the crew's day off, which had been scheduled for the day after the final EVA. But Mark Lee and Steve Smith now got the chance to experience one more day enjoying the glory of orbital space from the vantage point of Discovery's payload bay.

EVA 4 proceeded as planned on Flight Day 7. Harbaugh and Tanner installed a replacement Solar Array Drive Electronics module, provided by ESA. Then Steve Hawley drove the RMS arm to lift them upward to the top of the telescope. Their originally planned job was to replace the crude covers over the magnetometers that had been jury-rigged during SM1 with a proper set of permanent covers. In that location they also had easy access to the two large rips in the MLI. Using the spare material from the MLI repair kit, they wired the two patches securely into place (Figure 12.7). Tanner cut off a piece of torn MLI to be returned to the ground for later analysis.

Figure 12.7. Astronauts Greg Harbaugh and Joe Tanner wire a patch over one of two large tears in the MLI near the top of the telescope's light shield during EVA 4. Note the top of the second tear is visible at the bottom center. Both tears were patched. (Credit: NASA, Wes Ousley.)

After EVA 4 was completed, Doc Horowitz and Mark Lee worked for many hours fabricating the Bay 8 and 10 patches and the wire restraints for Bay 7 that would be installed the following day. As the work progressed, Horowitz periodically took pictures of what they were doing and sent them to Wes's team on the ground to have their work double-checked (Figure 12.8). Lee and Smith had little difficulty installing the finished products during EVA 5 on Flight Day 8 (Figure 12.9).

In future servicing missions in 1999 and 2002, the Hubble team continued to do imaging surveys to monitor the condition of the MLI and the patches installed during SM2 on the +V3 side of Hubble. Although some of the cracks and tears grew larger over the years, the degradation process more or less stabilized. Hubble never did fall apart or lose its skin, as the prophets of doom had proclaimed to the media it would during SM2.

Meanwhile Cepi's servicing team at Goddard designed and built sturdy stainless steel protective covers to be installed over each of the equipment bay doors that were exposed extensively to sunlight. They called them NOBLs or New Outer Blanket Layers. It was only toward the latter part of the 2000s, as SM4 approached, that the temperatures in Bays 5 and 8 reached a level that caused concern. Installing the NOBLs was typically designated as a low priority task. Three of them got installed during SM3a in 1999. But, it wasn't until the last EVA of the final servicing mission in 2009 that astronauts John Grunsfeld and Drew Feustel pressed at the very end to get the last of the NOBLs installed—the final bit of icing on the cake of Hubble servicing.

Discovery landed at the Kennedy Space Center at 3:35 am Eastern Standard Time on 1997 February 21, bringing the 11 day STS-82 mission to a highly successful conclusion. SM2 had dramatically advanced Hubble's scientific capabilities with the installation of the STIS and NICMOS instruments. It had introduced new technology and replaced hardware that was vulnerable to failure. It had also

Figure 12.8. Astronaut Scott Horowitz displays one of the patches he handcrafted for the Bay 8 door, based on the diagram sketched by Wes Ousley's team at JSC (Figure 12.6). (Credit: NASA, Wes Ousley.)

Figure 12.9. Patches installed at the top and bottom of Bay 8 door (right) and electrical wire used to flatten down and restrain original MLI covering on Bay 7 door (left) during EVA 5. (Credit: NASA, Wes Ousley.)

demonstrated that the seven astronauts in orbit and the hundreds of engineers, technicians, and managers on the ground could pull together as a team to come up with creative solutions to unexpected problems. It demonstrated the value of having thinking, adaptable human beings working in space.

AAS | IOP Astronomy

Life With Hubble
An insider's view of the world's most famous telescope
David S Leckrone

Chapter 13

Resurrection: The NICMOS Saga

The human eye is sensitive to light in the range of wavelengths from about 400 to about 700 nm. In terms of color vision, this is the range from violet to red. This makes sense. It corresponds to the colors of sunlight that have the maximum brightness at the surface of the Earth. The eyes of our primal ancestors evolved and adapted to the brightest daylight to give them the best opportunity to spot naturally growing food, or game, or threats from enemies—the best opportunity to survive under the conditions on this specific planet.

The universe is filled with electromagnetic radiation, from the shortest wavelengths of energetic gamma rays to the longest radio waves. Visible light is a tiny interval wedged into this broad spectrum (Figure 13.1). To study the cosmos astronomers use Hubble to expand their vision far beyond the part of the spectrum that they can see—with detectors mounted in scientific instruments that collect and measure light from the far ultraviolet (FUV) to the near infrared (NIR), that is with wavelengths from about 90 to about 2500 nm.

We can't see infrared light, but we can feel it as heat on our skin. The most sensitive infrared telescopes are cooled to very low temperatures so that the heat they naturally emit doesn't completely swamp the low-level infrared signals from distant astronomical sources. NASA's Spitzer Space Telescope was cooled with liquid helium so that its primary mirror stayed at a temperature of 5.5 K (−268°C; −450°F). Before its supply of liquid helium was exhausted in 2009, the Spitzer telescope was sensitive to infrared wavelengths from 3600 to 160,000 nm. The James Webb Space Telescope is designed to operate about one million miles from the Earth in the direction opposite the Sun. Its enormous primary mirror will be perpetually shielded from sunlight by a tennis-court-sized shade. The mirror and the telescope's four scientific instruments will be cooled to 50 K (−223°C; −369°F). The Webb telescope's instruments will be sensitive from 600 to 27,000 nm.

There are several major reasons why astronomers value infrared light. First, as photons of light travel across the vastnesses of space and time they are stretched by

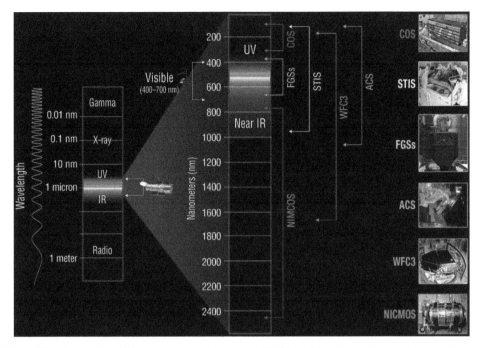

Figure 13.1. The portion of the electromagnetic spectrum Hubble and its scientific instruments can observe. The instruments labeled are those on the observatory after SM4 (2009). (Credit: NASA.)

the expansion of the universe. Even if they started their journey having been emitted by very distant stars, galaxies, or other objects as ultraviolet light, they will be red or infrared photons by the time we collect them in a telescope. If you want to study the deepest universe, as it appeared far back in the distant past, you need to observe in the infrared.

Second, the ability of small particles—gas molecules and dust—to scatter light depends on the size of the particles and the wavelength of the light. Short wavelength (ultraviolet or blue) light is more efficiently scattered by small particles than long wavelength (red or infrared) light. The sky overhead on a clear day looks blue because the air molecules (mostly N_2 and O_2) above us are small and scatter short (blue) wavelengths of light very efficiently in all directions. When the Sun is rising or setting, or is just below the horizon, its light is passing through a long path of air molecules. Much of the blue light is scattered away, while the red light is not. This leaves mostly red light to reflect off the clouds, and we marvel at the beauty of the sky at those times of day. Sunrises and sunsets can look even redder if the atmosphere is full of fine dust, for example from a volcanic eruption. Then even more of the blue light is scattered away by both air molecules and dust particles. What is left is red light.

The space between the stars in the Milky Way and in other galaxies is filled with a general distribution of gas (mostly hydrogen atoms and molecules) and fine dust particles—the interstellar medium. Starlight is skewed in color a bit to the red by the scattering away from our line of sight of photons of blue light from distant stars passing through this medium. Giant clouds of gas and dust, called gaseous nebulae, punctuate the interstellar medium. These are the nurseries within which new generations of stars form. To penetrate the dust and get a view of the fledgling stars within, astronomers must rely on infrared telescopes. The longer-wavelength infrared photons make it out of these clouds largely unhindered, while the shorter wavelength visible, blue, and ultraviolet photons get absorbed or scattered within. To our eyes, looking in visible light, the nebulae look opaque and usually quite beautiful. But through a telescope with light sensors designed to be sensitive in the infrared, our vision can penetrate the dust to see what's inside those nebulae.

Finally, one of the most intensively pursued topics in modern astronomy is the search for planets around other stars—exoplanets. The "holy grail" is to find exoplanets capable of supporting life. Astronomers are becoming adept at detecting and chemically analyzing the atmospheres of exoplanets. Hubble, working in tandem with the Spitzer Space Telescope, has been very successful in this kind of research. The near infrared range of wavelengths contains signatures of a number of important molecules that have been observed with the two telescopes in the atmospheres of exoplanets—water, methane, carbon dioxide, carbon monoxide, etc. Someday, we can hope that a combination of molecules indicative of the possible existence of life—water, oxygen, ozone, methane, etc—might be detected in the atmosphere of an Earth-like planet in the habitable zone around its parent star where water can exist in liquid form.

Hubble is a warm telescope. In general its internal temperature and that of its scientific instruments is maintained at about 20°C or so—i.e., room temperature. There are exceptions. The CCD visible light detectors in the original WFPC and WFPC2 had to be kept cold—at a temperature of about −70°C to −90°C. Small electrical devices called thermoelectric coolers chilled them and their heat was dumped into space via a large radiator panel protruding out the side of the spacecraft.

Although Hubble was not specifically designed to be a sensitive infrared telescope, the astronomers that monitored its design in the 1970s and early 80s pushed hard to extract as much near infrared capability as possible from such a warm system. John Bahcall, especially, was a champion of this cause. It became a "Core Requirement" that Hubble be capable of doing imaging and spectroscopy at near infrared (NIR) wavelengths.

An instrument designer could use cold baffles to block out the radiation coming from the warm sidewalls of the Hubble telescope. Other critical components such as optics, filters, and light detectors had to be cooled in some way to very low temperatures, so their own heat wouldn't be a source of unwanted background. But nothing could be done to subdue the heat coming directly from the telescope's mirrors. At wavelengths longer than about 2500 nm, the background of heat (that is, NIR radiation) given off by the telescope's primary and secondary mirrors simply would swamp even the most carefully designed infrared instrument. That was physics, and there was nothing further that could be done about it. However, even extending Hubble's long wavelength limit by a relatively small amount (by a factor

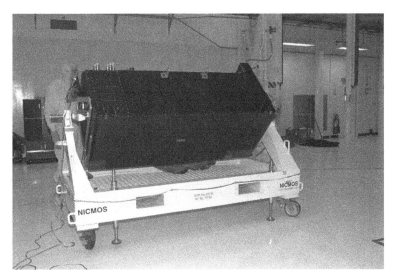

Figure 13.2. The completed NICMOS instrument in the clean room at Ball Aerospace. (Credit: NASA, Ball Aerospace.)

of 2.5—from 1000 nm up to 2500 nm) produced major gains in the telescope's scientific capabilities.

One major obstacle to enabling NIR astronomy on Hubble early on was the immature state of development of light detectors sensitive in that wavelength range. The detectors in instruments sensitive to visible light, such as the CCDs in WFPC and WFPC2, were decades ahead. They were based on silicon technology, similar to that used in microchips. They were already beginning to make inroads in commercial products like still cameras when NASA selected the first group of Hubble instruments in 1978. On the other hand no NIR instrument was selected at that time, because the expert reviewers evaluating those first instrument proposals did not believe the detector technology was ready yet.

The situation had improved by the time NICMOS (Figure 13.2) was proposed by Principal Investigator Rodger Thompson of the University of Arizona and his team in 1985. Its detectors were made from an alloy of the elements mercury (Hg), cadmium (Cd) and tellurium (Te). Unsurprisingly, they were called HgCdTe, or "Mer–Cad–Telluride" detectors. Although they couldn't compete with Hubble's CCDs in terms of size or image sharpness, they were plenty good enough to produce acceptable images for NIR astronomy. However, the rub was that they had to be very cold to work well—somewhere in the range 50–80 K (−223°C to −193°C; −369°F to −315°F).

The designers of NICMOS created a volume deep inside the instrument—the cold well—that would provide an ultra-cold environment for the detectors and cold optics. Five tanks or shells, nested together like a Russian Matryoshka doll, enveloped the cold well. Together they worked something like a very sophisticated thermos bottle. This assembly was called the Dewar or the cryostat (Figures 13.3 and 13.4). The innermost shell surrounding the cold well—the nitrogen tank—was

Figure 13.3. The NICMOS Dewar inside the instrument. (Credit: NASA, Ball Aerospace.)

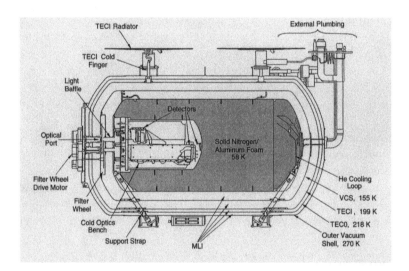

Figure 13.4. The interior structure of the NICMOS Dewar. The innermost volume, called the cold well, is surrounded by a block of solid nitrogen ice frozen into a matrix of aluminum foam. The cold well houses the three NICMOS near-infrared detectors and cold optics. (Credit: NASA, Ball Aerospace.)

filled with a 240 lb block of solid nitrogen ice that was intended to maintain the filters and detectors at a temperature of 58 K (−215°C; −355°F). As the ice gradually warmed up, it sublimated back into gaseous nitrogen, and that very cold gas was used to keep the next shell out in the "doll's nest" or Dewar, dubbed the vapor cooled shield, at 155 K. The third and fourth shells were kept cold with thermoelectric coolers at 199 K and 218 K, respectively. The outermost shell, the enclosure surrounding the entire apparatus, stayed at a relatively warm 270 K. The multiple shells were separated by a total of 124 layers of multi-layer insulation. That was a lot of insulation.

But how did they get a big, heavy block of solid nitrogen ice so deep inside the instrument? And how did they keep it frozen through the many months of testing, shipping, and preparation for launch? It was easier than it sounds.

The NICMOS Dewar had internal plumbing, three pipes that connected the nitrogen tank to the external world. Technicians used one of these pipes to fill the tank with liquid nitrogen. That pipe was also used to vent away excess nitrogen gas. Extremely cold helium gas was then circulated through the other two pipes into a coil of tubing that resided at the back end of the nitrogen tank. The helium gas was cold enough to freeze the liquid nitrogen into a solid block. To keep the solid nitrogen from sublimating away during the many months the instrument was on the ground being tested and prepared for launch, the NICMOS team purposely cooled it to temperatures as low as 35 K, well below the intended operating temperature of 58 K, but cold enough to assure that the nitrogen block would stay frozen and intact. When it warmed back up to about 58 K, they topped off the tank with liquid nitrogen and cooled it down again.

This was a clever and sophisticated design, intended to keep NICMOS working for up to five years—long enough for astronomers to use it for extensive research projects just as they would any of the other instruments on Hubble. But the design had a serious flaw, one that became apparent during testing at Ball Aerospace where the instrument was built.

NICMOS had three separate NIR detectors in three separate cameras to provide options for imaging wide, intermediate, or small fields on the sky with, respectively, low, intermediate, or high resolution. The optics were designed so that all three cameras were in sharp focus at the same time. Together the three would simultaneously provide clear images of three closely spaced fields of view on the sky.

However, as the instrument was run through the planned sequence of many warm up and cool down cycles during a year of testing, the situation eventually developed where the cameras did not all remain in focus. The Wide-Field Camera could be focused or the two narrower field cameras could be focused, but all three cameras could no longer be focused at the same time when the cameras were at their operating temperature, 58 K. However, when the nitrogen ice was cooled down to lower temperatures, the proper focus position was restored, or nearly so. This wasn't necessarily a big problem for astronomers, since the instrument had a mechanism that could bring any of the cameras into focus. But the phenomenon indicated that something was wrong internally within NICMOS.

An Anomaly Review Committee initially came up with a simple explanation for the focusing problem. When the nitrogen ice was cooled to well below its planned temperature, it shrank in volume—by a lot. When the ice warmed back up to its operational temperature, it expanded in volume—by a lot. A physicist would say it had a large coefficient of thermal expansion (CTE). The aluminum structures in the cold well and nitrogen tank had a much smaller CTE, by a factor of about 40. They didn't contract and expand nearly so much as the ice did, as the temperature changed. The expanding block of ice pushed on those structures, deforming them and moving the three NIR cameras forward and out of focus. When the block of ice was cooled to lower temperature, it contracted, relieving the pressure and allowing the aluminum to contract back almost (but not quite) to its original shape.

This state of affairs caused Cepi and his team great concern that the aluminum structures deep inside the doll's nest had been compromised. The question became, could we still have confidence that they would withstand the forces they would experience during launch without breaking? Cepi wanted to subject NICMOS to further testing, particularly a test in the huge centrifuge at Goddard to simulate the g-forces of launch and to additional acoustics testing to simulate the vibrations the instrument would feel during launch. The Principal Investigator, Rodger Thompson, strongly disagreed. He believed the instrument would be over-tested to levels of stress higher than it would actually see in flight. It might be broken by the tests themselves. This was a quandary. We couldn't launch NICMOS without assuring that it was safe to do so. But applying substantial forces during testing might make the situation even worse than it already was. Cepi won that argument with the top management of Goddard and NASA Headquarters standing behind him.

In 1996 mid-November NICMOS was accelerated in Goddard's centrifuge up to a level 25% higher than the acceleration it would experience during launch. A few days later it was subjected to a simulation of the launch acoustic vibration environment in the big acoustics chamber at Goddard. Afterward it was shipped back to Ball Aerospace in Colorado where it underwent a thorough performance check. The Ball engineers judged its performance to be "nominal and unchanged." The instrument had come through the dreaded stress tests unharmed and performing well. It was ready to be shipped to KSC to be launched on STS-82 on 1997 February 11.

On 1997 March 4, less than three weeks after NICMOS had been installed on Hubble, things started to go badly for the new NIR instrument. The temperature inside the nitrogen tank began to rise steadily above its nominal 58 K. The temperature of the next shell out from the nitrogen tank inside the Dewar, the vapor cooled shield that was supposed to be maintained at 155 K, began to drop. A thermal short had developed. As with an electrical short, two surfaces that were not supposed to touch had come together and made contact. Heat now had an unexpected path through which to flow from the warmer vapor cooled shield into the nitrogen tank. The extra heat input caused the solid nitrogen to sublimate into the gas phase and to be vented out of the instrument at a much faster rate than planned. It was predicted that NICMOS would stay cold enough to do its scientific work for less than two years, rather than the four or five years we had expected.

An independent team led by Ron Ross of JPL identified a more insidious problem with the block of solid nitrogen that had not been taken into account before launch and that could explain the thermal shifts inside the Dewar. As temperatures dropped in the nitrogen tank in the months of ground testing, and the solid nitrogen shrank in volume, voids were opened inside the tank. The nitrogen gas that had sublimated off the ice filled these empty spaces. During subsequent cool down cycles the gas refroze in places where it was not supposed to be, especially at the coldest place in the tank at the rear near its cooling coils. This process was called cryopumping. Over time the block changed in shape, becoming longer and longer. When the elongated block expanded as the nitrogen ice warmed up to operating temperature in orbit, it pushed hard against the rear end of the cold well, which in turn bent forward and made contact with the vapor cooled shield.

Of course we couldn't be certain of the conditions inside the Dewar. We could only infer those indirectly from a few measurements. But the situation clarified greatly when Richard Dame, a brilliant PhD structural engineer working at MEGA Engineering, under contract to the Project, used the CRAY Supercomputer at Goddard to create a remarkably detailed and precise model of the interior of the Dewar. His model was able to reproduce all the operational data about the conditions surrounding the development of the thermal short. For the first time we could visualize what had gone wrong, where in the Dewar the deformations of the aluminum structure had occurred, and where the thermal short was located.

We were the victims of our own ambition. We badly wanted NICMOS to work for a long time—up to five years. So, the scientists urged that the nitrogen tank be fully filled, leaving little extra space to accommodate the ice's expansion. When it did expand, the solid nitrogen put undue pressure on the aluminum structure of the tank, causing it to bend out of shape.

The Project and the STScI did all that could be done to extract as much NIR science from the new instrument as possible during its shortened lifetime. The STScI gave high priority to the observing programs already accepted for NICMOS and implemented a special call for new NIR proposals that would be evaluated on a fast track. So, a substantial fraction of the science astronomers had expected NICMOS to do got done. But it wasn't the same as with the other Hubble instruments, because it wouldn't be possible to follow up those first observations. New scientific questions would arise that could not be addressed later with the instrument. NICMOS would have stopped working in the meantime. NICMOS went dormant, its solid nitrogen depleted, on 1999 January 3, some 23 months after it had gone to orbit.

Meanwhile a miracle had happened, a technological miracle! Well, perhaps it was just dumb luck. But falling into our collective laps, out of nowhere, had come an entirely different approach to cooling NICMOS down to the deep cold state needed to continue its science program. Thanks to shuttle based servicing, the instrument would undergo a remarkable metamorphosis to be operated in a way that its original designers had never imagined. It would become a better, more capable tool for science. That was only possible because of a wild idea thrown out by Frank Cepollina at a monthly progress review at Ball Aerospace several years earlier.

My job as Senior Project Scientist for the Hubble Space Telescope was to assure that the observatory was meeting its scientific performance requirements, to assure that its scientific productivity remained at a high level, and to be the voice of science, the scientific conscience in the room, when the Program Manager made important decisions. His office was a short walk from mine. I was always available to him. But the domain of responsibility for the Hubble Project was vast. It was impossible for one person to keep track of everything it encompassed. So, I had help.

My Deputy Senior Project Scientist was a diligent and insightful astronomer out of Indiana University, Malcolm Niedner. There were two other Project Scientists, accomplished research astronomers, assigned to the two separate projects within the overall Hubble Program. Ken Carpenter, with a PhD from Ohio State, worked as lead operations scientist with a string of Hubble Flight Operations Project Managers over the years. Ed Cheng, whose degrees were from Princeton, was the lead scientist for the Flight Systems and Servicing Project, Frank Cepollina's outfit. All four Project Scientists were smart, well educated, experienced, and hard working.

Each Project Scientist brought his own unique knowledge and skills to his Project. However, Ed Cheng's particular talents took on special significance as time went on. He became Cepi's go-to problem solver. Ed was technically brilliant, ingenious, resourceful, and a natural leader of engineering teams, although he was an astrophysicist by training. He was a perfect match for Cepi.

As a teenager, a high school student in Old Tappan, New Jersey, Ed loved to tinker with electronics. He loved to build things. His dad, an electrical engineer, sometimes built things along with him. During his freshman year at Princeton, Ed met another student who also was very good at building things. He told Ed that he was working in the Physics Department as part of a work/study program. When this person dropped out of school at the start of Ed's sophomore year, he approached the Physics Department to see if they could use a replacement. That's where he met Dave Wilkinson, a professor of physics and renowned cosmologist, who was working on balloon experiments to observe the background of microwave radiation left over from the Big Bang. Ed impressed Wilkinson with his ability to make things work in the lab. He was a valuable guy to have around.

After completing work on his Bachelor's degree in Physics in 1977, Ed took a year off from his studies and continued working in Wilkinson's lab while he pondered what he wanted to do for the rest of his life. Wilkinson dissuaded him from going into medicine and convinced him to continue in physics. With Wilkinson as his advisor, Ed earned a Masters and a PhD in Physics from Princeton, the latter awarded in 1982.

Wilkinson was a member of the science team, led by John Mather at Goddard, developing NASA's Cosmic Background Explorer (COBE). That spacecraft and its instruments were being built to measure the microwave background radiation as a probe of the Big Bang and the very early universe. In the Princeton lab Ed had worked on prototypes of the microwave receivers to be used on COBE. In 1982 he moved to M.I.T. as a Post-doctoral Fellow to work with another renowned cosmologist, Rainer Weiss, who was also a member of the COBE science team. During those years, Ed also began collaborating with John Mather and his COBE colleagues at Goddard.

In 1989 Al Boggess, the Hubble Senior Project Scientist at that time, needed to fill the vacancy that opened when I left the Scientific Instruments Project Scientist position to become his deputy. There was great interest at Goddard in recruiting Ed Cheng for this position, to help oversee the development of the second generation

Hubble instruments, STIS and NICMOS. Ed was a natural choice because he would be able to continue working with the COBE team while also taking care of business on Hubble.

On more than one occasion Ed Cheng saved our bacon with respect to the Hubble scientific instruments. Because he brought such a diversity of innovative ideas to solve problems that arose on Hubble instruments over the years, I began referring to him as the Leonardo de Vinci of Hubble. He was a renaissance tinkerer.

After the completion of SM1 in December of 1993, it was full steam ahead on preparing for SM2. This included completion of STIS and NICMOS, both of which were designed and being built at Ball Aerospace Systems Division in Boulder, Colorado. Every month a contingent of us would visit Ball to check on their progress, to review their budget and schedule, and to confer with them on any technical problems they might be having. Cepi had overall responsibility within the Hubble Program for managing Ball's contract with NASA. But he especially resonated with the problem solving part.

Cepi was an accomplished engineer and inventor in his own right. When the Ball engineers identified a problem they were confronting, he simply couldn't resist trying to think of possible solutions. This could be frustrating for everyone else in attendance at a monthly review, as he would go off on some creative tangent for many minutes, while the rest of us were eager to get on with the meeting. Sometimes Cepi would become blazingly angry, usually directing an outburst at the Ball instrument managers.

Cepi hated negativity. He could not tolerate someone saying, "No," or "We can't do that," or "that won't work" without really proving the point. He respected a "can do" attitude, someone saying, "I'll give that a try." He was also a humane individual. He might yell at a man one day and take flowers to the man's wife in the hospital the next.

Cepi's approach might have been rough on occasion, but there can be no denying his accomplishments. The instruments and other modules built on his watch to service Hubble were of excellent quality, with innovations and improvements fostered by Cepi and implemented by his team. In the end he got people and companies to do their best work for Hubble. They all wanted to be seen as helping Hubble succeed. Cepi instilled that pride into them.

It was during one of Cepi's technical flights of imagination at a NICMOS review one month that he came up with an "outrageous" idea. We had been discussing how long NICMOS would last in orbit. Four years? Five years? He began musing about what could be done to extend its lifetime further by means of in-orbit servicing. "Can we refill its cryogen tank in orbit?" he wondered out loud. We all moaned! "Cepi, that's not possible. The liquid nitrogen has to be loaded into the instrument under gravity, in $1g$. It has to be frozen on the spot. There's no way to handle ultra-cold gases and liquids in orbit. It can't be done."

Perhaps surprisingly, this did not elicit anger from Cepi because of our "can't do" attitude. He simply sat quietly for a few seconds, grinned and said, "Indulge me. I want you to build valves connected to the nitrogen tank's plumbing that are accessible from the outside. Someone in the future could at least consider the possibility of replenishing the cryogen. Without valves like that there would be no way to do it."

At that stage in the construction of NICMOS, this was not a difficult thing for Ball to do. So, they indulged the creative genius in the audience and added three valves accessible to astronauts on a panel near the back end of the instrument. As on a number of other occasions, Cepi was prescient. His intuition about how to extend the instrument's life by replenishing the cryogen needed to keep its detectors cold indeed was not practical. But by adding the valves, by not foreclosing future options, he created a pathway for saving the instrument in another way several years later when it failed prematurely.

"It's broken!" It was Cepi's voice on the phone. "What's broken?" Ed Cheng replied. "NICMOS is broken!" This was standard procedure for Cepi in situations of dire technical emergency. It reminds me of Commissioner Gordon shining the bat signal spotlight onto the clouds to summon Batman in the movies. It was a call for help to his best trouble-shooter. It would be repeated several times over the course of the Project.

It was early 1997 March. The thermal short in NICMOS had just raised its ugly head. It had become clear that NICMOS was doomed to a severely truncated lifetime, probably 1.5–2 years. This was a challenge to Cepi, as well as being a matter of pride. An important piece of hardware coming out of his Project was compromised. He couldn't let that situation stand.

For several weeks Cepi held a seemingly endless series of meetings in his conference room, brainstorming about how to rejuvenate NICMOS. In addition to Ed Cheng, Goddard specialists in cryogenic cooling systems and other engineers attended these meetings. Cepi latched firmly onto his idea from years earlier to refill and refreeze the nitrogen inside NICMOS' Dewar in orbit. Some crazy ideas were kicked around. Since the tank had previously only been filled in $1g$, perhaps one could figure out how to rotate the entire Hubble structure in the shuttle payload bay to simulate gravity—an idea obviously destined to go nowhere. The cryogenic engineers continued to insist, it couldn't be done, it was just too difficult, it had never been done before, they couldn't figure out a way to do it, and in any event there was not enough time to come up with something totally new that would be feasible to implement. Of course these responses were the equivalent of waving a red cape at a hostile bull. "Can't do" was not in Cepi's vocabulary. "We pay you to come up with solutions," he barked. "Come up with a solution!"

It was in one of those meetings that a brilliant idea suddenly hit Ed Cheng. Suppose we ran this thing like a refrigerator. Instead of placing the cold source inside the instrument, why not generate ultra-cold gas outside the instrument in some kind of mechanical refrigerator and then run it into the NICMOS Dewar through the plumbing that was already there. Years before, Cepi had insisted, "Indulge me, I want to have access to the interior plumbing within the Dewar." Of course what he had in mind was the hopeless task of trying to refill the cryogen tank in orbit. But, Ed mused, couldn't we run ultra-cold gas from an external source through the cooling coils at the back end of the nitrogen tank? These were the same loops of piping that had been used on the ground to freeze the liquid nitrogen into a block of ice with cold helium gas. Even though the block of ice would no longer be there, the aluminum structures, including a sponge-like block of aluminum foam that had held the nitrogen ice in place, would still be there. Aluminum is a good heat conductor. So, there should be no problem for the heat in NICMOS' detectors to find a pathway back to the cooling coils, to be absorbed by the cold gas in the coils and carried outside the instrument. Ed did a few back of the envelope calculations and convinced himself that this just might work.

You could think of the original implementation of cryogenic cooling within NICMOS as being like an old-fashioned icebox, one like my Grandmother had when I was a small boy. Once a week a man would come to her house in a large truck, hoist a heavy block of ice over his shoulder with a big set of sharp steel ice tongs and carry it into the house, placing it into a compartment at the top of her icebox. That internal cooling source kept her food fresh for a week or so until the ice melted. Then along came the modern refrigerator—no internal block of ice. Rather a compressor and coils carrying a liquid refrigerant that could be turned into cold gas was housed in a separate compartment outside the food storage area. The cold gas was circulated through a heat exchanger inside the refrigerator, extracting the heat and dumping it into the outside air. Ed Cheng argued that we should completely change the operating principle for cooling the NICMOS detectors by switching from an icebox to a refrigerator. Voila.

There was a problem, however. Infrared astronomers observing at mountain top observatories or on airplanes had traditionally shied away from refrigerators, preferring cryogens like liquid nitrogen or liquid helium instead. That's because mechanical refrigerators were noisy. They vibrated loudly. Anything that vibrated was not welcome on Hubble. Mechanical vibrations could cause the spacecraft itself to jitter, degrading its very fine pointing stability and causing Hubble images to be smeared out.

So, the challenge was to find a small refrigerator of some kind that did not vibrate and that could be operated reliably in a zero-g environment, in a vacuum for at least five years. Moreover, it had to be ready to fly in the third servicing mission, SM3, then scheduled for late 1999 or early 2000, less than three years hence. Before such a new system that had never before been flown and operated in space could be accepted for Hubble, it would need to be thoroughly tested and verified, preferably in orbit on a precursor shuttle flight. There was very little time to put all of this together. It was a tall order.

At first the Goddard thermal experts were highly skeptical that this could be done. It was unrealistic, they argued, and in any event there would be so many ways that heat could leak into the system that it was doomed to fail. But Ed Cheng was persistent. In effect he said, ala Cepi, "Indulge me. What is the best mechanical cooler you know about?" The engineers knew about work being done for the Air Force by a small company in New Hampshire called Creare (pronounced Kree-air-ee).

The Goddard cryogenic engineers eventually came around, becoming convinced that this scheme might work and was at least worth a try. They worked with Ed and the Project to establish communications with Creare and to get them on board. The engineers at the company looked at what was needed for cooling the NICMOS

Figure 13.5. The rotating shaft and impeller assembly that was at the heart of the miniature turbines that moved, compressed, and expanded neon gas in the NICMOS cooling system. (Credit: Ed Cheng.)

detectors and said, "Yes, what you need is just on the edge of what we do." They had been working for about 10 years on a device that would suit the Hubble Project's needs very well.

In technical jargon the Creare device was a "Reverse-Brayton Cycle Cryocooler." What that meant was that it was not so much a refrigerator as a heat pump. Basically it extracted heat from a circulating flow of gas and dumped that heat into the external environment, leaving the gas colder than before. Repeating this cycle many times, the cryocooler produced gas that got progressively colder and colder.

At its heart was an amazing miniature turbine—emphasis on the word "miniature." An impeller, 0.6 inches in diameter, within the turbine was attached to a solid rotor shaft (Figure 13.5). The shaft was 0.25 inches in diameter and 2.5 inches long. The impeller plus rotor shaft weighed in at about 2 g. They were an integral part of an electric motor. When voltage was applied, the shaft, which contained a small magnet, rotated causing the impeller to push or pull gas through the attached plumbing. When in use, this tiny powerhouse rotated at up to 7300 revolutions per second (rps)! In the time it takes me to say, "one one-thousand," the shaft would have spun 7300 times. That high speed and low mass were very important. They meant that whatever tiny, high frequency vibrations might be created in the miniature turbine would impart a completely negligible amount of vibration energy into the rest of the spacecraft. Hubble's precise pointing would not be affected.

The rotator shaft was supported simply by gas pressure. While operating, it touched nothing solid. This meant it operated very quietly. It also meant that, because there was so little friction, there was no wear and tear.

Creare had already delivered an engineering model cryocooler of this type to the Air Force in 1994. It had operated for about 30,000 hr over a three year test run with no failures. In the process it cooled gas down to temperatures as low as 35 K (degrees Centigrade above absolute zero), working against heat loads similar to what would be encountered with NICMOS (about 7 watts). This sounded just like what the doctor ordered to cool the dormant instrument and bring it back to life.

In the spring of 1997 the Hubble Project and Creare began working in earnest to design and build the NICMOS Cooling System (NCS). By January of 1998 Creare had delivered the internal hardware for the NCS. That was a remarkably short time, especially for an emerging new cutting edge technology. It had not been easy. Major problems kept arising week-by-week. The turbines kept crashing, for example. Creare management was usually pessimistic at each monthly progress review. But Cepi just kept on pushing, and Ed Cheng kept working on the details of the problems and suggesting solutions. The Goddard Program Manager John Campbell felt certain that this thing was not going to succeed.

So, Ed Cheng had found a solution to the NICMOS cooling problem in a company the Hubble Project had not worked with before. That company just happened to have a new invention, the Reverse-Brayton Cycle Cryocooler, in an advanced stage of development that exactly met our needs and that could be delivered on a rapid schedule, compatible with the schedule for the next Hubble servicing mission. The juxtaposition of Cepi, several years earlier, insisting that Ball design valves that would allow future access to the interior of the NICMOS Dewar, coupled with Ed's concept for a new architectural approach to cooling NICMOS, and the Goddard engineers knowing about a remarkable new technology that seemed made to order for our needs and was essentially waiting for us to show up and claim it, stretched credulity. Was Hubble charmed? Did we have a guardian angel sent to make it up to us for the trauma we all had suffered due to spherical aberration? Or was it simply a matter of dumb luck? Ed's own explanation was that, if you always try to make good, prudent choices—to do the right things—eventually a few of those good choices might allow doors to unexpected opportunities to be opened. Cepi covered his bets by insisting on the "unneeded" valves. Ed pushed to go down the path of a refrigeration system rather than an icebox for NICMOS. Those were good choices.

The NCS (Figure 13.6) was designed to extract heat from the NICMOS Dewar and ultimately to dump that heat out into space. It did its job in multiple steps (Figure 13.7; Swift et al. 2009). One of the three Creare miniature turbines in the NCS, spinning at 1200 rps, circulated ultra-cold (70 K) neon gas through the cooling loop inside the back end of what had been the nitrogen tank inside NICMOS. This circulator gas absorbed the heat inside the Dewar and dumped that heat into the cryocooler through a heat exchanger. Using its own neon gas to absorb the heat, the cryocooler transported the heat-carrying gas around a closed loop of tubing. Downstream from the circulator, another miniature turbine, spinning at 7300 rps, compressed the warm gas, concentrating its heat energy within a smaller volume, thus raising the gas's temperature to nearly 300 K. The compressed hot gas very efficiently transferred its heat content into a third module, the capillary pumped loop (CPL), through a heat exchanger. The latter module's job was to dump the heat to a large (3 feet × 11 feet) aluminum radiator mounted outside the spacecraft (Figure 13.8). The circulation of neon gas through the cryocooler continued on to a third miniature turbine, rotating at 4500 rps. This unit, called the turboalternator,

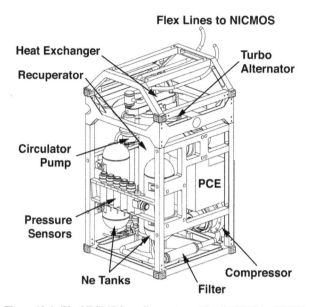

Figure 13.6. The NICMOS cooling system. (Credit: NASA, GSFC.)

forced the neon to expand dramatically in volume, diluting its heat energy and causing the gas's temperature to drop down to an ultra-cold 70 K. From there the 70 K neon continued on to the circulator, absorbing the heat coming from the Dewar and warming up in the process. This process of circulating the neon gas through a closed loop of plumbing, compressing and heating it, extracting heat out of it, expanding and cooling it and absorbing heat back into it would continue ad infinitum as long as the NCS was running. The heat from NICMOS' Dewar and its detectors and filters was, in this manner, continuously transferred out into space.

This was a great plan to restore NICMOS to science operation. But would the NCS actually work in orbit, in zero-g? It was critical to verify that it would before committing to actually installing the system on Hubble in SM3. Ed Cheng and his colleagues built a mock up of the system in a laboratory at Goddard to acquire some data about how the cooling system might perform in normal operations. But there was no substitute for proving it would work in orbit.

At 2:19 pm EST on 1998 October 29 space shuttle Discovery lifted off from pad 39B at the Kennedy Space Center to begin mission STS-95. Among the crew on board was the oldest person ever to fly in space up to that time, former Mercury astronaut and former U.S. Senator John Glenn. In the payload bay of Discovery was mounted, among other mission hardware, something called HOST, an acronym for Hubble Orbital Systems Test. The actual flight NCS hardware destined to be installed on Hubble was there on HOST getting a chance to prove that it could perform well in the microgravity of low earth orbit. A surrogate version of the NICMOS Dewar represented the heat source that the NCS would have to overcome.

The HOST mission also included flight tests of other key elements we were planning to install on SM3, including the new 486 class spacecraft computer, and a

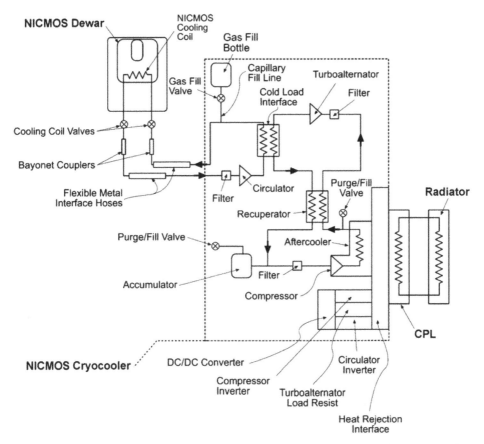

Figure 13.7. Schematic layout of the NICMOS cooling system. Triangles represent the Creare miniature turbines that drive the system. There are three of them—the circulator, the turboalternator, and the compressor, as described in the text. Sawtooth symbols represent heat exchangers where heat is passed from one part of the system to another. (Credit: W. Swift, F. Dolan, M. Zagarola, Creare Corp. Reprinted from Swift et al. 2007, with the permission of AIP Publishing.)

new solid state data recorder. Being on the same flight as John Glenn proved to be advantageous in an unexpected way for the Hubble Project. The old veteran astronaut drew all the media attention on this mission, leaving the Hubble testing work almost entirely unnoticed. It was a relief not to be under constant scrutiny.

The NCS system performed well, going through 185 hours of trouble-free operation. It achieved a minimum temperature of 72.65 K in the surrogate Dewar, achieving the cooling level that would be needed later with NICMOS.

The tests on the HOST mission, however, revealed a serious vulnerability. When the hardware was returned to Goddard afterward, the circulator that moves cold neon through the NICMOS cooling coil could not be started. An examination of the miniature turbine in the circulator revealed extensive damage to the tiny shaft. The culprit was moisture that had found its way into the plumbing and had frozen between the shaft and the surrounding housing. In response both the shafts and the

Figure 13.8. Radiator for the NICMOS cooling system, that was to be mounted by the astronauts on the exterior of the spacecraft, being tested in the clean room at Goddard. "CPL" refers to the capillary pumped loop that carried heat from the 300 K compressed gas to the outside radiator. (Credit: NASA, GSFC.)

housings of all the turbines were coated with a much harder, diamond-like material. And procedures for baking out any moisture that might be lurking in the system were strengthened. We were very thankful that this problem had been discovered during a test flight rather than after the NCS had been transported to Hubble.

With NICMOS brought back to life by the NCS, astronomers could expect better performance for a longer lifetime than ever could have been achieved with the instrument in its original, icebox configuration. The temperature of the block of nitrogen ice dictated the original temperature of the Dewar and detectors, 58 K. But a more optimal temperature in terms of detector sensitivity coupled with minimum background noise was around 70–75 K. The NCS allowed the cooling temperature to be set to the best value. And there was nothing limiting the instrument's operational lifetime, so long as it continued to produce excellent science. However, the infrared astronomers' gratification had to be delayed, while an emergency servicing mission (SM3a) was mounted in 1999 to replace Hubble's rapidly failing gyroscopes, essential for accurately pointing and stabilizing the telescope. The NCS would not be flown until Servicing Mission 3b in 2002.

Reference

Swift, W. L., Dolan, F. X., & Zagarola, M. V. 2009, AIPCP, 985, 799

AAS | IOP Astronomy

Life With Hubble
An insider's view of the world's most famous telescope
David S Leckrone

Chapter 14

Racing Y2K: Servicing Mission 3a

The Achilles Heel of the Hubble Observatory has always been *the Gyros*. There are six gyros on board, packaged in pairs in small boxes called Rate Sensing Units or RSUs (Figure 14.1). As the name implies, the gyros have the job of sensing the rate of rotation of the spacecraft around its three axes—pitch, roll, and yaw. The RSUs feed this information into the Pointing Control System software to assist in re-pointing or slewing the telescope toward a particular target in the sky and in holding it steady on that target until the fine guidance sensors can home in on guide stars and take over the pointing control duties.

The Hubble gyros are the best in the world. Hubble's requirements for incredible pointing accuracy (better than 0.030 s of arc) and stability (jitter less than 0.012 s of arc) demanded the best. Similar to the miniature turbines in the NICMOS Cooling System (Chapter 13), each gyro has a rotating wheel that is supported by gas pressure. They are called gas-bearing gyros. That makes their operation very smooth and quiet. They can detect extremely small angular motions.

There are numerous applications for gyros in airplanes, submarines, guided missiles, hand-held movie cameras, etc. These use gyros of various designs. None of them have the tight performance specifications of Hubble. The Hubble Project invested several millions of dollars, with matching funds from industry, attempting to develop other gyro technologies that could meet the observatory's stringent requirements. We needed choices. Those efforts were unsuccessful. So, in the end Hubble was married to a single, remarkably capable kind of gyro from a single manufacturer—Allied Signal (later called L-3 Communications).

At the time of Hubble's launch these gyros had a good reputation for longevity. Based on experience with similar gyros in other NASA and Department of Defense programs the Project expected them to last perhaps nine or ten years. A better way of saying this is that the gyros had a statistical mean time to failure of nine or ten years; half of a large sample of gyros would likely fail before nine or ten years and the remaining half after nine or ten years.

Figure 14.1. A rate sensing unit (RSU) containing two of Hubble's six gyroscopes. (Credit: NASA, L-3 Communications.)

This was critical to Hubble's scientific mission. Of the six gyros carried on the spacecraft, three had to be working at any given time during the early years of the mission to enable the telescope to be pointed at astronomical targets with the accuracy required for its scientific observations. The other three were spares, turned off and kept in reserve. If four or more gyros failed, the spacecraft would put itself into Zero Gyro Sun Point safe mode, a benign kind of hibernation, and wait for help from the engineers in the Operations Control Center at Goddard. Hubble would be safe, but its scientific work would cease.

In principle it was possible to carry out a more limited observing program with only two gyros or even one gyro in operation. But this required a lot more computing power than the primitive DF224 spacecraft computer could provide during the first decade of the mission. The Project planned to replace the DF224 with a more capable 486-class computer during Servicing Mission 3 (SM3), originally scheduled for a 1999 launch. That computer would be capable of enabling science observations in a two-gyro or one-gyro mode.

During the 44 months between Hubble's launch and the First Servicing Mission, four of the six gyros either misbehaved or failed entirely. These four were replaced during SM1. Three of them suffered from defective electronic parts. The parts problems were easily resolved. The fourth gyro revealed the first evidence of the problem that was to threaten Hubble science operations throughout the observatory's lifetime—a broken flex lead.

When I was a youngster, I used to play with a toy gyro—a metal wheel that I would set spinning by pulling a string that I had wound around its axle. Once set in motion, the rapidly spinning wheel would stay in the same orientation, balanced on the tip of a small wooden peg, as long as I didn't apply a force to it (and as long as it

Figure 14.2. A Hubble gyro assembly. Two gyros of this kind are housed inside three rate sensing units for a total of six gyros on board the spacecraft. (Credit: NASA, L-3 Communications.)

kept spinning rapidly). If there were no friction, it could keep spinning and stay fixed in space forever. When I did push its axle, the gyro would resist my efforts and in fact would tend to move at a right angle to the direction in which I was pushing. Its angular momentum was being conserved. This is called the gyroscopic effect.

By analogy the heart of a working Hubble gyro is a metal wheel, or rotor, spun up by an electric motor to about 19,200 revolutions per minute. The rotor is housed inside a small cylinder (the float assembly) with its axle at right angles to the long axis of the cylinder (Figures 14.2 and 14.3). As the observatory rotates in a particular direction, the axle of this spinning wheel resists the force trying to move it in that direction. Instead it moves at a right angle to the force, much like the toy gyro of my youth. This causes the float assembly to start to rotate around its long axis. The float assembly's rotation is sensed by a system of permanent magnets and electromagnets. The electromagnets are simple coils of electrical wire. The current sent through those coils exerts just the right amount of magnetic force to exactly offset the rotation of the float assembly cylinder. So, in fact the rotation is prevented. The amount of electrical current needed for this purpose is a measure of the motion of the spacecraft.

Each of the six gyro float assemblies and associated magnets is housed in a sealed cylinder filled with a dense liquid, about the consistency of 10W30 motor oil. Immersed in this liquid are seven tiny wires, about the thickness of a human hair. They are called "flex leads." Four of the flex leads carry electricity to the motor propelling the gyro rotor. The other three flex leads carry current to the electromagnetic coils. It was one of the four motor flex leads that had broken in the failed gyro, cutting off its electricity.

Here's how the problem arose. The flex leads were made of tiny nodules of silver bound together with copper. The heavy fluid in which they were immersed was forced into its housing with compressed air. The oxygen in the air reacted chemically with the fluid, releasing the chemical element bromine. Bromine corrodes copper. After a while, the copper in the flex leads became massively corroded and flaked

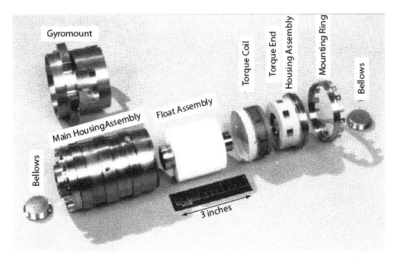

Figure 14.3. Exploded view of a Hubble gyro. The gyro's rotating wheel is housed in the cylinder labeled "float assembly." The wheel's axle is aligned perpendicular to the long axis of the float assembly, so that a rotation of the spacecraft, and the gyroscopic reaction of the wheel's axle, forces the float assembly to rotate about its long axis. This motion is exactly offset by a force applied by magnets to the right of the float assembly in the "torque coil" and "torque end." These components, together with seven thin electrical wires, called flex leads, all float in a thick liquid contained in the main housing assembly shown to the left. (Credit: NASA, L-3 Communications.)

away, leaving behind a very brittle chain of silver. If there were any tiny nicks, cuts, or other flaws in that chain, it could easily break at that location. With the copper corroded away the flex leads looked like Swiss Cheese.

In subsequent gyro assemblies, Allied Signal used compressed nitrogen gas instead of air to push the dense fluid into place. They redoubled efforts to avoid the tiny flaws in the flex leads. In later years when these measures had not proven totally successful, additional silver was used to coat the exterior of the tiny leads to strengthen and protect them. In the end gyros with silver-coated flex leads were the premium choice to go into the spacecraft during servicing missions. Unfortunately, a complete replacement of all of Hubble's gyros with these more reliable versions was never accomplished, and some gyros continued to fail because of flex-lead breaks.

Between SM1 in 1993 and SM2 in 1997 we experienced no gyro failures. All seemed well. We saw no need to replace any gyros in the second servicing mission. Statistical calculations, based on the assumption of a nine to ten year mean lifetime indicated that there was a 99.8% probability that the spacecraft would have at least three working gyros at the time of the next servicing mission, SM3, whose launch date had slipped by then to the year 2000 because of shuttle availability issues. The flaw in our thinking was that the statistics assumed that the gyro failures were random events. But there was a known physical failure process, and it was not completely random.

The consequences of our reliance on misleading statistical arguments began to bite us a little over two weeks after Discovery returned from orbit at the end of SM2 on 1997 February 21. Gyro 4 failed on 1997 April 9. Gyro 6 failed on 1998 October 22. Gyro 3 began behaving erratically in 1999 January and failed completely on 1999 April 20. All three failures apparently resulted from broken flex leads. That left the observatory in great scientific jeopardy. All three spare gyros had been brought into service. There were no spares remaining. If any of the three functioning gyros failed after that, Hubble would stop doing science and go into safe mode.

It was in this time frame that the Shuttle Fleet was being pressed into service to construct the International Space Station (ISS). On 1998 November 20 the Russians launched the first ISS module into orbit. Two weeks later the first US unit was carried to orbit on Shuttle Endeavour during STS-88. A proton rocket orbited a second Russian module in July of 2000, while the next US component went up on Atlantis on STS-98 in early February of 2001.

As a result, Hubble servicing and other missions unrelated to ISS construction had serious competition for use of the Shuttle Fleet. Three shuttle orbiters—Discovery, Atlantis, and Endeavour—were outfitted with new external airlocks that could be mated to the ISS. Hubble had used one of these, Discovery, flying for the first time with the new airlock in 1997 February for SM2 (Chapter 12). The oldest active orbiter, Columbia, was unsuited for the external airlock and thus could not participate in ISS construction. It was relegated to other duties as assigned.

NASA's plan was to launch the flagship observatory, Advanced X-Ray Astronomy Facility—AXAF, later renamed Chandra—on Columbia. Next in line for the old orbiter would be Hubble's third servicing mission. The originally scheduled launch date for AXAF was 1998 August. Hubble's SM3 was scheduled to follow in 1999 December. However, the AXAF Project ran into a number of technical problems and kept slipping to later and later launch dates. Each time the AXAF launch date slipped, Hubble's SM3 launch had to slip accordingly. Ultimately AXAF was scheduled to fly in 1999 July. That caused SM3 to be rescheduled to 2000 June.

With Hubble's last spare gyro going into operation in 1999 April, we faced the prospect that the three operating gyros had to last for more than a year. Given the recent history of gyro failures due to broken flex leads, there was great concern in the Project that observatory science operations could go down at any time. Hubble could be off-line, unavailable for astronomers to use, for many months.

The impacts of Hubble down time were considerable. Hubble observing time was very valuable, both in terms of delayed or lost opportunities for scientific discoveries and the considerable financial value of lost time. The amortized cost of one month of Hubble observing time was about $20 million in 1999 dollars. Moreover, it would be a public relations nightmare for NASA, if Hubble became unusable for many months because of technical glitches with AXAF, particularly given Hubble's earlier PR problems with spherical aberration.

Hubble Program Manager, John Campbell, was ahead of the game on this problem. In 1998 November he began lobbying NASA Headquarters for a change in the shuttle orbiter assigned to SM3 from Columbia to one of the vehicles with the external airlock. Headquarters requested that the feasibility of doing SM3, given the presence of an external airlock, be checked with trial runs in the Neutral Buoyancy

Lab at Johnson. Those tests were concluded during the first week of 1999 January and the results were favorable.

On 1999 January 13 Campbell formally requested a change of orbiter assignment for SM3 to completely decouple *Hubble* from any further launch delays of AXAF. If the change were approved promptly, he would set the Project's launch readiness date for 2000 June 1.

In early February of 1999 Hubble engineers declared Gyro 3 to be unusable. Though it would be about two more months before the gyro totally failed, the writing was on the wall. We were about to go into a period of uncertain length with only three operating gyros. This was shaping up to be an untenable situation. Drastic action needed to be taken.

In a presentation at NASA Headquarters on 1999 February 19, Campbell laid out the rationale for flying an emergency repair mission to Hubble the following October, eight months later, based on the high probability of a prolonged period of Hubble down time if nothing were done. Accepting Campbell's recommendation would undoubtedly complicate the lives of the managers at Headquarters, both in Space Science and in Human Space Flight. But to their credit Ed Weiler, who at that time had taken over as NASA Associate Administrator for Space Science, Ken Ledbetter, then Division Director for Mission and Payload Development under Ed, and Joe Rothenberg, then Associate Administrator for Human Spaceflight and famed for his earlier management of SM1 at Goddard, picked up the ball and ran with it. Of course this was a receptive audience for Campbell's recommendation. All three of them had invested significant fractions of their careers in the Hubble Program.

The plan was to split SM3 as originally conceived into two parts—SM3a and SM3b. SM3a would be the emergency "call-up" mission to replace all of Hubble's gyros. It would also include some of the other tasks originally planned for SM3, those that could be fully tested and ready to go on such short notice. SM3a was originally manifested for a launch as STS-103 on shuttle Discovery on 1999 October 14.

SM3b would be delayed to sometime in 2001. That would allow an orderly completion of our wonderful new camera, the Advanced Camera for Surveys, and some other payload components that could profit from a more relaxed schedule. As currently laid out, before the split into two missions, SM3 was becoming more and more complex. It would have been the first Hubble mission to require six spacewalks to fit in all the servicing work that needed to be done. So, the split of the mission into two parts simplified the job SM3 was intended to accomplish and increased the probability of success for both missions.

In the 1960s and 1970s the large mainframe computer became a major tool in manufacturing, commerce, public utilities, government agencies, national defense, and scientific research. Airlines used them to schedule flights. Banks used them to maintain account records and calculate interest. Automated assembly lines were operated with them. NASA used them to control the launch countdowns of rockets, to calculate the complicated trajectories of interplanetary probes, and to plan and schedule satellite operations.

The data storage and memory in those early main frames were limited in size and were expensive. So, programmers cut whatever corners they could to limit the volume of data going into the computations. Some software required them to enter calendar dates. Suppose a mortgage lending company needed to calculate the interest on a loan payment due on a specific date in a specific year. One shortcut was to specify the year with only two digits. The day, month and year for the date 1972 June 30 became 30/06/72, for example. It was understood in the software that the leading two digits, "19" were supposed to be there for the year number. The "19" didn't have to be repeated over and over, using up precious computer memory and data storage space. Why use four digits when two would do the job?

The people who wrote these computer programs probably didn't expect them to be used for more than a few years up to perhaps a decade. After all, computers would steadily grow in capability and speed, data storage and memory would become cheaper, and stating a year number with the full four digits instead of two, would become standard procedure. What they didn't reckon on was that good computer codes that worked well might continue to be used for a long time into the future.

In the mid-1980s a few computer experts began calling attention to what they called the year 2000 problem. In the mid-1990s it was dubbed the Y2K problem. The letter "Y" stood for "year," the letter "K" was the designation in the metric system for "kilo" or one thousand. So, Y2K was shorthand for "The Year 2000."

Simply stated, if your computer program designated the year 1999 as 99, then what would happen to the program's computations at the stroke of midnight on December 31 of 1999? The calendar would turn over to 2000 January 1. The program would be asked to recognize the year "00." But what would that mean—1900, perhaps? Or 19100? What havoc would these strange values wreak on whatever it was that was being calculated?

It wasn't until 1995 or so that the world started to take serious notice of the Y2K problem. A sense of urgency about the problem set in around the globe. There were simple tests one could do on a computer program, for example by artificially setting the year designation ahead to "00" and then observing what happened. But a degree of paranoia took hold about what was being missed. What have we overlooked? Where were the hidden year designations lurking in all the computers that basically were in control of *everything*?

NASA began working on the Y2K issue in 1996, checking thousands of hardware and software systems associated with every flight program and ground-based activity and verifying that they were immune to the problem, or if necessary repairing or replacing them. In total the Agency invested $60–70 million insuring that there would be no critical computer failures. In addition, virtually all missions were to be put into a quiescent state, cease nearly all activity, as the clock moved from midnight of 1999 into the early moments of 2000. This did not include orbiting satellites and planetary probes, as they did not specify time in days, months, and years in their on-board computers. Typically they used Mission Elapsed Time, starting the clock at the moment of liftoff or at some other well-defined mission milestone.

The Shuttle Fleet was thoroughly checked. Any needed modifications were made and the Fleet was declared impervious to Y2K problems. Nevertheless, out of extreme caution, Shuttle Program managers required that STS-103 carrying Hubble's SM3a be completed and back on the ground before the calendar turned over from 1999 into 2000. That wasn't a serious concern for us. We were going to fly on Discovery in October. All would be safe and well long before Y2K. We had no idea what was in store for Hubble as the end of 1999 approached.

After many delays Chandra/AXAF finally made it off the ground on 1999 July 23, carried to orbit by Columbia on STS-93. This was an historic mission not least because the mission Commander was Eileen Collins, who, on this flight, became the first woman astronaut to lead a shuttle mission in that role. During launch the Shuttle encountered several problems, one of which would directly affect the schedule for Hubble's SM3a.

Five seconds after liftoff an electrical short disabled two digital control units that operated two of the orbiter's main engines. Normally, the loss of two engines would require a Return to Launch Site abort of the mission—a very risky proposition that had never been done before. However, redundant units in both engines took control and the launch continued into the proper orbit.

After Columbia landed back at KSC on July 28, its wiring was given a thorough inspection. An electrical short was found in poorly routed wiring that had rubbed against an exposed screw head, so that its insulation was worn off. With that discovery the entire shuttle fleet was grounded until each orbiter's wiring could be inspected and repaired. Discovery was rescheduled to carry SM3a into orbit no earlier than October 29. However, as it turned out, it's wiring required extensive repairs. The launch date slipped to no earlier than November 19.

On November 13 one of Hubble's three remaining operating gyros failed. The spacecraft placed itself into Zero Gyro Safe Mode, oriented its solar arrays toward the Sun to assure it had plenty of electrical power and waited for help from controllers on the ground. When Hubble entered safe mode, it would shut off its transmitters to save power. The spacecraft in effect went dark. Program Manager John Campbell's heart raced as he waited in panic while the STOCC team tried to restore communications with the telescope. Relief came only when we received confirmation that Hubble was talking to us again. But we all realized on this November day in 1999 that scientific observations of the Universe with Hubble had stopped and could not be resumed until the suite of gyros was replaced in orbit. *We were dead in the water.*

Things continued to go downhill for Discovery, and consequently for SM3a, after that. A further delay of the launch date to December 2 resulted from a longer than expected time to complete the wiring repairs. Then a wayward drill bit was found, lodged in engine number 3. December 6 became the new launch date. Discovery was transported to Pad 39A, and engine number 3 was replaced on the pad.

Another round of wiring damage, this time in the umbilical connecting Discovery to the external propellant tank, put another dent in the launch schedule—this time to December 9. Now Thanksgiving was upon us. The hard working shuttle crew at

KSC deserved the holiday off to be with their families. So, we agreed to a slip to December 11.

A closeout inspection of Discovery's engine compartment revealed a dent in a line that carried liquid hydrogen to the main engines during launch. A launch date no earlier than December 16 was planned. The damaged plumbing was replaced, tested, and verified to be in good order on December 13. The shuttle team verified that December 16 was still a valid launch date. The countdown to launch was initiated by the KSC Launch Director at 1:03 am on Tuesday, December 14—right on time, I sarcastically thought.

I flew into Orlando on Tuesday afternoon, 1999 December 14. As usual I rented a car at the airport and drove the 50 min down the Beeline Expressway to Cocoa Beach. This was the fifth time I would be a witness to a shuttle launch. From prior experience I knew that the visceral thrill of seeing (and feeling) a shuttle roaring off the pad, on top of the prospect of Hubble being made whole again, would be indescribably thrilling.

The following afternoon I made my way to the KSC Media Center. That's where I normally watched shuttle launches. It was the best place for viewing, closer to the pad than any of the other guest viewing locations. I had appointments for interviews with reporters there.

I stood by the countdown clock near the turnaround pond. It said—13 hr, 44 min, 44 s, deceptive because of the many built-in holds. The American Flag next to the clock was at half-staff, marking the 200th anniversary of the death of George Washington. Only 200 yr between General Washington and a shuttle launch to the Hubble Space Telescope. Amazing!

Across the road was the Vertical Assembly Building (VAB) where, a few weeks before, Discovery had been mated with its external tank and solid rocket boosters. The complete shuttle stack had then been gently placed upright on the gargantuan 5.5 million pound crawler vehicle that transported it to Pad 39B at a breathless 2 miles per hour. The VAB looked like a huge monolith 600 feet high. Dozens of vultures rode the updrafts created by the building, as though its walls were cliffs in a mountain range. Watching them circle lazily around the VAB, I was reminded once again that this place was first a primeval jungle before it was a rocket port.

The atmosphere at the Press Site that afternoon was surprisingly quiet. I recalled the first servicing mission in December of 1993. Then the place was buzzing with activity. Back then, this close to the time of launch, the parking lot was full of reporter's cars. It was nearly impossible to find a parking place. On this afternoon six years later, however, the parking lot was perhaps 25% full. The huts housing reporters were still there, bearing signs designating NBC News, CBS News, ABC News, AP, and Reuters. From their rooftops television cameras would capture the launch when it happened.

The apparent lack of public interest in SM3a, as compared with SM1 and SM2, was a bit deflating. With no new instruments flying on SM3a, it was perhaps a less interesting and certainly a less glamorous mission, less likely to capture the broad attention of the worldwide media. As an astronomical observatory, Hubble had been asleep, out of commission, for a month. It was our job to wake it back up. That's all.

Unsurprisingly, Ed Weiler didn't plan to attend the launch as he had for the prior two servicing missions. He had been promoted to the Associate Administrator position at NASA Headquarters. His purview was now much broader than just Hubble. In his place at KSC and JSC throughout the mission would be Anne Kinney, Ed's replacement as Director of the Astronomy and Physics Division in the Office of Space Science.

At 2:45 in the afternoon of 1999 December 15, at the KSC Media Site, a cool breeze blew off the water. There were occasional clouds in the sky but also plenty of blue. Probably good enough weather for a launch. "I wish we could go today," I thought to myself.

After my interviews were finished, I drove to a small building on the outskirts of the KSC campus, the Vertical Processing Facility (VPF), where Cepi and his HST servicing team were ensconced. Their critical job there was to receive and inspect all of the mission payload hardware, to run tests to make sure it all had survived the long trip to the Cape, and to get the payload fully integrated into Discovery's payload bay, ready to launch.

I checked in with Mike Kienlen, Cepi's deputy, and asked him how everything was going. "Bad news," he said. "We've been delayed another day, to the 17th." The story he then told me was of the hair-pulling-out variety. It seems that somehow the JSC Shuttle Program Manager, Ron Dittemore, and his staff had gotten word of a potential problem with welded joints on the shuttle's external fuel tanks (ETs). Apparently, welding rods made from the wrong material had been used in the welding of an external tank at the facility where the ETs were manufactured at NASA's Michoud Assembly Facility near New Orleans. This did not refer to our ET, the one in the shuttle stack standing at that moment on Pad 39B, but to some other ET that had been assembled at Michoud sometime in the past. Our ET had already passed all of its welding inspections and had been judged fit to fly.

However, the same people who did the ET welding also had done welding of structural components within the engine compartments of the shuttle orbiters back in the 1970s. Questions were now being raised about their quality control processes. Dittemore asked that the launch of STS-103 be delayed for 24 additional hr to give JSC time to review the original quality control paperwork from the 1970s. Discovery had flown successfully 26 times previously without a problem. However, it didn't seem right to second-guess the JSC team's desire to be absolutely safe, to leave no stone unturned. But Hubble was having problems of its own that day in orbit and further delay exacerbated our own serious worries about its safety.

Hubble had been off-line in Zero Gyro Safe Mode for a little over a month. This was by far the longest period of science down time for the observatory since it was launched. In safe mode it was healthy and more or less stable. When it passed from daylight into darkness once per orbit it went into an uncontrolled free drift. As it came back out of darkness into daylight it drifted to off-nominal orientations that usually deviated by less than 30° from its correct orientation. In daylight the spacecraft's coarse sun sensors locked onto the Sun, causing it to re-orient itself with the solar arrays face-on to the Sun. The electricity generated by the arrays recharged its batteries, keeping the spacecraft power-positive and well protected.

The Hubble Project team, however, was growing concerned because recently Hubble had shown a tendency to emerge out of the darkness of orbital night drifting to abnormally large off-nominal orientations—sometimes by as much as 90°. So far, even with these surprisingly large excursions, it had been successful always locking onto the Sun. But the situation was worrisome. It was ironic then that today, the day the shuttle team had delayed our launch yet again for reasons I felt were beyond conservative, Hubble experienced by far its largest excursion, its largest instability in Zero Gyro Safe Mode. As it was coming out of the darkness into daylight, it had drifted so much that one of my colleagues said, "It nearly flipped over." We were growing anxious to get to orbit and remedy the situation.

The mission of STS-103 originally had a planned duration of ten days. This included two days to catch up with Hubble in orbit, to grab it with the remote arm and to dock it to the Flight Support System platform in Discovery's payload bay. Four days were allocated for four spacewalks to replace Hubble's six gyros and to install other equipment in the spacecraft. One more day in orbit provided additional time for an unscheduled EVA, in case one were needed to complete any remaining work or to handle problems that might arise in deploying Hubble back into its own orbit. The following flight day was a "crew day" for the shuttle crew to rest, handle personal tasks, and prepare the orbiter for the return to Earth. Two additional days in orbit had to be allocated as possible "wave-off" days in case bad weather materialized that would prevent a safe landing of the orbiter at KSC. If necessary, it could land instead on the Dry Lake bed at Edwards Air Force Base in California. A launch on December 17 implied a potential landing as late as December 27. Y2K was growing brighter on our radar screen. There was little time to spare. We needed to get this thing off the ground.

That evening I strolled from my hotel to a small nearby Greek restaurant, Zachary's, for dinner. The waitress who served me was amazing. She was very chatty, and also clearly very bright. Probably in her early 60s, I guessed. She asked me if I was in the Space Program. I replied in the affirmative, but did not tell her exactly what my job was. Then she asked, "What do you think about the delay due to the welds?" I told her I wasn't very happy about it. She replied, "Oh yes, I know, I know; it just seems to me like they're so skittish they're almost afraid to launch. It's a shame because you may lose the last EVA and not be able to put on all the gyros." I corrected her misimpression about that, telling her we would change out all the gyros in the first EVA and that the last EVA had less critical things in it.

She sat down across the table from me. Her skin was leathery and aged, a pattern of intersecting crevices—too much time in the Florida sun I reckoned. She smelled of cigarette smoke. She told me how much she really loved those Hubble images. She said that late at night before she went to bed she would get on the web, bring up the Hubble web site and just enjoy looking at Hubble images. In order to get a clearer view on her computer, she had gone out and bought a higher quality monitor. After that the images looked really great.

I was a bit stunned by this conversation. Here was Ms Every Person, a middle-aged waitress in a modest restaurant, who was knowledgeable about Hubble, loved what Hubble did and worried about its wellbeing. "My God," I thought. "We have an incredible responsibility. People are counting on us. They really care." The encounter brought to mind many other such conversations I had had in the past. When people heard that I worked on Hubble, they wanted to talk about it. Whether it was the person sitting next to me on an airplane, or my barber, or the guy who serviced my car at the local dealership, they had questions. These people were incredibly supportive. But I had never encountered someone in everyday life before who seemed so personally invested in Hubble as the waitress at Zachary's.

The next morning I sat in on the daily press briefing given by Ron Dittemore and others from JSC and KSC. The questions focused on the welding concern and the delay it had caused. At one point a reporter who knew me turned to me and asked directly, "Does the Hubble Project agree with this decision?" I didn't say what I was actually thinking. Putting on a brave face, I replied, "I'm very frustrated every day Hubble isn't working. On the other hand I want to make certain that we have a high probability of full mission success. The shuttle people have to do what they have to do to assure it's successful. So, yes, we support them doing what they believe is necessary." But I was deeply frustrated, and so was John Campbell, that we had to press ever closer to the Y2K cutoff for the mission, so that people could go research quality control paperwork that was over 20 yr old. Perhaps their extraordinary caution was justified, but the logic of it evaded me.

At this same press briefing, it was announced that the probability of launch tomorrow, December 17, or rather the probability of a launch scrub tomorrow due to bad weather had increased to 60%. I believe the probability of a launch scrub today had been about 20%. So, in the course of being extraordinarily cautious and conservative, the powers that be had lost another day in the race with Y2K. It was looking more and more like we weren't going to fly until January or later.

On Friday, December 17, I was once again at the KSC Media Center. It was a little past noon. My primary job during these servicing mission launches was to be available to consult with the Hubble Program Manager—in this case John Campbell—as issues arose. Beyond that my fellow Project Scientists, Mal Niedner and Ken Carpenter, and I, together with Anne Kinney, were spending time at the Media Center taking requests for interviews. There were many such requests from newspapers, magazines, national TV and radio networks, local radio and TV stations from all over the country, and from media around the world. It kept the four of us well occupied as we waited for STS-103 to launch.

The whole business of the welding rod quality control issue and the search for the applicable decades-old paperwork had been cleared up. It was no longer an issue. The company involved had a superb record over the years. They had never had parts failures that weren't caught and corrected. This particular welding rod issue was discovered by one of the company's own inspectors in their own plant and was caught before the external tank that was being welded left the plant. There was no reason to suspect or doubt that the welds they had been doing for decades on NASA hardware were anything but excellent.

As of that moment, early Friday afternoon, it was 80% "no-go" for a launch that day. Earlier in the morning there had been some sunshine and blue sky. The KSC Launch Director had given the go-ahead to begin filling Discovery's external tank with liquid hydrogen and liquid oxygen, the process called "tanking." As the afternoon wore on, rain continued to fall, sometimes very hard. After the Sun set the precipitation lightened to a sprinkle. We grew a bit more optimistic.

The launch window would close nominally at 9:29 pm It was now 8:15 pm There was a lengthy list of weather-related criteria that had to be met before the launch could proceed. I checked the board: lightning, green; cumulus clouds, red; anvil clouds, green; debris clouds, green; disturbed weather, red; thick cloud layers, red; smoke plume, green; field mill, green; flight precipitation, red; launch wind, green; temperature, green; ceiling visibility, red. The countdown clock was at $T - 9$ and holding.

At 8:51 the launch window was extended by 4 min. That step would result in a one-day delay in the rendezvous with Hubble, but at least it would get Discovery off the ground. Then a voice on the public address system said, "Okay, we've polled the weather community and we've got several constraints out there with the weather. We've got some showers down south of us. There's just not enough clearing time. So, it appears we need to scrub for the day and recycle for 24 hr turnaround."

The next day, Saturday, December 18, launch was scrubbed early in the morning. It was continuing to rain. It appeared that it was going to keep raining all day, and it did—thick clouds and a steady rain with no break in sight.

Nominally that Saturday would have been the last launch opportunity for the year. A few days had to be reserved at the end of the mission as a concession to Y2K. The shuttle team needed sufficient time, with margin, to make sure all systems were safe and in a quiescent state at the stroke of midnight on December 31.

However, consideration was now being given to the possibility of launching the following day, the 19th. For this to happen, the shuttle team would have to give up one of their two weather wave-off days, thereby increasing the risk of having to land at Edwards rather than back here at Kennedy. The Hubble Project would have to give up the fourth and final EVA day, leaving only three EVAs to do all the servicing that could be fit in. The fourth EVA had been earmarked for the installation of fresh insulation to cover the degraded MLI found during SM2 on the forward shell and light shield—the tube structures surrounding Hubble's primary optics. The replacement insulation had been dubbed the SSRFs (pronounced "serfs"), which stood for shell/shield replacement fabric.

I left KSC for the evening not knowing whether there would be a launch attempt the next day, Sunday, or not. The last weather information I heard before leaving indicated a 60/40 chance of favorable conditions for a launch tomorrow. The KSC and JSC shuttle managers, along with their Headquarters counterparts, were also discussing how they were going to handle the eventuality that Discovery might have to land at Edwards and not get back here to Kennedy in time for its button-down for the Y2K transition—a big concern for them.

That evening, while I was at dinner, the pager on by belt buzzed. It was John Campbell. I called him back. He said that another factor was being introduced into the decision as to whether or not to launch tomorrow. Joe Rothenberg, the Administrator of the Human Spaceflight Program (Code M in Headquarters parlance) and his people were arguing that they didn't want to fly unless they could do all four EVAs. The fourth EVA had been an add-on late in the mission planning process. I considered it a bonus, not a necessity. There was no immediate urgency in installing the SSRFs. That was something we could postpone until the next servicing mission.

John wanted to hear my opinion about this, and I told him just that. I viewed the fourth EVA as a bonus and I was unwilling to trade the SSRFs for another month or more of down time, another month of lost Hubble science observing time. Apparently several other key Hubble Project people—Dave Scheve, John's Deputy, Preston Burch, Project Manager for Operations, and Cepi—had all been agreeing with Rothenberg that we should wait until January when we could do four EVAs. John, at least up to that point, had been alone in his opinion that we needed to get on with it and get the gyros changed. John's opinion was also heavily weighted by the spacecraft's recent bad behavior in safe mode, deviating by large angles from its proper orientation as it came out of darkness into daylight. How much worse might it get before we could get back up in January? And how much risk did those large excursions hold for the safety of the observatory? I also strongly agreed with that concern.

So, I retired for the evening not knowing whether or not there would be a launch attempt the following day. What would the decision be? I felt we were losing the race with Y2K. I was thankful for John Campbell's priorities. He wanted to get the scientific perspective and that's mainly what I was there for. A meeting had been scheduled for 8:30 in the morning where all of this would be debated. As I went to bed that night, I did not know how it was going to turn out. It was a mystery.

The next morning, Sunday morning, I drove to the VPF to check in with the Hubble team. The sky was very blue with just a few wispy clouds floating by. The sunshine was brilliant. Last night's mystery had been resolved. The Air Force weather guys were saying we had an 80% "go" probability for a launch at about 7:50 that evening.

A short walk down the narrow road from the VPF was a large drainage pond. I enjoyed watching the alligators there (Figure 14.4). Today two large ones were sunning on the bank, napping. They had been deprived of sunlight for the prior several days. Now the cold-blooded creatures could warm up a bit. When they're in the water, floating on the surface, one could only see eyes, snout, and a little bit of their horny greenish-black backs showing. There were a lot of shore birds around. I imagined that's one reason the alligators looked so well fed.

One evening prior to one of our servicing missions—I don't recall which one—I was watching four of the gators swimming in the pond. Each made a long narrow bow wave in the water. Standing on the shore I mentally lined up the central axis of each of the four bow waves. "Hmm," I thought. "That's interesting. If I eyeball the directions toward which the four axes are aimed, they all seem to be converging on one spot on … on me!" My first hypothesis about this observation was that people frequently must come to that pond for an early evening nature walk and throw food

Figure 14.4. One of our neighbors in a drainage pond a short walk from the building at KSC where the Hubble Team was busy preparing for the launch of SM3a. (Credit: the author.)

into the water to watch the gators feed. It was their dinner. My second hypothesis was that the gators had eyed me, standing on the shore close to the water, and surmised that *I* was their dinner. Pondering a choice between these two possibilities, I decided to retreat back to the VPF.

At around 4:00 that Sunday afternoon I began to drive toward the Media Site to be there for the launch that evening. At the security gate I was stopped by the guard and told to pull off the road, because the astronauts were right behind me. The guard invited me to get out of my car and watch as they drove past. I saw a security helicopter flying directly in front of them with heavily armed people partially hanging out of the open doors—watching out for terrorists, I supposed. Then came a convoy of police cars and civilian cars, followed by the astronauts' legendary silver colored van—the Astrovan. It looked a bit like a small house trailer. A large, vividly colored NASA meatball was emblazoned on its side. The van was equipped with ventilators that blew dry, cool air through the astronauts' orange launch and entry suits—a real blessing in the hot Florida sun. This parade of vehicles went zipping by at high speed, about 60–70 mph I guessed, with lights flashing. They had an appointment at the launch pad and didn't want to be late.

Early that evening, as the Sun dropped lower in the west and shadows grew long, I was enveloped by the beauty of it all—the beauty of the natural setting in the evening light, the beauty of the incredible machine standing at the ready on Pad 39B, and, in my memory, the silver and gold beauty of the Hubble Space Telescope, soon to be coupled with it. The sky was deep blue, accented with some low haze and high cirrus. The moon was three quarters of a bright disk, waxing gibbous phase.

Bill Harwood, one of America's best space reporters, had kindly invited newcomer Anne Kinney and me to watch the launch from on top of the CBS news hut. That was an offer not to be refused, viewing from a prime location.

Figure 14.5. Launch of STS-103 on Discovery at 7:50 pm EST on 1999 December 19—the very last possible launch date before the Y2K shutdown. (Credit: NASA.)

The countdown was ticked away by a sometimes unintelligible announcer over the PA system: "T minus one minute and counting; T minus 12, 11, 10, 9, 8, ..., 3, 2, 1; we have booster ignition and liftoff of Space Shuttle Discovery on a mission to repair the Hubble Space Telescope as we venture into the twenty first century..." (Figure 14.5). "All right, all right, oh my goodness!" "HA!" I exclaimed. Finally a deep rumble reached us—a crackling staccato. "Discovery Houston, go at throttle up." "Discovery copy, go at throttle up." From the erratic PA I caught some of what was being said. ".... solid rocket boosters burnout and separation from the orbiter has been confirmed...." "Discovery, Houston ... forty miles altitude fifty five miles down range from the Kennedy Space Center, traveling at 3500 miles per hour." Then a short time later, ".... negative return...."

It was a beautiful, flawless launch on the very last day we could have done it without violating the Y2K strictures. Now, it was on to Houston.

It was about 11:00 pm CST in Houston on Monday, December 20. We were into Flight Day 2 of STS-103. As usual, I was sitting in the Customer Support Room

(CSR) in Building 30 at Johnson Space Center with other members of the Hubble Management Team. Everything about the mission was going by the book—smoothly, no problems. The orbiter's payload bay doors were opened about 2 hr after launch. The Flight Support System (FSS) had been oriented so that its "lazy Susan" platform was horizontal (that is, parallel to the shuttle's wings), its latches open, ready to clamp onto the docking rails on Hubble's aft bulkhead. Monet might have painted a picture of the scene—the gold foil insulation covering the FSS, projected against the vivid white skin of Discovery's tail, projected against the blue Earth with its white clouds in the background, and beyond, the blackness of space. It had been a quiet and contemplative day.

The next day, Tuesday, December 21, at about 1:00 in the afternoon the STOCC operations team at Goddard put Hubble into its deepest form of safe mode, hardware sunpoint. This was an essential step so that the spacecraft would remain in a fixed orientation, allowing Jean-Francois Clervoy, the remote arm operator, to grapple it and swing it down onto the FSS in the payload bay. The Hubble team had been a bit nervous about this procedure. It required us to turn on temporarily a set of gyros called the retrieval mode gyros (RMGs). These were not the high precision gyros that had failed and were to be replaced. Rather, the RMGs were coarse mechanical gyros, not intended to be operated continuously, but good enough to stabilize Hubble during the rendezvous and docking process. The worry was, if one of the RMGs failed, when we had no other gyros working, Hubble would be sent spinning. It was our job to worry about improbable failures that could have catastrophic consequences. So, we worried.

Later that evening, after completing a thirty-orbit chase to rendezvous with Hubble, Mission Commander Curt Brown and Pilot Scott Kelly gently guided Discovery upward under the observatory to a point where Clervoy could attach the end effector of the arm to the grappling fixture on the side of the telescope. As I sat in the CSR following what was going on, one of the JSC people commented that Commander Brown had more experience as a rendezvous pilot of the shuttle than any other astronaut. He had done what he was then doing, safely flying the shuttle to get in close proximity to another satellite, more times and was more skilled than anyone else. Our observatory was in the best of hands. A short time later, Hubble was safely locked into the latches on the FSS, ready to be serviced.

On the ground we were unable to see the rendezvous and grappling on video because the tail of the orbiter was blocking the line of sight to the TDRSS communications satellite. At that very moment, however, Hubble and Discovery were actually passing over Houston. Some of us went outside and looked up. The two magnificent machines, reflecting sunlight, looked like a single bright planet, moving rapidly across the sky. We couldn't resolve the two of them, but intellectually we were aware that we were seeing their moment of union with our naked eyes.

The following day, Wednesday, December 22, was the most important day of the mission. On that day, during the first spacewalk of SM3a, astronauts John Grunsfeld and Steve Smith were to install six new gyros, housed in three RSUs, into Hubble. That's what had to be done to restore the observatory's ability to do astronomy again. It's why we had mounted this emergency mission. Except for a momentary problem getting the first RSU bolted down, the change out went smoothly (Figure 14.6). We all sat nervously watching the telemetry pages on the monitors in the CSR, as each new gyro was turned on for its aliveness test. They all spun up beautifully, although it seemed to take a long time. It was like watching the pot of water that refused to come to a boil. There were sighs of relief and general elation throughout the room.

Then things became a little strange. There was not enough EVA time available on SM3a to install the new NICMOS Cooling System (NCS, see Chapter 13). It would have to wait for SM3b in 2001 or 2002. But there was one task that needed to be done on SM3a to prepare for the installation of the NCS a few years hence.

We had learned from the test flight of the NCS on STS-95 in October of 1998—the HOST mission—that there was one potentially fatal problem with the NCS design that had to be avoided at any cost. If any water got trapped inside one of the miniature turbines and froze there, it would jam the rotation of the turbine, causing it to fail. In that scenario the NCS simply wouldn't work and NICMOS could not be brought back into operation.

Figure 14.6. Having completed installation of three new rate sensing units (the three silver colored boxes near the top of the open bay), astronauts Grunsfeld and Smith work to open two valves and to remove two covers from inlet and outlet ports on the exterior of NICMOS to their right (not visible in this picture), preparing the way for the installation of the NICMOS Cooling System in SM3b. (Credit: NASA.)

The internal dryness and cleanliness of the NCS system could be verified on the ground prior to its launch. The only part of the NCS that still might be susceptible to contamination would be the circulator loop that would circulate ultra-cold neon gas through the plumbing—the cooling coil—at the rear of the nitrogen tank inside the NICMOS Dewar. That was the only part of the NCS that would directly "see" the interior of the NICMOS cooling coil and any residual water that might linger there. The gas was circulated through the coil by one of the miniature turbines. That turbine could be jammed if any ice crystals found their way into the circulator.

To greatly reduce this risk, it was proposed that the cooling coil be opened to the vacuum of space well in advance of SM3b. Any residual water trapped there would evaporate away, leaving a pristine set of plumbing to which the NCS would later be attached. That was a small task for Grunsfeld and Smith to do after they had installed the three new RSUs.

With the −V3 axial bay doors on Hubble already opened for the RSU installation, they could see and reach the exterior of NICMOS in the adjacent axial bay to their right (Figure 14.6). They needed to do two straightforward things. First, unscrew and remove two covers, one on the inlet port and one on the outlet port, to which the two tubes of the NCS circulator would be attached. Through these ports neon gas would flow into and out of the cooling coil. Second, open the two valves that enabled the neon gas to flow (Cepi's famous "indulge me" valves that he had insisted on many years before).

The astronauts began by trying to remove the inlet and outlet port covers by hand. They could get one of them partially turned, but the other one wouldn't turn at all. When this was communicated to the ground, it mistakenly came across that it was the valves that wouldn't open. The flight rules required that, if the valves didn't open, the astronauts should not force them, but simply close out the work site and move on. That would, however, have left us poorly prepared to install the NCS in the next mission. Following the incorrect flight rule, Grunsfeld and Smith pulled away from NICMOS and began the process of closing the aft shroud doors.

We were sitting in the CSR monitoring all of this on video, when the door tore open and in ran Darrell Zimbelman, the systems engineer responsible for the NCS. He was very distraught, practically in tears. "They're not doing it right! They're not doing it right!" he exclaimed. "The NCS is not going to work!" He had correctly understood that it was the inlet and outlet port covers that would not come off, not that the valves were stuck. In that situation the applicable flight rule was, if the bayonet fitting covers resisted being removed, the crew was supposed to get additional tools and apply more torque, forcing them off.

Meanwhile, Project Scientist Ed Cheng, the prime mover behind the NCS effort, was back in his hotel room helplessly watching this all unfold on NASA TV. Ed had worked the planning shift and was off duty. He should have been asleep. But he also could tell that the crew wasn't doing the right things. He made separate calls to two colleagues, Hubble engineers, back in Building 30, telling them something was wrong. Apparently they didn't fully appreciate the gravity of the situation. In both cases they were busy, distracted, and impatient to hang up the phone and get on with whatever they had been doing. Later they apologized to Ed for not grasping what he had been trying to tell them.

As on other occasions, John Campbell was exasperated. His major complaint during the missions in which he led the Goddard Hubble Management Team, was the lousy communications between the Hubble Team and the Shuttle Mission Team. Only a hundred feet or so of hallway separated their respective control rooms, but it might as well have been a mile. In this case he tried to activate the usual, serpentine communications path into shuttle management. But the astronauts in orbit were proceeding to close the aft shroud doors, leaving the job undone. He felt helpless. Time was running out.

There was one remaining servicing task to be done during that EVA—installation of voltage improvement kits (VIKs) on Hubble's six batteries. It had high priority. Throwing his hands up, Campbell concluded that the astronauts should move on to the VIKs, lest they run out of time to complete that work.

Then suddenly, impulsive, won't take "no" for an answer Cepi, leapt up and marched resolutely down the corridor to the Mission Control Room. He yanked the door open, walked in and spoke face-to-face to Lead Flight Director, Linda Ham. Soon word went up to the EVA astronauts that they needed to re-open the doors and give the NICMOS job another try. There had been a miscommunication between them and the ground. And so Smith and Grunsfeld retrieved a wrench and used it successfully to open the recalcitrant inlet and outlet covers. They then proceeded to manually turn the valves, which opened properly.

The NICMOS confusion had cost significant EVA time. In real time the two astronauts improvised a shortcut that would allow them to install the VIKs in less time than had originally been allocated. Human ingenuity triumphed once again in orbit.

At the end of the EVA Smith and Grunsfeld returned to a problem that had come up as they tried to stow one of the old RSUs they had just removed from Hubble. It was supposed to fit into a box-shaped protective enclosure for its return to the ground. However, they were unable to get it all the way into the enclosure because of interference from a rubber bumper, a cushioning material that was attached to the RSUs exterior. Exposed to the vacuum of space the material had swelled up in volume. The RSU would have been safe to return it as it was—partially stowed, half in and half out of the protective enclosure. So, they had left it stuck there while they proceeded on with the rest of the EVA. At the end of the EVA they returned to the RSU determined to finish the job. Grunsfeld aligned himself horizontally over the RSU, and grabbed its handrail. Smith then floated above him, holding onto Grunsfeld's backpack, and began bouncing up and down, trying to push the RSU the rest of the way into the protective enclosure. This was amusing to watch—comic relief at the end of a very long spacewalk. And they did succeed in fully stowing the RSU into its protective enclosure and closing the door.

EVA 1 turned out to be the second longest spacewalk in the history of the Shuttle Program, and perhaps in the history of manned space flight—8 hr, 15 min.

Claude Nicollier was a citizen of Switzerland. He had earned a Bachelor's degree in Physics in 1970 from the University of Lausanne and a Master's in Astrophysics from the University of Geneva in 1975. In parallel he had been a fighter pilot in the Swiss Air Force and later an airline pilot for Swissair. He graduated from training as a test pilot in England in 1988. In 1978 Claude was selected by the European Space Agency to be a member of the first class of ESA astronauts. Under an agreement between ESA and NASA, he was accepted in 1980 for astronaut training in Houston as a Shuttle Mission Specialist.

In 1993 on STS-61 Claude had done a masterful job operating the Canadian Remote Arm in Hubble's first servicing mission. STS-103 was to be his fourth and final shuttle flight. At age 55 he was one of the older astronauts.

Having been assigned to duties inside the Crew Cabin on previous missions, Claude never thought he would get to do an EVA, though he had always wanted to. With his last shuttle mission looming, he summoned his resolve and asked to be assigned as an EVA astronaut on STS-103. His request was accepted. He would be doing the second and fourth spacewalks working on Hubble during SM3a. Sadly, because of the Y2K problem, the mission had been abridged to include only three EVAs.

EVA 2 would be Claude Nicollier's first and last spacewalk. He and English astronaut Mike Foale exited Discovery's airlock at about 1:06 pm CST on December 23 to commence the EVA. Their job was to install Hubble's new, more powerful 486-class computer—20 times faster and with six times more memory than the old DF224 it would replace. We called this "Hubble's Brain Transplant."

They would also replace one of the original fine guidance sensors with the FGS that had been removed and returned to the ground during SM2, STS-82. It had been refurbished and adapted to correct for spherical aberration before being returned to orbit on this mission.

All went well with both tasks. Nicollier encountered some difficulty taking the old FGS out and inserting the refurbished replacement unit. Aligning the bulky FGS's with the guide rails on which they had to slide proved difficult. The body of each FGS blocked Claude's vision, so it was difficult for him to see what he was doing. In the end, however, he was successful in getting the job done (Figure 14.7).

EVA 2 was, once again, a very long spacewalk—8 hr, 10 min. It seemed at the end of the day that Nicollier and Foale were reluctant to come back inside Discovery— especially Claude. This would be his only experience working freely outside in the vacuum and darkness of space, the beautiful Earth drifting beneath him (Figure 14.8). It was a pinnacle of his career. It appeared he just wanted to savor the moment a little bit longer. He took his time finishing up and going in to dinner.

At the end of EVA 2, with the successful completion of the aliveness and functional tests of the new computer and the refurbished FGS, we could declare the mission fully successful. All the criteria had been met.

EVA 3 the next day, December 24, was a bonus. Grunsfeld and Smith replaced a failed S-Band radio transmitter with a new one, installed a new solid state data recorder to replace a ten year old reel-to-reel tape deck, and ended the EVA by replacing damaged insulation on some of the equipment bays with New Outer Blanket Layer (NOBL) insulation panels. This EVA went smoothly, but was still quite lengthy—8 hr, 8 min.

Figure 14.7. ESA astronaut Claude Nicollier installing a refurbished fine guidance sensor during EVA 2 on SM3a. English astronaut Mike Foale is seen to the left. (Credit: NASA.)

Figure 14.8. The Sun glinting off the surface of the Hubble Space Telescope with the crescent Earth in the background. It is possibly the indescribable beauty of scenes such as this that make EVA astronauts reluctant go back into the shuttle at the end of their shifts outside. (Credit: NASA.)

It was Christmas Eve. Nat King Cole singing, "I'll Be Home For Christmas," kept looping through my brain. For now, home had to be in Building 30 at JSC. When I was a very small boy I was in a production being put on at the local high school in my small hometown of Salem, Illinois. It was Dickens' "A Christmas Carol." I played Tiny Tim. I had only one line to speak, "God bless us every one." On this night, 1999 December 24, I said it to myself with not a little emotion—"God bless us every one."

The next day, Christmas Day, Hubble was unberthed. It was on the remote arm, hoisted several feet above the Flight Support System, and was being oriented to its release position. The aperture door had not yet been opened. As always it was a spectacular vision projected against the blue Earth, almost ready to go back to work exploring the cosmos.

A contingent of Hubble team members working at JSC during the mission, organized by Administrative Assistants Mindy Deyarmin and Coleen Ponton, did a wonderful deed for all of us, boosting everyone's morale on that Christmas Day when we were so far from home and families. They arranged to have a large quantity of food delivered, sufficient for a complete Christmas dinner for all the workers on both shifts—the orbit (EVA) shift and the planning shift. The food arrived on the afternoon of December 24, all of it uncooked and most of it frozen. Fortunately, some of the Hubble people were staying at the Candlewood Suites or at the Residence Inn, where they had full kitchens with ovens adjacent to their bedrooms. Everyone else had a microwave oven in his or her hotel room. Mindy, Coleen, and their squad collected room keys from 12 members of the Hubble team, and used all the available ovens and microwaves to cook turkeys, hams, and side dishes. The men who had donated their room keys requested that the women who were cooking in their rooms not answer the telephone if it should happen to ring. If their wives were calling they were worried about what reaction it would provoke back home if a strange woman answered the phone in Houston. About ten volunteers worked 16–18 hr that Christmas Eve and Christmas Day in order to have food for both shifts to eat. Everyone was treated to a fine holiday meal.

There were over a hundred professionals from Goddard, NASA Headquarters, the Space Telescope Science Institute and numerous contractors working on SM3a at JSC. Every one of them made significant personal sacrifices to be in Houston over the holidays. I overheard one of the women in our group comment that she had been doing just fine until her husband told her on the telephone that the kids had started opening their presents. Hearing that, she almost broke down.

Every one of those people, from secretaries and administrative assistants to the highest-level engineer, scientist or manager, were passionate about and totally dedicated to the Hubble Space Telescope. That's why they were there. But we were all grateful to Mindy's volunteers for picking up our spirits, and filling our bellies at a time of year when it was painful to be away from home.

Hubble was deployed that evening a little before 5:00 pm. I went back to my hotel. At about 7:15 pm Hubble was supposed to pass overhead once again over Houston. I went out behind the hotel into the dark soccer field of a neighboring high school where the glare from the streetlamps was a lot dimmer. I looked up, expecting to see a single, unresolved point of light go over—Hubble and the Shuttle still close together. I was not prepared for what I actually saw—two beautifully separated bright points of light, perhaps a third of a degree apart, one trailing the other in tandem across the sky. Shuttle Discovery, and the Hubble Space Telescope, now deployed, drifting apart from each other, each a separate free-flying satellite once again. With averted vision I was able to see what looked like a comet tail moving away from the leading point of light. I took that to be the wastewater dump that had just taken place from Discovery a few minutes before.

As I walked back to the hotel, John Campbell and Cepi drove up. They asked if I wanted a ride to dinner. "Sure, thanks," I replied. About 16 of us Hubble Huggers met at the Steak and Ale. Campbell was able to have his first beer in several weeks. It wasn't his favorite kind, but it did the job. We toasted the best Christmas present of all—a working Hubble Space Telescope.

AAS | IOP Astronomy

Life With Hubble
An insider's view of the world's most famous telescope
David S Leckrone

Chapter 15

Dangerous Liaison: Servicing Mission 3b

It had been five years since new scientific instruments—the essential tools astronomers use to study the cosmos—had been installed in Hubble. The spacecraft had slots for six instruments, including the Fine Guidance Sensor used for astrometry. All six slots were filled at the time of Hubble's deployment in 1990. During the First Servicing Mission in 1993 one instrument had been removed to clear space for COSTAR. COSTAR was an ancillary device designed to correct for spherical aberration in three of the instruments, and not a scientific instrument in its own right. By the late 1990s the Faint Object Camera had become obsolete and was in little demand. It was decommissioned. The FOC was the last instrument to make use of COSTAR.

Both of the new instruments installed during SM2 in 1997 had been immensely productive, but both had run into technical problems. NICMOS was dormant; its solid nitrogen coolant was prematurely depleted in January 1999. At that point three of the observatory's six instrument slots were no longer in active service.

In May 2001 STIS suffered a short circuit in its Side 1 power supply resulting in a blown fuse and loss of functionality in the Side 1 electronics. The instrument had been built including a redundant set of electronics—Side 2. The Hubble Operations Team commanded STIS to switch to Side 2, and it continued to produce outstanding scientific data as normal. However, its redundancy was gone. Any future failure of the electronics on Side 2 would result in complete loss of the instrument.

Having been installed in 1993, WFPC2 was the oldest active instrument on Hubble. It was beginning to show its age. Its CCD detectors were continuously bombarded by the cosmic rays they encountered in orbit day in and day out. The radiation exposure was taking its toll.

Servicing Mission 3a had been purely an engineering campaign, an urgent mission called up on short notice to replace Hubble's failing gyros. Now, as we looked ahead to 2002 and the launch of SM3b, there was an urgent need to restore Hubble to scientific health. Only two primary instruments remained available to astronomers—STIS and

WFPC2. Both were very powerful and, together, satisfied the demands of most research programs. But both were also battle scarred and vulnerable.

Our hopes for the observatory's scientific future rested on the promise of a new advanced-technology camera, the Advanced Camera for Surveys (ACS), selected by NASA for development following an Announcement of Opportunity in 1994. The Principal Investigator, Holland Ford of Johns Hopkins University, was a seasoned Hubble veteran, having been the Deputy P.I. on the Faint Object Spectrograph and the Instrument Scientist assigned by NASA to watch over the development of COSTAR.

The most experienced builder of Hubble instruments, Ball Aerospace, had won the job of designing and building the ACS. Ball had also produced the first-generation Goddard High Resolution Spectrograph and the second-generation instruments, STIS and NICMOS, as well as COSTAR. Holland asked the Hubble Project for assistance in managing the Ball effort and Cepi was happy to oblige by assigning a first-rate young Goddard engineer, Carolyn Krebs, to the job. So, taken all together, this was an excellent scientific and technical team. We had high confidence in the instrument they were building to fly on SM3b.

The ACS was substantially more capable then WFPC2. The WFPC2 in its wide-field mode was a 2 megapixel CCD camera. The Wide Field Camera in ACS had two large CCDs butted together to create a 16 megapixel camera. Each pixel in the big ACS camera sampled an area on the sky only one quarter the size of the area covered by a WFPC2 pixel in its three wide field CCDs. So the ACS had approximately twice the angular resolution of WFPC2 and could bring out finer detail in its images of planets, nebulae, and galaxies. The field of view of ACS was nearly a factor of two larger than that of WFPC2. The sensitivity of the ACS Wide Field Camera, summed up over all visible wavelengths of light, exceeded that of WFPC2 by about a factor of five. It could detect five photons of visible light in the same amount of time that WFPC2 could only detect one (Figure 15.1).

No wonder Ford's science team named its proposed instrument the Advanced Camera for Surveys. A deep core sample through the universe looking far back in time, the original Hubble Deep Field, required about ten days with WFPC2. The same observation could be completed in less than three days with the ACS and at the same time cover a larger area on the sky with finer angular resolution. Of course, with longer observing campaigns ACS could (and ultimately did) go much deeper than WFPC2, probing closer to the time of the Big Bang in the Ultra-Deep Field (see Chapter 11).

WFPC2 did retain some unique capabilities relative to ACS. The filter sets of the two cameras were somewhat different, for example. Also, WFPC2 would provide a back up to ACS, should the latter encounter technical problems. But it seemed likely that ACS was destined to become Hubble's workhorse camera.

We also had high hopes for the new NICMOS cooling system (NCS). It had been successfully tested on the HOST mission, part of STS-95 in 1998. During SM3a in 1999, the external valves and ports on the NICMOS instrument had been opened to expose its internal plumbing to the vacuum of space, making it ready to interface with the NCS during SM3b in 2002. The cryocooler unit was to be installed inside

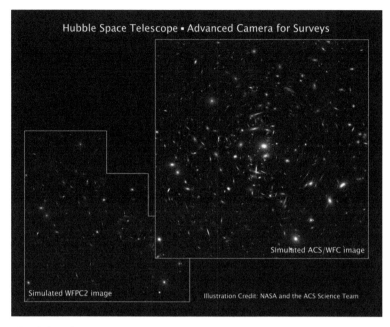

Figure 15.1. Simulation of a WFPC2 and an ACS image of the same field in the sky, a rich cluster of galaxies observed for the same length of time, through the same set of filters. Note the ACS image is larger, sharper, and deeper than that of WFPC2. This simulation was created just before SM3b was launched to demonstrate the improvements that would come from the new ACS. (Credit: NASA, the ACS Team, STScI 2002-06.)

the spacecraft's aft shroud and hooked up directly to NICMOS. It would also be connected to the large NCS radiator to be mounted outside the spacecraft. This attempt to resuscitate NICMOS with a mechanical cooling system was advertised as an "experiment" with no guarantee that it would actually be able to bring NICMOS back into full operation.

Adding ACS to Hubble's arsenal of scientific instruments and restoring NICMOS to operating status during SM3b would go a long way toward securing the outlook for the observatory's future with fine imaging capability spanning from the ultraviolet into the near-infrared.

The spacecraft itself was due for a major upgrade in the form of new and more capable solar arrays. We were grateful to the European Space Agency for providing the two sets of arrays—the original ones that were attached to Hubble when it was launched in 1990, and an improved set that was installed during SM1 in 1993. But they had lost efficiency over the years. Now it was time to move on, to assure that we had an adequate supply of electrical power to meet the future needs of our new scientific instruments and of the spacecraft as a whole.

Unlike the ESA solar arrays, which were flexible and rolled up like window shades, the new arrays were flat, rigid, and folded on hinges like a book. Although the new arrays were one third smaller in area than the old ESA arrays, they would provide Hubble with about 20% more power. That's because the new solar cells were made from the compound gallium arsenide (GaAs), which was more efficient in converting sunlight to electric current than the silicon cells that had been used for the earlier Hubble arrays.

The extra power would be sufficient to allow all of our scientific instruments to remain powered on at all times either in "hold" or in "operate" mode, so that they each could be brought up to full operation relatively quickly. This would give the astronomers planning the observations a lot of flexibility and allow Hubble's precious observing time to be used more efficiently.

The new arrays would not flex or "twang" as the old arrays did when Hubble moved into and out of orbital night. They would not impart any significant vibrations to the spacecraft that would cause it to jitter as it pointed toward astronomical targets. Being smaller, the new arrays would encounter less atmospheric drag, reducing the rate at which Hubble's orbit decayed.

Cepi and his team saved the Project a substantial amount of money in purchasing the new solar arrays. Iridium Communications was in the business of building and deploying a constellation of 66 active satellites, plus some spares, in low Earth orbit to enable a worldwide system of satellite phone and pager coverage. The company needed a large number of solar arrays to power all of those satellites. The Hubble Program bought solar array panels straight off the iridium production lines, and realized the economy of scale that came from iridium's purchasing in large quantities.

At Goddard, technicians in our shops built rigid support frames on which the new solar panels, eight of them in all, would be mounted. The support frames were made of an alloy of aluminum and lithium. They were lighter and stronger than the aluminum structures commonly used in spacecraft construction. In orbit the EVA astronauts would be able to fold up the two solar array wings and move them out of the way, making them easier and less dangerous to work around than the big old flexible arrays. There was everything to like about the new arrays—more power, less jitter, less drag, and safer to work around during spacewalks.

The scary 900 pound gorilla on the SM3b manifest was the power control unit or PCU. As designed, it was one of the simplest boxes on the spacecraft—mostly packed with relays (electrically operated switches) and cables. I thought of it in analogy to the circuit breaker box in the basement of my house. The main source of electricity coming into the house entered by way of that box. The box then delivered electricity to a large number of separate circuits that wended throughout my house, delivering electricity to the outlets in every room. Similarly the PCU received the electric current put out by the arrays of solar cells as they were exposed to sunlight. It routed the incoming electricity both to power all the systems and instruments on Hubble and to charge the spacecraft's six nickel–hydrogen (NiH_2) batteries during orbital daytime. During orbital night every system and instrument received its power from the re-charged batteries, via the PCU.

The Project began pondering the need to replace the PCU in 2000, shortly after the conclusion of SM3a. The original PCU had received heavy usage since it was launched in 1990. It was showing signs of wear. Several of the relays had failed, cutting off the flow of electricity from some of the solar array sections. Additional relay failures in the future would potentially result in insufficient power for full-up operations of the scientific instruments. Working around the failed relays was possible, but would require the rapid development of new hardware.

The big driver of the decision to replace the PCU, however, was an intermittent electrical fault that apparently resulted from a loose joint in a bus bar that carried current to various locations within the box. The fault had occurred twice since Hubble's original deployment—once in 1993 and once in 1999. The loose joint resisted the flow of electricity. Both times the problem had been relatively small in magnitude. However, our electrical engineers projected that, if it increased by about a factor of three or more, Hubble would not be able to maintain normal science operations. It would produce an imbalance in the amount of current going to charge each battery, threatening that some of the batteries might overheat. That could potentially be a dangerous situation.

The PCU was not designed to be an in-orbit replaceable unit. How could such a simple device ever fail, after all? It was rather large (4 ft × 2 ft × 1 ft) and bulky. To remove and replace it would require astronauts to disconnect and reconnect 36 cables, which protruded from the side of the box (Figure 15.2). The connectors for these cables were crowded together in a restricted space and would be difficult to see through the fish-bowl helmet that came with a spacesuit. They would also be difficult to reach and unfasten while wearing those bulky spacesuit gloves. These issues were not showstoppers, however. Cepi's engineers had become masters of creating specialized tools to allow the astronauts to perform almost any task, no matter how awkward. The EVA astronauts had demonstrated on earlier Hubble servicing missions that they could handle such difficult challenges.

The bigger problem, the kind of problem that caused us to lose sleep and have a perpetually dry mouth, was the fact that, in order to replace the PCU, the entire Hubble spacecraft had to be turned off. No electricity could be allowed to be present anywhere for a period of up to eight hours or so.

Hubble had been powered up prior to its 1990 deployment while still in the womb of Endeavour's payload bay. The crew on STS-31 oriented the shuttle so that Hubble's unfurled solar arrays faced the sun, sending a life-giving flow of electrons through the PCU and into Hubble's batteries. The Operations Team in the STOCC in 1990 had systematically turned on all of the electrical components in the spacecraft, following a carefully planned script, until the observatory was completely operational and ready to be sent along on its journey to investigate how the Universe works. On subsequent servicing missions new systems and instruments installed by EVA astronauts started out with no power. But they were powered up and tested by the STOCC operations team soon after their installation. Never since its launch had Hubble's electricity been completely turned off.

Common wisdom held that, once a spacecraft had been turned on in orbit, no one in their right mind would turn it completely off again. It just wasn't done. It was presenting a really serious temptation to fate. The threats were numerous. Once a box was powered down, there was no guarantee that it would turn back on again on command. Power surges during turn on could cause damage to basic electronic parts. While a spacecraft had no power, its active thermal control systems—heaters and such—would be unavailable to keep all systems warm. Components that got too

Figure 15.2. Replacement power control unit (PCU) installed in the VEST test facility in the large clean room at Goddard. Note the long column of cable connectors on the left side of the box. Disconnecting and reconnecting these 36 connectors while wearing spacesuit gloves and a large helmet was a difficult challenge for the astronauts. Specialized tools were designed to help them. (Credit: NASA, GSFC.)

cold could break. Different components would grow cold at different rates. The internal parts within a box would shrink and expand at different rates as the temperature changed, causing mechanical stress and possible breakage. It took sophisticated thermal models running on powerful computers to calculate those rates and to predict which components were most susceptible to failure. Also, with the power off, there was no telemetry. Our engineers would be totally in the dark, not knowing how cold things were getting in the spacecraft and its myriad boxes and instruments. It would be like performing heart surgery without being able to monitor the patient's vital signs.

As the Project grappled with the difficult decision of whether or not to remove and replace the PCU, I openly expressed my own reservations. The electrical anomaly at the bus bar joint had only occurred twice in six years. Each time its magnitude was small and did not directly threaten any of the spacecraft systems. It was only the worry that the problem might return at a much higher level sometime in the future that was causing us to consider replacing the PCU. But in replacing the PCU we might well cause irreparable damage to other parts of the spacecraft, especially if thermal control got…well…out of control.

On the other hand, like everyone else, I wanted Hubble to have a long and happy life to 2010 and beyond. We needed a healthy PCU for that to be the case. In the end I voted to go ahead with the change out. But it was a close decision.

Of course, we on the Project were not the only ones with significant worries about the riskiest servicing task ever planned for Hubble. The NASA Administrator was as worried by all of this as we were. Astronauts John Grunsfeld and Rick Linnehan would be doing the spacewalk to change out the PCU during SM3b. Grunsfeld described a meeting he had, together with Ed Weiler, in Dan Goldin's office (Grunsfeld 2014):

"During the lead-up to servicing mission 3b, one of my more personal moments with NASA Administrator Dan Goldin occurred when he called the space-walking team up to his office in Washington. I thought, 'Well, this is a reasonable thing. He wants to have eye-to-eye contact with the team that is going to change out this PCU.' The administrator had also had a discussion with Ed Weiler, another Hubble Hugger, who had to make the call as to whether we were going to change out this box. The PCU had a grounding strap that was connected to a bus bar with a bolt, washers and a nut. Something to do with the washers and bolt had been creating some electrical resistance—measured in just tens of milliohms—between the cable and the bus bar. This was tiny resistance, but over a number of years it eventually would [might, if it had grown to a higher value,] have caused the observatory to fail. However, if we tried to replace it and failed, we would lose the observatory right away. So, there was a fair amount of pressure in making that kind of decision. Ed Weiler went to the administrator and suggested that we change it out to give Hubble the opportunity for a successful, longer scientific lifetime. The administrator went with that but with the caveat that he wanted to look us in the eye first. He said, and I quote, 'I don't mean to put a turd in your punchbowl, but for the sake of the agency, you have to fix this.'"

No pressure, right? As though Grunsfeld, Linnehan and all the rest of us Hubble Huggers would not have been feeling any pressure "to successfully fix the world's most fantastic scientific instrument," and not to break it.

And so, the PCU change out was scheduled for EVA 3 on SM3b. The first two EVAs would be dedicated to installing the new solar arrays. These arrays were designed so that they could be "dead faced," that is electrically disconnected from the spacecraft circuits without having to be physically de-mated, unlike the ESA arrays they were replacing. That would allow the PCU change out to be completed in less time. The PCU work would take a long time—about 8.5 hours in all. It could not be fit totally into one EVA. There were, however, some short segments during EVA 1 and EVA 2—20 min here, 30 min there—when some preparatory work could be fit in to get a head start on the PCU change out. Similarly, there was time on EVA 4 to tidy up some PCU odds and ends left over from the prior day. In this way the main body of work on the PCU change out could be fit into one 6.5 hr spacewalk.

As always, putting together the EVA schedule was a challenge, but it was more so than usual for SM3b. When a list of EVA tasks was made up of a large number of relatively short jobs, there was a lot of flexibility in how the EVA planners could sequence those jobs. However, in this case five lengthy jobs needed to be completed— installation of two new solar array wings, the PCU replacement, the Advanced Camera for Surveys, and the NICMOS Cooling System. Each of these required approximately one EVA day. In fact the only order that allowed the work to be finished in just five EVAs was: first solar array wing, second solar array wing, PCU, ACS, NICMOS cooling system. That was how it transpired that the two major science upgrades on SM3b had to wait until the last two space-walking days. If needed, a sixth, unsubscribed EVA opportunity was held in reserve by the Flight Director for use only if some of the tasks on the first five spacewalks had not been completed or if other problems had arisen that required astronaut intervention.

The servicing team was very resourceful in creating ways to reduce the risk of removing and replacing the PCU. The planned timeline for EVA 3 included two breakout points. When the spacewalkers reached either of those points the Flight Director in Houston would assess whether or not they were seriously behind schedule. If they were, they would need to break out of the task and perform contingency procedures to ensure that Hubble would stay warm enough to safely make it through the overnight period and to the next EVA. To be safe overnight, the crew had to make sure at least nine specific cables (out of the 36) were attached to the PCU. For Hubble to be safe to operate autonomously in orbit, should the Shuttle have an emergency that forced it to return to the ground on short notice, a specific set of 23 cables had to be attached to the PCU.

Manually turning off every piece of equipment on the spacecraft safely would be a prohibitively long process. Some boxes would grow cold unnecessarily as other boxes were individually powered down by commands sent from the ground. There was also the risk of getting something wrong during such a lengthy and complex procedure.

This led spacecraft operations engineer Mike Wenz and his colleagues to create a "Superproc," a big package of pre-programmed computer subroutines that would implement the detailed power-down procedures moment-by-moment, step-by-step, component-by-component. In other words they created an autopilot that would control the entire process with precise timing and without much in the way of human intervention to get the observatory powered down as safely and as quickly as possible. The beauty of the Superproc was that it could be fully tested well in advance of the mission. If bugs were found, there was time to correct them. And there was time to refine and optimize this new software.

The physically demanding process of EVA astronauts, wearing bulky spacesuits, trying to disconnect and reconnect 36 difficult-to-reach cable connectors was especially concerning. Cepi's team had to design, build and refine a set of tools that would simplify the job as much as possible. And then the crew had to practice, practice, and practice some more. The Goddard team built a full-scale mock up of the PCU that provided a precise representation of the location, feel, and tightness of the 36 cables and connectors. The astronauts had this transported to the large

Shuttle Simulator Building at JSC. Every night before going home from work, John Grunsfeld and Rick Linnehan would use the PCU mock up and the specially designed tools to go through the complete process of removing and reattaching the 36 cables. Again and again they practiced this until they each had completed the task hundreds of times. They developed terrific muscle memory.

The Hubble spacecraft had no on-board propulsion—no rockets, no gas jets, no propellant. It could not be commanded to move from place to place. It could only rotate about its center of gravity, pointing at targets distributed all over the celestial sphere. To drive its rotation, the spacecraft used reaction wheels packaged in reaction wheel assemblies (RWAs). Four RWAs ran in normal operations, but the Pointing Control System (PCS) required a minimum of three to properly control the rotation of the spacecraft about three axes. These were relatively large and heavy flywheels—2 feet in diameter, 100 lbs in mass—spun up by electric motors to as much as 3000 rpm. Operating in concert, the angular momentum of each individual wheel contributed to a net angular momentum for the whole spacecraft, causing it to rotate in a particular direction with a particular speed. Four metal rods on the exterior of Hubble, each 8 feet long, and wound with electrical wire, operated as electromagnets, pushing against the Earth's magnetic field to slow the spin of the reaction wheels and the rotation of the spacecraft. This was called "dumping momentum" in spacecraft parlance.

On 2001 November 10, less than four months before the scheduled launch of STS-109, one of Hubble's four reaction wheels stopped rotating for a time. The STOCC Operations Team was able to re-start the wheel. But that anomaly gave the Project pause. If that RWA failed, Hubble would be left with three working RWAs —enough to continue normal operations, but with no redundancy. As a precaution, NASA decided to delay SM3b to evaluate the situation. An RWA installation would take about one hour of EVA time. There was time to do it near the end of EVA 2. We had a spare RWA at Goddard. Cepi's team prepared it for flight, tested it, installed it into a protective enclosure and shipped it to the Cape. There it was integrated with the rest of the payload onto a carrier at the rear of the payload bay.

It was cold at Cape Canaveral, at least by Florida standards—low temperatures in the low-to-mid-30s Fahrenheit. The launch of Columbia on STS-109 had to occur around sunrise—the coldest time of the morning—to properly phase it to rendezvous with Hubble two days later. NASA and the nation had paid a dear price—the loss of Challenger and its crew in January 1986—the first time they had attempted to launch a shuttle mission under very cold conditions. The Shuttle Program was now very conservative about cutting things too close with regard to launch conditions. And so, they postponed the start of SM3b from Wednesday, 2002 February 27, to the following Friday, March 1. The forecast for the ambient temperature at liftoff at 6:22:02 am on Friday was 55°F—much better. Launch occurred right on time at the opening of the launch window.

By this time I had been on the scene at KSC to witness six Hubble-related shuttle launches. Each was uniquely beautiful—a work of art painted by humans onto a canvas provided by nature. In this case we were looking east in the early dawn light from the Media Center, across the Turning Pond, toward Pad 39A. Shortly after

Figure 15.3. Columbia rises through a thin layer of clouds at dawn, carrying STS-109, SM3b toward orbit. (Credit: NASA.)

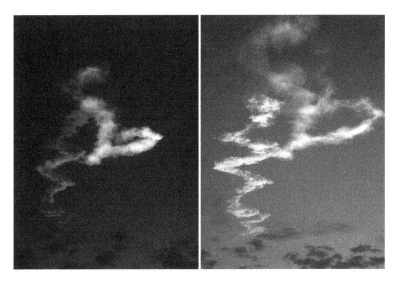

Figure 15.4. Exhaust plume left behind by the launch of STS-109 dissipates as the Sun comes up. (Credit: the author.)

Columbia lifted off, it entered a thin bank of clouds (Figure 15.3), and penetrated through them shortly thereafter. The sun, lurking just below the horizon, illuminated the trail of rocket exhaust left behind from the launch. Colors varied along the plume, depending on its altitude as it was exposed to the rising sun (Figure 15.4).

Shortly after launch we had to hustle to Houston to pick up the mission timeline in Building 30 at JSC. We had a new Program Manager leading the Goddard team, Preston Burch. Preston was a large man, outgoing, and affable, with a hearty laugh, a large inventory of funny quips, and the best long-term memory I had ever encountered in another human being. He was also a highly experienced engineer. Preston had previously served as the Hubble Project Manager for Flight Operations and Ground Systems. He was selected to take over the entire Hubble Program replacing John Campbell in 2001. John had been promoted to the position of Director of Flight Projects at Goddard in 2000.

The Program office at Goddard had arranged for a number of us to fly from Kennedy directly to Ellington Air Force Base near JSC on one of NASA's Agency aircraft, a Gulfstream II. Preston was on board, as well as Chris Scolese, Ed Weiler's Deputy for Engineering, Keith Kalinowsky, our lead Systems Engineer, Frank Cepollina, myself, and a few others. The flight was smooth and uneventful until the plane was on final approach to land at Ellington. We were passing through a low bank of clouds. As the Gulfstream dropped out of the clouds it was immediately evident that we were not lined up with the runway. It was off to our right some distance. I assumed the pilot would go around again to get lined up properly for a landing. Suddenly the plane banked hard to the right, followed by a very tight bank to the left. "This pilot must think he's flying an F-16," I thought.

I looked out my window and saw the ground, seemingly only a short distance from the Gulfstream's wing tip. At that moment I was positive we were going to die that day. I tried to avoid fear, but thought to myself, "so this is how it's going to end." The plane completed the sharp second leg of this S-shaped maneuver and then almost immediately rolled to horizontal and touched down only a few seconds later. There were a lot of pale faces and sweaty palms in that cabin. Normally, after these flights the pilot and first officer leave the cockpit and mingle a bit with the passengers. This time they did not emerge. The cockpit door remained closed. We disembarked and proceeded on to pick up rental cars and check in at our hotel. At that time, Keith Kalinowsky's wife was a high-ranking executive at the FAA. I heard a rumor that the incident had been reported through her, but am not certain that anything came of it. I was just glad to be alive.

Preston and I arrived in the Customer Service Room in Building 30 at about the same time that night. Having just settled into his station, Preston was approached by the JSC Mission Manager, Phil Englehauf. It was Phil's job on STS-109 to serve as the main point of contact between the Hubble Management Team and Shuttle Program management. He asked Preston if they could have a side conversation about the mission. Having gotten a brief impression of what the conversation was going to be about, Preston asked if I could join them. "No problem," Phil replied.

"I wanted to give you guys a heads up that we've had a problem come up with the orbiter and there's a chance we may have to terminate the mission and bring it back home." Phil drew a sketch on a yellow pad to illustrate the problem. "There are two Freon cooling loops in the payload bay, one on the port side and one on the starboard side. These keep the shuttle's electronics cool. Shortly after liftoff our guys noticed that the flow rate of the Freon through the port side cooling system was low.

It's not clear if that is a problem with the sensors that measure the flow rate, or an actual reduced rate of flow. The port side cooling loop performance is within what the flight rules require, but just barely. The starboard side cooling system is working perfectly. But the question is, could the port side provide an adequate back up, in case a problem came up with the starboard side system? We need at least one working system. We need that back up. Our guys are going to sharpen their pencils and see if they can find some extra margin for the port side system. There will be a meeting of the shuttle managers this evening (Friday) to determine what to do. Stand by."

I felt some of the wind go out of my sails. We really needed this mission to be successful. The Hubble team had spent many years preparing for it. The ACS and NICMOS were critically important to Hubble's scientific future. I tried to bury my worries in a side compartment within my brain, and get on with the work at hand.

At the conclusion of their evening meeting, the shuttle managers gave the STS-109 crew the go ahead to proceed with their previously planned tasks. They were to fine tune their orbit and prepare for the rendezvous and docking with Hubble around 3:30 am Sunday morning. The EVA crewmembers were to check out their spacesuits, tools, and the airlock to get ready for their spacewalks, with the first scheduled to begin early Monday morning.

The shuttle managers left their meeting expressing confidence that the mission would continue to its planned conclusion. Nevertheless, they agreed to meet again the following day (Saturday) to review the situation. At that meeting they concluded that the faulty cooling system, though not working at full capacity, did provide adequate cooling for shuttle equipment and did not threaten mission safety or operational capabilities. SM3a should proceed, following the original plan—a great relief to all of us Hubble Huggers there and in the STOCC at Goddard.

Mission Commander Scott ("Scooter") Altman was a distinguished naval aviator and test pilot who had flown on two prior shuttle missions. He specialized in flying F-14 Tomcats. What made Scooter's naval career especially interesting to those of us who were meeting him for the first time was that he had played a role in the 1986 Tom Cruise movie, "Top Gun." He stood in for Tom Cruise flying stunts in his F-14. In the film it was Scooter who buzzed the tower and Scooter whose middle finger was visible as he flew upside down over the enemy Mig pilot. He was a graduate of the University of Illinois and received an MS degree in aeronautical engineering from the Naval Postgraduate School.

Pilot Duane ("Digger") Carey was a rookie, flying his first shuttle mission on STS-109. Just before coming to NASA for astronaut training, he had served as a test pilot, flying F-16's at Edwards Air Force Base. Digger had an MS degree in Aerospace Engineering from the University of Minnesota.

Altman and Carey maneuvered Columbia to the vicinity of Hubble with several burns of the orbiter's engines during the first two days of the mission. One of these was the longest rendezvous engine firing in the history of the shuttle program up to that time—three and a half minutes with the orbiter's two large orbital engines—to slow Columbia's rate of approach to Hubble. Altman and Carey took over manual control at about 2:30 am Sunday with the orbiter about one half mile below the observatory. They guided the shuttle gently to within 35 feet of Hubble, close enough to allow Flight Engineer Nancy Currie to seize hold of its grappling fixture with Columbia's Remote Manipulator System (RMS) arm. Hubble was captured at 3:31 am Central Standard Time (Houston time) on Sunday, 2002 March 3.

About half past midnight CST on Monday, March 4, astronauts John Grunsfeld and Rick Linnehan emerged from Columbia's airlock to begin EVA 1. Their job was to remove the old ESA solar array wing from the starboard (−V2) side of Hubble and replace it with a new rigid array wing and its associated electronics.

This mission was Grunsfeld's second tour of duty servicing Hubble. He had also performed two spacewalks on STS-103, SM3a in 1999. John was an astrophysicist by profession, specializing in high-energy phenomena observed in the universe in X-rays and gamma-rays. He received his PhD in Physics from the University of Chicago in 1988. Prior to servicing Hubble, John had flown on two other shuttle missions, one devoted to ultraviolet astronomy, and one that entailed a five day stay docked to the Russian Mir space station.

Rick Linnehan was, perhaps surprisingly, a veterinarian. He specialized in large animal medicine, including zoo animals. He received his Doctor of Veterinary Medicine degree from Ohio State in 1985 and spent two years on an internship in exotic animal medicine at the Baltimore Zoo. From 1989 to 1992 Rick was chief clinical veterinarian for the US Navy's Marine Mammal Program, serving in California, Florida, and Hawaii. He worked with whales, seals, sea lions, and walruses. Prior to STS-109, Rick had flown on two shuttle missions. Both of these entailed extensive life sciences research in Spacelab modules in the shuttle payload bay.

In the 2013 film, "Gravity," Sandra Bullock plays Dr Ryan Stone, a medical doctor, chosen as a Mission Specialist to service the Hubble Space Telescope on an imaginary future mission, STS-157. (The screenplay writer destroys Hubble with space debris early in the film. I had never considered that the Hubble observatory would meet such an end!) Popular science communicator and astrophysicist Neil deGrasse Tyson wrote, in numerous Twitter messages, a widely read critique of the film. Many of his comments were spot on. But I took issue with one: "Mysteries of #Gravity: Why Bullock, a medical Doctor, is servicing the Hubble Space Telescope." I sent a reply to Tyson, noting that Rick Linnehan, a large animal veterinarian had once serviced Hubble. There was a delightful irony in the idea that a large animal veterinarian had been called upon to perform surgery on the largest telescope ever flown in space.

The third critical participant in all the EVAs, though working from inside the crew cabin, was Nancy Currie. Nancy held a PhD in industrial engineering from the University of Houston. She was a Colonel in the US Army and an instructor pilot at the U.S. Army Aviation School, logging over 3900 hr in rotary-wing and fixed-wing aircraft. In three shuttle missions prior to STS-109 she established a reputation as a highly skilled operator of the shuttle's robotic arm. And that was her role in SM3b, to move the EVA astronauts attached to the RMS to their various worksites.

Except for a communications problem with Grunsfeld's space suit, EVA 1 transpired uneventfully. At the end of the day Hubble had half a set of new solar arrays (Figure 15.5).

Next up for EVA 2 were Jim Newman and Mike Massimino. Their job early that Tuesday morning was to replace the remaining ESA solar array, the one on the port (+V2) side of the spacecraft, with the second new solar array wing plus its electronics. They also had to swap out the reaction wheel assembly that had been so worrisome to the Hubble Project with a replacement unit.

Jim Newman was a physicist with a PhD from Rice University awarded in 1984. Newman had flown on three prior shuttle missions. The most recent of these was the first International Space Station assembly mission. On that mission Jim performed three spacewalks to hook up power and data cables between the Russian and American modules, among other tasks. He had also spacewalked once during his first shuttle flight, STS-51. So, today on EVA 2 of STS-109, Jim would be in Columbia's payload bay performing his fifth spacewalk.

In contrast STS-109 was Mike Massimino's first shuttle mission and today would be his first spacewalk. He was a pure rookie. His colleagues called him by the nickname, "Mass." This seemed appropriate because Mike was a large man. He had

Figure 15.5. Astronaut Rick Linnehan prepares to install a new rigid solar array panel on the starboard (−V2) side of the spacecraft. In this view, the array wing has been folded closed like a book. (Credit: NASA.)

a wonderful sense of humor. He was also highly experienced in the study of human–machine systems and remote human operator control of robotic systems in space. Mike earned his PhD in Mechanical Engineering from M.I.T. in 1992.

The EVA went smoothly with Nancy Currie once again transporting the two spacewalkers around the payload bay and to their worksites at the telescope. Jim Newman and Mike Massimino concluded the day with some preparations for the next day's EVA task, including installing a thermal blanket, doorstops and foot restraints to give Grunsfeld and Linnehan a head start on their PCU change out work.

Two relatively straightforward EVAs had been completed. Now it was time for sweaty palms. Late that night (Tuesday), the STOCC would command all power to Hubble to be shut off using the Superproc. We would all wait for many hours in the blind while the spacecraft and all its contents grew colder. It was indeed a really big deal—a dangerous liaison with fate, the future of the world's most famous telescope hanging in the balance.

Growing up in a small town in central Massachusetts, Christine Cottingham was in elementary school when space shuttles were just beginning to be launched. Her teachers would take the students out into the hall to watch the launches on TV. It was so exciting to Christine. She knew then that she wanted to work on the space program. She read every book she could find on space. That passion led her to a Bachelor's degree in Mechanical Engineering at Penn State and a Master's degree in Mechanical Engineering with a specialty in Thermal Fluids at Johns Hopkins. At about the time she was preparing to graduate from Penn State, she read an article in National Geographic about the first Hubble Deep Field. The sight of all those galaxies amazed her. She knew she was getting into the right job when she was hired by Lockheed Martin as a contractor to NASA to work on the Hubble Space Telescope Project at Goddard. "And to think," she mused, "In my new job I could be a part of something like that—producing the Hubble Deep Field! When a Hubble discovery or image appeared on the news, I could say to my family, with a lot of satisfaction, that I was part of that."

As a thermal engineer on Cepi's team, Christine worked on the NICMOS Cooling System from the beginning, analyzing and testing its cooling ability, participating in the HOST mission (STS-95 in 1998) and planning for the NCS installation on Hubble in SM3b. Her thermal analyses also informed the planning of EVA timelines. How long could a component like a Reaction Wheel Assembly be out of its protective enclosure before it started to get too cold? In other words how much time did the astronauts have to get the system installed in Hubble so that the STOCC could power it up? This was very important input to planning upcoming servicing missions. But it paled in comparison to what was looming as discussions of changing out the PCU revved up.

Work became really intense, as the Project asked for Christine and her colleagues to study the feasibility of doing the PCU change out. This was risky business. They had to get it right. How could we turn the power completely off on Hubble for a period of many hours without it getting to the thermal red line—to the point where components were so cold that they would break?

For over two years, Christine and her colleague, Melissa Fassold ran computer model after computer model, calculating cooling curves for every component and system on Hubble—graphs that showed how quickly the temperature would drop when no electrical power was present to keep the system warm. When it reached a temperature colder than it had ever been before, that established a yellow or caution limit. When it reached a temperature where physics and laboratory data said components would degrade or break, that established a red limit. Don't get to a red limit, ever!

The thermal team also came up with strategies to mitigate the risks as much as possible. What systems were slow to cool down? Those were likely to be the ones you would turn off first. What systems were at greatest risk? The WFPC2 radiator, for example, protruding outside the spacecraft, would start already at a cold temperature. It would reach its limits quickly. So, the team asked that an MLI insulation blanket be fabricated for the astronauts to put over the most critical part of the radiator—the central section that was connected to WFPC2's electronics—to slow down the rate at which they would grow cold. The Reaction Wheel Assemblies would cool down quickly. But when they were spinning at 3000 rpm, they put out a lot of heat. Running all four reaction wheels at full speed in advance of the EVA would not only warm them up before their power was cut, it would also warm several of the adjacent equipment bays in the spacecraft. Virtually every component in the spacecraft would be pre-heated to as high a temperature as it could safely accommodate, to lengthen its cool-down time as much as possible.

Christine was assigned to cover the EVA shifts in the STOCC throughout the mission, together with two other Hubble thermal engineers, Jorge Piquero and Teri Gregory. Jorge was the team lead and made any big decisions. Teri would be monitoring telemetry, as long as they had it. Christine, sitting at a console with a headset and microphone on, talked on the loops to communicate for the team. Her call sign was "Thermal."

When Christine sat down at her station on the evening of 2002 March 5, ready to monitor EVA 3, she took confidence in the certainty that the team had done everything they could think of to prepare for the seven or so anxious hours to come. Christine and Melissa (who was covering another shift that day) had modeled every part of the observatory's thermal properties exhaustively. They knew the systems inside and out, and how they should perform during the PCU change out. They had documented the thermal analyses and all the cool-down curves in a folder that they named, "The Thermal Bible." They had documented every contingency that they could think of—every problem, every anomaly that might come up—and had a plan for handling every one of those. They had practiced responding to contingencies in what seemed like an endless series of practice runs during local simulations (SIMs) at Goddard and Joint Integrated Simulations (JISs), full dress rehearsals managed by JSC. There was nothing more they could think of to do except monitor the air-to-ground communications loop as Grunsfeld and Linnehan proceeded with their spacewalk to replace the PCU.

Mike Wenz's Superproc, designed to power down the entire Hubble spacecraft safely and rapidly, was initiated at about 11:30 pm Goddard time or 10:30 pm in

Houston. The idea was to time the power-down so that it would just be completed and Hubble would have no internal electrical power, precisely at the moment the two astronauts arrived at the door to Equipment Bay 4, the bay housing the PCU. In that way minimal time would be lost and Hubble would have gotten no colder than necessary when the astronauts started disconnecting the old PCU.

The two EVA astronauts, assisted by Jim Newman and Mike Massimino, entered Columbia's airlock to complete the process of donning their spacesuits at about midnight Houston time. Previously, in the middeck area of Columbia's cabin, Grunsfeld and Linnehan had put on several flexible garments, including long johns and a Liquid Cooling and Ventilation Garment. The latter contained plastic tubing through which chilled water flowed to keep the astronaut's body at a healthy temperature. They climbed into the flexible bottom portion of the spacesuit, the Lower Torso Assembly, just prior to entering the airlock.

The upper part of the spacesuit, the Hard Upper Torso, including the arms, gloves, helmet, and a large backpack, called the Primary Life Support System, was latched onto an Airlock Adapter Plate rigidly mounted to the wall of the airlock. Massimino assisted Linnehan in climbing upward into his upper torso, while Newman gave Grunsfeld a hand. Having sealed and locked the lower and upper sections of the spacesuit together, Newman reached around Grunsfeld to unlatch him from the adapter plate. That's when he felt moisture on the suit, saturating the back. It appeared that there was a water leak somewhere in Grunsfeld's suit, probably in the plumbing that circulated water through the cooling undergarment. Newman told Grunsfeld that his suit was wet. Newman grabbed a towel and started mopping up water (Figure 15.6). Someone in the crew asked, "Are we going to be able to go out?" Newman shot back, "Not in this suit he's not!"

Grunsfeld had not felt the moisture. He didn't know his suit was wet. He didn't realize at that moment that he was potentially in danger. Fortunately for everyone

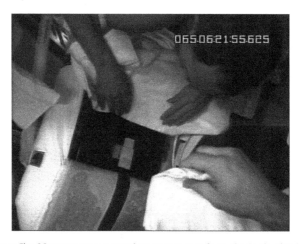

Figure 15.6. Astronaut Jim Newman uses a towel to mop water from the back of John Grunsfeld's leaky spacesuit. (Credit: NASA.)

Newman found the problem well before the airlock was de-pressurized and the outside door was opened to the vacuum of space.

As Grunsfeld told the story to a large and appreciative audience at Goddard years later: "If you have water in the back of the spacesuit and you go outside, as soon as you get to low pressure, that water is going to flash freeze. If there's any water in the electrical connecters and plumbing, [because] water expands a little bit as it freezes, it would have broken the spacesuit, and that would have been the end of me.... and then we wouldn't have finished the EVA"—the latter comment drawing laughter from the audience.

Of course John was speaking ironically, as well he could after the fact. But I couldn't help thinking that, with all our worries about the threat to Hubble's survival presented by turning the spacecraft's power off, we were blindsided by the possibility that a completely unrelated malfunction in a spacesuit might have led to a tragic end. In space danger can come out of nowhere.

The immediate response of Columbia's crew to this sudden contingency was to start mopping water with paper towels, to get Grunsfeld out of the wet upper torso of the spacesuit, and to begin the process of getting him outfitted with alternate attire. The shuttle did not carry one suit per crewmember. Instead the suits were designed to be re-sized in real time to fit more than one astronaut. In this case, the crew worked on Jim Newman's suit to adapt it to Grunsfeld's dimensions (Figure 15.7). This was a bit of a challenge. Newman was 6 feet tall; Grunsfeld was 5 feet 8 inches.

Re-sizing Newman's suit was going to take some time, perhaps an hour or more. And the clock in Mike Wenz's Superproc was ticking, methodically turning Hubble's electricity off. The telescope, the instruments, and the other spacecraft systems were getting colder. Normally, spacesuits were re-sized overnight—time pressure usually wasn't an issue. But now, the crew on Columbia had to work quickly.

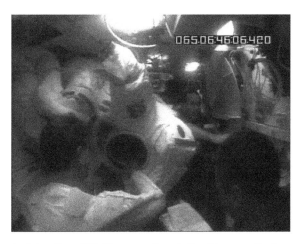

Figure 15.7. Astronauts John Grunsfeld (foreground) and Jim Newman (background) work to re-size Newman's spacesuit to fit Grunsfeld following the water leak in Grunsfeld's suit. (Credit: NASA.)

Back in the STOCC Christine Cottingham was taking all of this in while monitoring the air-to-ground loop. When she heard that there was water in the suit, she thought, "That's not good." But at first she didn't fully appreciate how bad the situation was. Then she started hearing conversations over various Hubble management loops. Al Vernacchio, the Mission Operations Manager (MOM) called out, "Stop the commanding. You need to figure out what we should do at this point."

Both in the STOCC and in Houston the Hubble Management Teams were beginning to brainstorm about aborting the PCU replacement today and doing something less time consuming instead. "Is it still possible to do the PCU change out today? Should we switch EVA plans, do the Advanced Camera installation now and do PCU tomorrow or Friday?"

Christine and her thermal engineer colleagues did not like that option. It would put the installation of the brand new ACS *before* the PCU job. The ACS would have to go through the trauma of being powered off and cooled down like the rest of the instruments and spacecraft systems when they tried to perform the PCU replacement later. She had run thermal models for the ACS but felt it was not a chance worth taking.

In the Customer Support Room at JSC, most of the rest of us didn't like that idea either. It would mean switching around the pairs of EVA astronauts, each of whom was highly trained to do the various installations. The crew was cross-trained to do each other's jobs but not nearly as much. There was a lot of conversation with the shuttle team about how long the crew was going to need to get Grunsfeld's new suit configured.

Back in the STOCC, Bill Crabb, "HST Systems" on the loop, took control (Figure 15.8). "We've got to turn Hubble back on. What do we need to turn back on

Figure 15.8. Two views from the Space Telescope Operations Control Center at Goddard during EVA 3, the change out of the power control unit (PCU). At the left is the "Thermal" team, led by Jorge Piquero (on the phone). To his right is thermal engineer Teri Gregory. Behind them are thermal engineer, Christine Cottingham and, to her right, Greg Johnson, a thermal analyst from Ball Aerospace. The picture on the right is of Bill Crabb, "HST Systems," who took charge of coordinating the overall STOCC response to the spacesuit water leak crisis and the subsequent improvised powering up and powering back down of the Hubble spacecraft. (Credit: NASA, GSFC.)

first?" That was a question directed squarely at "Thermal." They had worked through planning and training for many contingencies prior to the mission. But, "Water in the astronaut's suit," was not one of them. Their response had to be improvised, "seat-of-the-pants."

The Superproc was pretty much complete at this point. Almost everything had been powered down. All three thermal engineers pulled out their copies of "The Thermal Bible." They began going through all the cooling curves to see what systems had the most margin left before they reached their red limit. When had they been turned off? How fast had they been predicted to cool down? What temperature would you expect them to be at right now? Starting at that temperature, how much more time might they have before they got to the yellow limit and then to the red limit?

"Thermal" identified the half dozen or so systems that had the least temperature margin left, that were closest to the danger point. "Turn those on now in this order," Christine relayed to Bill Crabb. Bill immediately called up the engineers responsible for those specific systems. Those were the experts who knew how to run the command procedures to power up their systems. And they were locked, loaded, and ready to issue those commands. Had they not been well prepared, had they needed time to look up the commands and execute them, it might have been disastrous. But that was not the case. Each of them responded quickly. The STOCC was a well-oiled machine.

They turned on the Fine Guidance Electronics, the system that controls the Fine Guidance Sensors. They turned on electronics that controlled the pointing of the High Gain Antennas. They turned on some of the optical telescope's heaters. They turned on the computer that handled commands and data for the scientific instruments. They turned on WFPC2.

They had to put Hubble back into a safe and stable configuration until the spacesuit issue was resolved. It wasn't a certainty that the PCU change out would still be done during that EVA, but "Thermal" wanted to have Hubble in a state that would make it possible to continue, if that's what management elected to do. They didn't turn everything back on, just the most vulnerable systems. The question was, could today's big and risky job—the one hundreds of people had been dreading for a long time—be salvaged?

For a brief time during the emergency power-up, Christine and her colleagues had telemetry again for the systems that were back on. They could see how cold those systems actually had gotten. That gave them a new starting point to use on their cooling curves, a more accurate reading of where they stood, if the EVA continued. So far everything looked okay.

During this intense period, everyone in the STOCC, under the direction of "HST Systems," kept their cool and focused intently on what they had to do. Art Whipple, the Anomaly Response Manager, drew laughter and lightened the mood when he said on the loop, "This is like a really bad SIM!"

Christine was of two minds. On the one hand she kept the panic and paralysis that might have set in at bay. Higher brain functions took over. Her mind was clear. There was a moment when she had inhaled deeply. A lot was going on. But she knew what to do. She had pulled out the "Thermal Bible" and started working the problem.

However, in another compartment of her brain she felt emotion, concern, and some level of fear. She was very anxious because she knew that a lot of people were counting on her to get this right; a lot was riding on what she did. She thought, "This will be the news story tonight on TV. I don't want to mess up." But she had lived with the thermal analysis of Hubble's power-down for the last two years. She had comfort in knowing that she really understood it, knew its limitations, understood that the models were conservative and had a lot of margin built in. That level of preparation made the present situation easier to handle.

The entire team in the STOCC had gotten Hubble into a safe state. Now all they could do is wait to see if the crew in orbit could do a re-start of the EVA and proceed with the originally planned work. Later analysis showed that, had the engineers not gotten those systems powered back up, had they continued to cool down during the two hours or so that the start of the EVA had been delayed, it probably would have been a very bad day. Some of them, particularly WFPC2, would likely have hit their red temperature limit. But the team's training and professionalism took over, and they saved the Hubble Space Telescope from likely disaster. This episode, as related by the people who experienced it, including Christine Cottingham, may be seen online on YouTube in the video, "Hubble Memorable Moments: Powering Down."[1]

Back on Columbia it took over an hour, but the crew finally had a replacement suit configured for John Grunsfeld. They went back into the airlock to get him into his Hard Upper Torso. There they discovered that they had forgotten about Rick Linnehan. He was still in his own Hard Upper Torso, still latched firmly to the airlock wall, sound asleep.

Grunsfeld and Linnehan egressed from the airlock into Columbia's payload bay at 2:28 Houston time on Wednesday morning, two hours later than planned. There was relief that everyone had agreed to press on with the PCU change out for which these two astronauts had been intensively training for the past two years. A few minutes before the EVA crew's egress, the STOCC began an expedited power down of the systems that previously had been turned back on. The spacecraft started to cool down once again. The power down was completed at 3:37 am, just before Rick Linnehan opened the Bay 4 door to begin removal of the old PCU.

From this point until the end of the EVA, Christine and her "Thermal" colleagues had no telemetry, no insight into what was going on. Telemetry was the lifeblood of spacecraft operations. Engineers wanted to see voltages and temperatures. But they had none of that. Waiting those nearly seven hours during the EVA was like a parent sitting through their child's surgery, not knowing how they are doing, and not having any control over what was going on.

Nancy Currie drove Linnehan on the remote arm up to Bays 2 and 3 where he disconnected all the batteries. Meanwhile Grunsfeld completed the installation of MLI covers at several locations to slow down the cooling of Fixed Head Star

---

[1] https://www.youtube.com/playlist?list=PLiuUQ9asub3Ta8mqP5LNiOhOygRzue8kN.

Trackers, the WFPC2 radiator and other components. Linnehan proceeded to open the door to Bay 4 and started the process of disconnecting the first 30 of the 36 cable connectors on the left-hand side of the old PCU. Afterward he and Grunsfeld traded places. Linnehan moved to the PCU's protective enclosure and began preparing the new unit for installation. Grunsfeld, attached to the end of the arm, removed the remaining six connectors on the old PCU, unbolted the box and brought it down to the payload bay. He stored it in the protective enclosure for the ride back to Earth. Linnehan handed the new PCU box to Grunsfeld, who carried it back up to Bay 4, bolted it in place and reattached the 36 cable connectors (Figure 15.9).

After closing out their work sites, the two EVA astronauts ingressed back into the airlock at 9:16 am. In all, the spacewalk had taken 6 hr, 48 min. That turned out to be the shortest EVA of the mission—absolutely remarkable, considering the nearly calamitous way in which it had begun and considering that the PCU change out was the most difficult task of the mission. Possibly the saga of the wet spacesuit, and the two hour delay had gotten the adrenaline flowing in Grunsfeld and Linnehan. They realized they were getting off to a bad start and worked at peak efficiency to get the job done expeditiously.

Back in the STOCC, after the PCU had passed its aliveness and functional tests, the time had come to restore electrical power to the spacecraft. The STOCC couldn't turn all systems back on at the same time because that would pull too much power all at once. They had to be patient as each box and each instrument was powered on carefully, one at a time. "Thermal" chose to turn WFPC2 back on first. Even with the MLI blanket on its radiator, it had gotten very cold, probably colder than it had ever been. It exceeded its yellow limit, and was probably an hour away from the red limit threshold. It turned on easily; the yellow light on Christine's screen turned to green as it warmed up. The instrument was in fine shape after its ordeal, as were all

Figure 15.9. Astronaut John Grunsfeld holding one of the tools used to disconnect and reconnect the 36 cables on the power control unit. (Credit: NASA.)

the other systems. There was a flood of telemetry. There were green lights all across the monitor screens in the STOCC. Hubble had come fully back to life. Nothing had broken.

Next up, on Thursday, March 7 was EVA 4, featuring spacewalkers Jim Newman and Mike Massimino. The previous three EVAs resulted in major improvements to Hubble's electrical power systems, assuring a bright future for the spacecraft. Now, at last, the time had come to perform a much-needed major upgrade of Hubble's scientific capabilities by installing the most powerful camera the observatory had ever carried, the Advanced Camera for Surveys (ACS).

The EVA crew egressed the airlock at 3:00 am Houston time, and began the spacewalk with the usual setup procedures. With Newman on the remote arm and Massimino the free-flyer, the spacewalkers then opened the aft shroud doors, unplugged and removed the decommissioned ESA FOC (Figure 15.10), and stowed it temporarily on the sill of the payload bay.

The astronauts removed the ACS from its protective enclosure. Newman, with Nancy Currie piloting the remote arm, transported the new camera to the empty bay that awaited it (Figure 15.11). With Massimino's assistance he slid it onto the rails that guided it into its position inside the aft shroud (Figure 15.12), and connected its communications and power cables.

The FOC had resided in one of the two coldest locations in the aft shroud, an axial bay on a quadrant of the −V2 side, which usually received little direct sunlight. NICMOS occupied the other cold location in a similar position on the +V3 side. It was imperative that ACS remain as cold as possible, because it's large CCD detectors needed to operate at about −80°C. The heat was extracted from the CCD chips by thermal electric coolers and was dumped directly into the aft shroud.

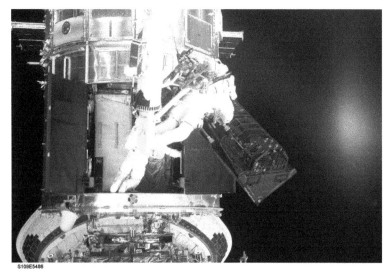

Figure 15.10. Jim Newman and remote arm operator Nancy Currie maneuver The Faint Object Camera to the sill of the Payload Bay. (Credit: NASA.)

Figure 15.11. Jim Newman carries the Advanced Camera for Surveys to the aft shroud of Hubble for installation in the axial bay formerly occupied by the FOC. (Credit: NASA.)

There was no external radiator built into the ACS structure, directly viewing space, as there was for WFPC2. As a result this new source of heat would be expected to warm up the ambient environment experienced by all the instruments in the aft shroud.

The Project had developed a new Aft Shroud Cooling System (ASCS), with a plan to install it along with ACS during SM3b. The ASCS included its own external radiator, similar to the one being installed on the outside of the spacecraft on this mission for the NICMOS Cooling System. However, SM3b became oversubscribed. There was simply no EVA time available to install the ASCS. It was left on the ground with the expectation that it would be on the manifest for the final servicing mission, SM4, sometime in the future.

After the ACS installation was finished, Massimino took Newman's place on the remote arm. Together they retrieved the electronics support module from its carrier and installed it on the floor of the axial bay just in front of the ACS. This box of electronics was designed to operate the NICMOS Cooling System that would be installed tomorrow on EVA 5. Since the NICMOS cryocooler was destined for the axial bay that already housed NICMOS, on the opposite side of the aft shroud, why would its control electronics be situated so far away?

The answer was an example of opportunistic engineering by the Hubble team. The FOC had been the last of the original Hubble instruments that needed COSTAR. It was now gone and COSTAR was no longer active. That meant that COSTAR's cable hookups for electrical power, commanding and telemetry had become available for the NCS to use.

Massimino and Newman disconnected COSTAR's cables and re-connected them to the electronics support module. They also connected another set of cables to the

Figure 15.12. Jim Newman slides the ACS onto its guide rails, assisted by Mike Massimino. Note the Faint Object Camera temporarily stowed on a fixture attached to the sill of the Payload Bay. (Credit: NASA.)

electronics module and routed it across the aft shroud to the location where the NICMOS cryocooler would be installed by Grunsfeld and Linnehan the next day. They then closed the −V2 doors, tidied up the Payload Bay and called it a day, ingressing at 10:30 am.

EVA 5, the final one of the mission, was complicated, involving work both on the inside of Hubble and on its exterior. This was the day the new-technology NICMOS Cryocooler was slated to be installed. Success would restore Hubble's near infrared eyesight. But NICMOS was not designed to be cooled by a mechanical refrigerator. This was pure innovation, perhaps a long shot, but the team had to give it a try.

Grunsfeld and Linnehan were out for their third spacewalk at 2:46 am. Houston time Friday morning. Nancy Currie gave Linnehan a ride on the remote arm to the +V2 doors of the aft shroud. After opening the two doors, Linnehan rode back down into the payload bay to retrieve the new mechanical cooler. Together, he and Grunsfeld installed it adjacent to NICMOS and attached the cables from the

Figure 15.13. John Grunsfeld, with Nancy Currie driving the remote arm, transports the 3 × 11 foot NCS radiator to the installation site where Rick Linnehan waits to assist. Note the yellow, curved handrails at the top and bottom of the aft shroud to which the radiator will be latched. (Credit: NASA.)

electronics support module that had been routed there by Newman and Massimino during yesterday's EVA 4.

Next the two spacewalkers removed the NCS radiator from its carrier and began installing it on the exterior of the aft shroud, next to the bay where NICMOS resided. The design of the radiator required that it be latched onto two handrails, one at the top of the aft shroud and one at the bottom (Figure 15.13). Grunsfeld was to close the top two latches, while Linnehan attended to the single latch at the bottom. There was some awkwardness here, as the two struggled to get the radiator locked in place. It seemed as though it didn't quite fit. Trying to force it into place was physically demanding to the point that the two astronauts almost gave up. But with one last forceful push, they managed to seat the latches onto the handrails and close them. They were not able to align the radiator top to bottom, so in its final orientation it appeared askew (Figure 15.14). But it was mechanically secure, and so they left it as it was.

Linnehan ran a conduit from the radiator through an opening in the bottom bulkhead of the aft shroud. The conduit would carry ammonia that had been vaporized by the heat being transferred out of the NCS cooling loop. That heat would be radiated into empty space by the NCS radiator, causing the ammonia gas to cool and liquefy, after which it was circulated back into a heat exchanger, and the cycle of extracting heat from the NICMOS Dewar would continue. Inside the aft shroud Grunsfeld attached the conduit to the cryocooler. He also ran the neon inlet and outlet tubing from the cryocooler into the corresponding ports on NICMOS and completed the connections.

Figure 15.14. The NCS radiator mounted on handrails at the top and bottom of the aft shroud. The astronauts had difficulty installing the radiator. It is securely latched in place, but slightly misaligned top to bottom. (Credit: NASA.)

There were numerous other small, detailed steps in installing and setting up the NCS that I won't describe here. But in broad outline this is how EVA 5 proceeded. Grunsfeld and Linnehan concluded their spacewalk at 10:06 am. This EVA had lasted for 7 hr, 20 min. In total the five EVAs of STS-109 lasted 35 hr, 55 min—a record up to that time for EVAs during a single shuttle mission.

Shortly after EVA 4 the day before, the ACS had passed its aliveness and functional tests conducted from the STOCC. The process of bringing the new NCS up to an operational state and cooling down the NICMOS detectors to the desired temperature, about 77 K, would require weeks. We would have to wait patiently to find out if this "experiment" had succeeded.

Columbia landed back at the Kennedy Space Center at 4:33 am. Eastern Standard Time on 2002 March 12 after a highly successful STS-109 mission to service Hubble. Tragically, that would be the last time the old and revered space ship would land on Earth. On its very next flight, STS-107, it broke apart upon re-entry into the atmosphere on 2003 February 1. All seven crewmembers perished. Once again the entire shuttle fleet, or what was left of it, would be grounded, this time for 2.5 years.

The newly installed ACS started producing some amazing images and scientific discoveries within a month after the end of SM3b. NASA released Early Release Observations (EROs) from the ACS at a press briefing on April 30 (Figure 15.15).

But NICMOS, now being cooled by the NCS, was another matter. The NCS was functioning, but it seemed to be taking forever to cool the instrument down. We waited and waited, growing more frustrated by the day. The NCS was turned on by the STOCC on March 18, less than a week after Columbia landed. At first the temperature dropped rapidly, by 78 K in only one day. But then the rate of cooling slowed dramatically. Six days later, on March 24, the temperature had only dropped

Figure 15.15. ACS Early Release Observation of the "Tadpole Galaxy." The galaxy has been pulled apart by a collision with a smaller galaxy seen as the blue clump at the upper left. (Credit: NASA, H. Ford, the ACS Team, STScI 2002-11.)

an additional 39 K. The engineers tried reorienting Hubble to a cooler attitude. That helped but only by a small amount; the temperature continued to drop steadily, but very slowly. On March 29, they turned NICMOS off, so that the NCS would not be fighting the heat generated by the electronics in the instrument it was supposed to cool.

Finally, on April 5, after the gas in the cryocooler had passed through a rocky, turbulent period called surging—which had been expected—and began to flow smoothly again, the team revved up the speed on one of the tiny turbines. At that point, the temperature finally fell to about 75 K. We were there!!

NICMOS was turned back on. By adjusting the speeds of the turbines, that low temperature could not only be maintained, but could be adjusted up or down as desired to optimize the performance of NICMOS's detectors. With the old cooling method that used a block of solid nitrogen ice, there had been no choice. The temperature of the ice dictated the temperature of the detectors and optics, and that was too low by about 20 K for optimum performance of the instrument.

Figure 15.16. Early Release Observations taken with NICMOS after the NICMOS Cooling System had restored it to operation. The top frame is a visible light image taken with WFPC2 of the edge-on spiral galaxy NGC 4013, showing the galaxy center as completely obscured by dust. The middle frame is a three-color image taken with NICMOS, showing how the dust is penetrated by infrared light. The bottom frame is a NICMOS image taken in the light of hydrogen. It reveals a thin disk of stars circling the center of the galaxy. (Credit: NASA, R. Thompson, the NICMOS Team, STScI 2002-13.)

The experiment had worked. NICMOS was now being cooled in a manner its original designers had never envisioned. And the instrument performed better than ever before. NASA released EROs taken by the newly revived NICMOS on 2002 June 5 (Figure 15.16). NICMOS kept operating in this new way and producing excellent infrared images of the universe for another six years.

Reference

Grunsfeld, J. M. 2014, Hubble's Legacy, ed. R. D. Launius, & D. H. DeVorkin (Washington, DC: Smithsonian Institution Scholarly Press), 61

AAS | IOP Astronomy

Life With Hubble
An insider's view of the world's most famous telescope
David S Leckrone

Chapter 16

Stormy Weather: Hubble Under Siege

NASA's Space Astronomy missions have historically followed a particular life cycle, but with variations. First comes a concept for one or more new missions or new instruments, coupled with a scientific justification for why they would be important to do. These are developed by groups of advocates and sometimes marketed in advance within the community or to committees advising various branches of the Science Mission Directorate (previously called the Office of Space Science) at NASA Headquarters. Proposals for the bigger flagship missions have traditionally welled up from within the scientific community. This was the case for Hubble and for Chandra, the big orbiting X-ray telescope. The James Webb Space Telescope described later in this chapter also originated in this way. Their proponents seek the endorsement of the committees carrying out the decadal surveys, which recommend new initiatives to NASA once every ten years. The National Academies of Sciences, Engineering, and Medicine conduct the decadal surveys. On a more or less regular basis NASA Headquarters issues announcements of opportunity for new missions—typically medium or small in scope—or for new science instruments to populate a mission platform, such as that of a major flagship observatory. Proposals are written and submitted by science and engineering teams. A peer-reviewed competition and selection process follows.

Next comes the preliminary design phase, followed by a detailed critical design period. Each of those phases culminates in a design review by a panel of independent experts. If a mission or instrument passes its Preliminary and Critical Design Reviews, and if the funding is available, the project can go forward into manufacturing, testing, and preparation for launch. Development of an operations control center and creation of a ground system of computers and software are also a big part of this process. The development of a new mission is spread out over a period of years. The number of years depends on the scope of the mission, the skill of the development team in bringing it to fruition and, in most cases, a healthy dose of good luck.

That completes the mission development phase. At that point the usually very large development budget phases down to a low level, or to nothing at all. What then follows is the operations period, supported by its own line of funding. During operations, an observing plan is generated by the scientists for whom the data are meant, or by their surrogates at a place such as the Space Telescope Science Institute (STScI). That plan is converted with computer software into a sequence of commands that are transmitted to the spacecraft's on-board computer for execution on orbit. After the programmed observations are executed, the data acquired are transmitted to the ground and ultimately disseminated to the observers, who process, analyze, and interpret them. They then communicate the results to the broader community via scholarly papers published in professional journals. They may also describe their results to the news media in cases where particularly important results are obtained. The spacecraft and instrument functions are monitored, and anomalies are addressed and resolved to the extent that commands from the ground allow.

After a number of years, typically five to ten years or perhaps a bit longer, the aging spacecraft and its instruments will begin to fail. Engineers will become very clever at working around problems to keep the mission going for as long as they can. But sooner or later either the orbiting systems will fail, or the decision will be made simply to stop collecting data, turn the spacecraft off and dispose of it, hopefully in a controlled manner. That ends the mission operations phase.

The yearly mission budget for operations is usually much lower than that allocated for development. As development phases down and operations pick up, funding in the year-to-year flow of the Space Science budget becomes available to begin the life cycles of new missions. Of course reality can become quite complex when the NASA Science Mission Directorate is trying to manage the budgets and schedules of a couple of dozen worthy missions at the same time in various phases of their life cycles.

The Hubble Space Telescope Program was different. It was an anomaly among Space Science missions. Because it could regularly be refurbished and updated with new technology, new spacecraft systems and new scientific instruments by means of space shuttle servicing, its mission development phase was never-ending. Hubble was kept perpetually young and at the cutting edge of astronomy and astrophysics. It was the equivalent of a long-lived mountaintop observatory in space, or at least that was the original idea. Its development budget could have gone on for as long as there was a space shuttle to service the observatory. In return for that investment, Hubble continued as one of the most important NASA missions ever and one of the most important scientific endeavors of our time. Also, it became very familiar to and much beloved by the taxpaying public.

For better or worse, Hubble stood out like a sore thumb within the Space Science budget. "It is just too expensive." Or "Shuttle servicing is just too expensive," were common complaints. The Space Science leadership at NASA Headquarters had to deal with the conundrum of having to nurture their most productive and famous mission, while at the same time responding to other scientific constituencies that also had valid claims to the limited overall funding available.

This situation created schisms within NASA between the Hubble Program team at Goddard and the Office of Space Science at HQ. The former was totally dedicated to the dream of maintaining Hubble as the world's greatest and most productive observatory for as long as possible. The latter had to face a wider gamut of budgetary pressures.

It also created schisms within the astronomical community between those who used and cherished Hubble, and had their research grants funded by the Hubble Program, and those who were eager for Hubble to get out of the way in NASA's budget so that they could go on to the next big thing. Sometimes those were the same people.

In the hiatus following the loss of Challenger in 1986, while Hubble's future and that of the Shuttle Program remained uncertain, STScI Director Riccardo Giacconi assigned a new task to Deputy Director Garth Illingworth and Research Branch Head Pete Stockman—to begin thinking about what should be the next major astronomy mission after Hubble. Riccardo was justifiably concerned about the long time it takes to give birth to a major new space astronomy mission—usually decades. So, it was never too early to start thinking about it. As Garth recounts the story (Illingworth 2016), he thought, "This was somewhat crazy, given the work remaining on Hubble." After all, Hubble hadn't even been launched yet. But there was no use resisting their boss, and so the two of them, together with astronomical engineer Pierre Bely, started down this conceptual path. What would be a worthy successor to Hubble?

At that time Bely and other astronomers had already been working on design concepts for very large aperture (8–16 m), passively cooled space telescopes that would be especially sensitive to infrared radiation, the wavelengths of light that one would detect from objects in the high redshift universe. Such a telescope would be perfectly suited to the study of the very earliest epochs of star and galaxy formation, when the universe was very young.

The STScI group produced a succession of studies, workshops, and scholarly papers, exploring this idea. In 1995 October a small engineering feasibility study was initiated at Goddard, funded by Ed Weiler to the tune of a few tens of thousands of dollars, and supported by John Campbell, the Hubble Project Manager. Engineering manager Bernie Seery and astrophysicist John Mather were asked to lead the Goddard effort. They called the object of these studies the Next Generation Space Telescope, NGST.

Meanwhile, AURA, the parent corporation of the STScI, had commissioned a separate study called "HST and Beyond," by a distinguished group of astronomers chaired by Alan Dressler. Their 2.5 year effort culminated in a report dated 1996 May 15.[1] The Dressler Committee's recommendations included the following:
1. "The HST should be operated beyond its currently scheduled termination date of 2005 in a no-repair, no-upgrade mode at approximately 20% of the current cost of operation and maintenance."

---

[1] "Exploration and the Search for Origins: A Vision for Ultraviolet-Optical-Infrared Space Astronomy," Alan Dressler ed. (Washington, DC: Association of Universities for Research in Astronomy).

2. "NASA should develop a space observatory of aperture 4 m or larger, optimized for imaging and spectroscopy over the wavelength range 1–5 μm. Extension shortward to about 0.5 μm and longward to at least 20 μm would greatly increase its versatility and productivity. We believe that a 4 m or larger version could be built for well under $1 billion, and have set an approximate cost of $500 million as a desirable goal."

For reference, the cost of Hubble's development up to the time of its launch was about $1.8 billion in 1990 dollars. Shuttle launches for Hubble servicing missions came essentially free of charge to Ed Weiler and the Office of Space Science. All of the shuttle costs were borne by the Office of Space Flight, which included Human Space Flight—a real bargain for the Hubble Program.

To provide a rationale for the much lower cost estimate for a much larger space telescope, Dressler cited a recent study from the Space Sciences Board of the National Academy of Sciences (NAS), entitled "A Scientific Study of a New Technology Orbital Telescope." Quoting from the Dressler Committee report, "In this study U.S. industry showed the scientific community that it is technically feasible to develop and launch a passively-cooled, filled aperture 4 m space telescope within the next decade for a cost well below $1 billion, and possibly as low as $500 million." Specifically, the NAS study referred to telescope designs being considered for the U.S. Defense Department's "Star Wars" anti-missile defense program. These were not immediately suitable for the kind of infrared astronomy the Dressler Committee had in mind. But, nevertheless, industry was claiming such relatively inexpensive missions could be done.

I had several reactions when I first read the "HST and Beyond" report:

First, I was grateful that it supported a five year extension of Hubble's lifetime. The original 15 years, set long before Hubble was launched, was a purely arbitrary construct intended to facilitate budget estimates, and not based on any scientific considerations at all. Surely, if Hubble were still going strong at the 15 year mark, NASA and the scientific community would be very reluctant to shut it down.

Second, I was unhappy that the report attempted to set budgetary limits a decade into the future on the scope of Hubble's new instruments, technology upgrades, and in-orbit servicing. Surely such decisions should be science-driven and made much closer to the time when they were relevant to the actual situation at hand. In 1996 Hubble had just begun to scratch the surface of what it was capable of doing. We hadn't even flown SM2 yet, which was schedule for the following year. That servicing mission would be the first to introduce advanced technology scientific instruments, STIS and NICMOS. Of course at the time, I didn't know what the future held for Hubble science. Neither did the Dressler Committee. But by proposing to end future servicing missions and to drastically limit Hubble's budget during those later years, 2005–2010, their intent clearly was to free up funds to go into their new telescope, while abandoning the original concept of a "permanent" (or at least very long-lived) observatory in space, regardless of the remaining scientific potential of the latter.

Third, I applauded the idea that the successor to Hubble should be a large-aperture infrared telescope—a big step up from Hubble's capabilities. That made perfect sense. Hubble had been conceived originally as a deep-space "time machine." Its successor should push those boundaries further, as well as advancing many other areas in which Hubble had blazed trails. Hubble would still provide extremely valuable and unique access to ultraviolet and optical wavelengths, sparsely covered or not covered at all by the infrared telescope Dressler had in mind.

However, when I saw the cost estimates, $500 million to $1 billion, I asked myself, "What have these guys been smoking?" I was incredulous. That seemed so naïve. It would be wonderful if true. Reducing the costs of large, high-tech, state-of-the-art space science missions, capable of doing frontier-expanding research, could only lead to more opportunities for exploration and discovery. But, at least in the mid-1990s, it seemed more like fantasy to me—or at least more like a good sales pitch by for-profit aerospace companies. A very large, passively cooled telescope and associated instruments had to be technologically challenging. Such low cost estimates did not seem credible. But breaking the "Hubble Cost Paradigm" became the mantra of the team that had begun working on Hubble's successor.

At the January 1996 meeting of the American Astronomical Society (AAS) held in San Antonio, NASA Administrator Dan Goldin gave an invited speech, which he titled "NASA in the Next Millennium." Goldin was an intense and intimidating personality. He was almost impossible to resist. His speech to the AAS matched his personality. It was impassioned and inspiring. He was an evangelist, preaching to a congregation about the gospel of new technologies and how they could lead to less expensive and more capable space science missions. It was a persuasive vision of what the next 25 years could hold.

Goldin praised Hubble, showing the recent WFPC2 image of the Eagle Nebula (the "Pillars of Creation") and talking about the media excitement produced by the Hubble Deep Field, which had been made public for the first time at that very conference. But he also disparaged Hubble saying, "But my concern is we're still flying ground-based technology in space. The telescope that flies in space is designed for one g, so is it any wonder that the Hubble is 25,000 pounds? There is no vision of where we need to be."

Earlier in the talk, Goldin made another pronouncement that disturbed many of us who had been working on Hubble—Lyman Spitzer's dream—for decades: "There are those in the science community who like to talk of the Hubble Space Telescope operating for another two decades. We cannot do that either. The Hubble is wonderful. It's terrific, but at a quarter of a billion a year, we must plan a replacement to the Hubble, and we cannot say, because the budget's declining, we're going to put it off indefinitely. We have to begin now. We have to work on the technology. But the next mission doesn't have to cost two billion dollars to build and four billion dollars to operate. It doesn't have to weigh 25,000 pounds."[2]

---

[2] Goldin's "four billion dollars to operate" evidently is a rounded-up value of 15 years-worth of Hubble's total yearly budget $250 million.

This was a far cry from the early NASA vision in that 1972 memo from John Naugle to George Low (see Chapter 2) about what would in the future be called the Hubble Space Telescope: "... as an essentially permanent observatory in space through the marriage of automated spacecraft technology and the unique capabilities of the Shuttle transportation and maintenance systems." Now, it felt like the NASA Administrator was calling the undertaker to come and start taking measurements for Hubble's coffin.

That quarter billion dollars per year to which Goldin referred was split up with roughly 30% going to maintaining Hubble servicing capabilities and preparations for upcoming missions; 20% allocated to the development of the new, next-generation instruments and other technology upgrades; 20% going to spacecraft operations at Goddard; 20% going to AURA for the Space Telescope Science Institute; and 10% for research grant funding for Hubble observers. The total yearly cost of Hubble upgrades and servicing was $125 million. The $125 million balance of the annual budget went to spacecraft operations, science operations, and scientific research with Hubble data. What Goldin and NASA were getting for this $250 million annual investment was a long list of important benefits to the Agency and to Humanity:

Hubble was a "bird in the hand," as opposed to a paper concept. It was real, already in orbit and operational, refreshed by SM1 with its vision fully corrected—one of NASA's most valuable assets and one of its most important missions ever. Hubble was producing a continuous stream of extraordinary new data about the universe. It was already proving to be one of the most productive and high-impact scientific tools of our age, revolutionizing many areas of astronomy and astrophysics.

The marriage between Hubble and the Shuttle Program provided scientifically important work for NASA's Human Spaceflight Program. It unified the separate "stovepipes" of NASA as no other mission ever had.

Hubble servicing had demonstrated the feasibility of space station construction. Tools and spacewalking techniques used for Hubble were later adapted to building and maintaining the ISS.

The development of new scientific instruments and new spacecraft systems entailed the development and application of new technologies. For example, the new focal plane light sensors, the NICMOS cryocooler, the precise figuring of the complex small optical surfaces needed to correct for Hubble's spherical aberration, etc, appeared to be exactly the kind of new technology development Goldin was calling for in his speech.

Not least, Hubble generated immense good will for NASA from the media, from the American taxpayer, from Congress, and from people all over the world. You couldn't buy the kind of favorable publicity Hubble brought to NASA. It was earned.

Goldin's enthusiasm became more expansive with another comment. "I see Alan Dressler here. All he wants is a four-meter optic that goes from a half micron to 20 microns. And I said to him, 'Why do you ask for such a modest thing? Why not go after six or seven meters?'" (Audience laughter) And later in the talk, "Again I'm picking on Alan Dressler, and I shouldn't, but he came to me and said, 'All I want is a four-meter telescope.' And I said, 'What beyond that?' And he said, 'Get me a four-meter telescope.' **That's not good enough**," Goldin concluded.

The approximately 4 m telescope recommended by the Dressler committee was the largest with a monolithic primary mirror (i.e., made from a single piece of glass or other material) that would fit into the protective structure or fairing at the top of the largest commercial rocket then available. A larger telescope, say an 8 m, would require a segmented mirror with multiple smaller individual mirrors hinged together in some way. It would have to be folded up to fit inside the launch vehicle, together with the instruments and spacecraft.

Many astronomers had concluded that a new 4 m telescope, as the Dressler Committee had recommended, would only provide a small, incremental gain in capabilities over Hubble. But now the NASA Administrator had given the astronomical community permission to think big—very big. Moreover he had practically directed them to get on with it as quickly as possible. With new technologies, it shouldn't be necessary for the price tag to go as high as $2 billion, or at least that was Goldin's assertion. Implicit in this was the necessity that, once started, the project had to be completed relatively quickly. The longer any project dragged on the more costly it became.

Of course this was an offer no astronomer could refuse. Dream freely about the optimum large-aperture space telescope, sensitive to near- and mid-infrared wavelengths. It would be the perfect tool for exploring the early universe, studying planets in orbit around other stars, and much, much more. It would be so much more powerful than Hubble, at least for those areas of research that needed a very large light collecting area and infrared capability.

I was as excited by this prospect as any other astronomer. But it also created an intellectual schism for me. A large number of people had devoted decades of their lives, and NASA had devoted a large amount of taxpayer money to bringing the Hubble Space Telescope—the crown jewel of astronomy at that time—to fruition, to realizing the dreams of many astronomers going back decades. Wouldn't it be better for now to devote our energy to preserving and enhancing the asset we had just been given a few years before, to see where it would take us scientifically, to extract its full value? Rather than an arbitrary 15 years, Hubble's "natural lifetime" should correspond to the period during which it continued to produce cutting-edge, boundary-breaking science, with continued servicing and technological upgrades for as long as the shuttle could support it. Give Hubble its full moment in history with the NGST to have its time later, hopefully with a period of overlap between the two. Or at least that was how I felt about the matter.

Of course work in the background on Hubble's ultimate successor was a prudent thing to do. The new technologies had to be advanced for fabricating large, precisely ground and polished segmented mirrors; for maintaining the segments in proper alignment; for cooling large optics and instruments to extremely low temperatures; and for sensing infrared light with large detector arrays.

But suddenly it seemed like rapid development of the NGST to meet a 2007 or 2008 launch date became the primary focus of many in the community. Was Hubble already "old hat?" Sustaining Hubble as a "permanent" ultraviolet, optical, and near-infrared observatory in space and regularly upgrading its technology with periodic shuttle servicing calls required a steady allocation of about $125 million per year, half of Hubble's annual budget. Each servicing mission left Hubble essentially as a fresh, new observatory with greater scientific capabilities than ever before. I strongly believed that was a real bargain. But suddenly there was a chimera competing for the attention of the community—a much bigger space telescope with intriguing capabilities that would cost relatively little as major space projects go and that would be ready to fly in a decade or so. And Hubble, or more precisely the Hubble budget, was viewed as an impediment in its way.

In March of 1996 the NGST Study Office at Goddard conducted a competition for independent outside teams to define a concept for and assess the feasibility of an NGST mission. Two aerospace companies with associated partners were selected—TRW and Lockheed Martin. In addition Goddard itself was to do an in-house concept and feasibility study to establish a baseline for what NASA considered possible. For several months during the spring and summer of 1996 the three teams laid out the case for three different design concepts. They reported their results back to NASA in 1996 August.

In a 1997 document entitled "Next Generation Space Telescope: Visiting a Time When Galaxies Were Young,"[3] editor Pete Stockman pulled together a complete story of where the NGST concept stood as of that date—its scientific rationale, design considerations, launch and candidate orbit options, relevant technologies, etc. Included was a summary of each of the three design concept and feasibility studies carried out by TRW, Lockheed Martin and Goddard.

The Lockheed Martin-led team investigated a design that incorporated a 6 m monolithic glass primary mirror, while both the TRW team and the Goddard team considered 8 m segmented telescopes. All three studies emphasized keeping the observatory much lighter in weight than Hubble so that it could be launched in a medium-sized launch vehicle; all focused on using already existing or promising new technologies that would require a minimum of innovations; all emphasized keeping the telescope optics, instruments, and spacecraft as simple as possible. All three studies priced a realistic program as costing $175 million upfront money for design studies and technology development, plus $500 million for construction of the flight observatory and development of the operations ground system.

These conclusions were right in line with the $0.5–1.0 billion price range specified by the Dressler Committee. They were also consistent with Dan Goldin's assertion that a faster, better, cheaper modern observatory should cost far less than Hubble's price tag.

On 1999 March 8 Ed Weiler signed a "Formulation Authorization," that essentially was a formal kickoff for the NGST mission. In that same year NASA selected Lockheed Martin and TRW for preliminary concept studies. The launch date was set for 2007 or 2008. In 2002, NASA awarded an $825 million contract to Northrop Grumman Space Technology to serve as the prime contractor for the new

---

[3] June 1997 (Washington, DC: Associated Universities for Research in Astronomy).

mission. Northrop Grumman had acquired TRW in 2002. The mission's name was changed from NGST (to avoid confusion with the name of the prime contractor) to the James Webb Space Telescope (JWST) in honor of NASA's visionary third Administrator, who had done much to promote space science as well as human spaceflight.

On 1999 June 3 at another meeting of the American Astronomical Society in Chicago, Dan Goldin doubled down on his desire to rein in the Hubble mission. He said that people who were arguing for a fifth servicing mission to extend Hubble's life were "Hubble Huggers." "Don't be a Hubble Hugger!" he said, derisively.

That term quickly caught on and became a rallying cry for all of us who strongly supported the Hubble Space Telescope. Back at Goddard, Frank Cepollina (Cepi), who had been actively campaigning for an additional servicing mission, SM5, for several years, had "Hubble Hugger" t-shirts made up, which we all wore with pride.

In the spring of 1997 Associate Administrator Wes Huntress had formally extended Hubble's operational lifetime by five years to 2010, but with no additional servicing beyond SM4. This was in keeping with the recommendation of the Dressler Committee that Hubble Operations be continued at very low cost for an additional five years.

When SM3b had been delayed until March of 2002, SM4 was penciled in on the schedule for March of 2004. Cepi and others in the Project believed that the spacecraft wouldn't last from a final servicing mission in 2004 until 2010 without an intervening maintenance visit in 2007 or 2008. Not only were the gyros prone to failure, but also other systems were vulnerable—the Fine Guidance sensors and their control electronics, reaction wheels, the data management unit, etc.

In 2002 Helen Wong from the Aerospace Corporation performed a detailed statistical analysis of Hubble failure probabilities under various assumptions. She concluded that Hubble, at the completion of a fully successful servicing mission, had a formal probability of continuing to provide science data for six years thereafter of about 18%. For seven years of science the probability dropped to about 9%.

Helen's calculations were based on conservative assumptions. She assumed that the failure of any individual part in a system would lead to the failure of the entire unit, which might not be the case; a system might be degraded without failing entirely. Second, there was no way the calculations could account for the ingenuity of Hubble's engineers in finding ways to work around hardware failures. It was not possible in advance of a failure to anticipate what could be done to mitigate that failure. Consequently, we in the Hubble Project viewed her calculations as providing a worst-case scenario. We might get lucky and be able to keep Hubble going in spite of such failures. But clearly, continuing the paradigm of regular servicing and upgrades was the best way to keep Hubble operating at its full capabilities.

Cepi had assembled a world-class team of engineers who had developed a deep expertize in all things technical about Hubble and Hubble servicing. Once the last servicing mission was completed, that team would quickly disband and its members would be scooped up by other organizations. "So, be careful," Cepi argued. "Once you stop using the servicing capability, you will very quickly lose it."

As late as 2003, NASA's plan for Hubble's end of life was to fly one final shuttle mission in 2010 to retrieve the telescope and return it to Earth for display in the National Air and Space Museum. Detailed studies of what such a mission would entail, however, revealed that it would have to be the equivalent of a full up and complex servicing mission. With the new rigid solar arrays and the external radiator for the NICMOS cooler recently installed in SM2, the spacecraft would no longer fit in the shuttle's payload bay. It would require several spacewalks for astronauts to remove and dispose of these appendages. Moreover, such a mission could not utilize the normal carriers and the Flight Support System flown in prior servicing calls to Hubble. There would be no room in the payload bay for anything other than Hubble itself. So, the retrieval mission would of necessity have to be of an unusual and previously untested design.

John Campbell and other Hubble engineers also argued that the internal structures of the Hubble telescope and spacecraft might not tolerate the landing forces when the shuttle touched down on the runway. They might break, endangering the shuttle and its crew.

The final straw came from the astronauts themselves. They had been more than willing to fly servicing missions to maintain and enhance Hubble's scientific capabilities and productivity. In so doing they were risking their lives, but in a great cause—expanding the frontiers of science. However, they were unwilling to risk their lives simply for the purpose of creating a museum display. As John Grunsfeld put it, "If there were to be a mission after the SM4 for the purpose of returning Hubble to Earth in the Shuttle Payload Bay, the Astronaut Office would have reservations supporting the mission ... In a sense this mission would be risking human lives, and a unique national resource (the Space Shuttle), for the purpose of disabling great science, albeit due to necessity at end-of-life (Foust 2003)."

So NASA changed the end-of-life plan. Sometime in the future a propulsion module would be attached to the aft end of the spacecraft that could be used to de-orbit Hubble in a controlled fashion into an unpopulated area of the open ocean. This could be done in a final shuttle mission in 2010. That would also present the opportunity for at least a minimal amount of servicing to allow Hubble to continue doing science well into the second decade of the century. Alternatively, some kind of unmanned mission might be flown to install the propulsion module robotically when it became necessary to do so.

A controlled re-entry for Hubble was essential. At the altitude of Hubble's orbit, the Earth's atmosphere is extremely rarified. But the air still has enough density, and creates enough friction, to cause the spacecraft's orbit to decay slowly. Over time, without the orbital re-boosts provided by the shuttle during servicing missions, Hubble would gradually drop in altitude, with the decay accelerating as denser air was encountered. At some point the atmospheric drag on Hubble would make the spacecraft very difficult to control and the fine pointing needed for science observations would no longer be feasible. As Hubble fell to Earth, most of the spacecraft would break apart and burn up due to atmospheric friction. However, heavier components like the primary mirror and its support structure would likely survive all the way to the ground. It is possible (though highly improbable) that debris might fall in a populated area, potentially endangering unlucky people in the impact area.

The situation would be exacerbated during periods of maximum solar activity when the Sun's radiation causes Earth's atmosphere to swell and the air density hundreds of miles up to increase. Looking toward the future from 2002 or 2003, the strength, starting date, and duration of the next solar maximum could only be roughly predicted. Various computer models of solar activity made contradictory predictions. "Let's play it safe," Cepi argued, "by giving Hubble another re-boost in 2007 or so."

Ed Weiler was undoubtedly in an awkward position. He had been playing a major leadership role as Hubble Program Scientist at NASA Headquarters for two decades. But now his boss was insisting that NASA move beyond Hubble—support Hubble within the schedule and budget boundaries originally defined for planning purposes years before, perhaps extend its life a bit if that could be done inexpensively—but don't do anything more than that. There was to be no grand, long-term vision for keeping Hubble operating as a long-term national asset in space as long as it continued to produce cutting edge science. Hubble was to be terminated in 2010 to be replaced by JWST—and that was that.

Another challenge for Weiler was where to find the money to pay for JWST development. He and John Campbell reached an agreement on this. The roughly $250 million per year budget line that had been allocated for Hubble would now be split between Hubble and JWST. As the Hubble mission phased down after the final servicing mission in 2004, the JWST development budget would rapidly phase up. The $250 million per year would continue, essentially in perpetuity, until the $1 billion dollar successor to Hubble was completed and launched, no later than 2009 or 2010.

So, from the late 1990s on, Hubble and NGST/JWST became budgetary competitors. There was pressure to get the final Hubble servicing mission, SM4, flown as quickly as possible. There was immense pressure from some parts of the astronomical community to get the cost of Hubble operations down as low as possible, as quickly as possible. The Hubble Operations Project worked diligently to do just that. The cost of Hubble operations at Goddard steadily dropped after 2000, until it reached a point where few spacecraft controllers could be found in the STOCC. Instead, if problems arose on the spacecraft, the Lockheed and NASA engineers would receive electronic pages (later text messages) alerting them to come in to the STOCC to address the issues at hand.

We Hubble Huggers in the HST Project Office at Goddard were also in an awkward position. After all, the new study project for NGST had been established at Goddard. It had the support of our Hubble Program Manager, John Campbell. Some of the Hubble team members were already beginning to migrate onto the NGST study. Similarly, our colleagues at the STScI had been assigned the responsibility of operating the new telescope. So, that institution began bifurcating into a Hubble office and an NGST office.

My Project Scientist colleagues and I were always careful not to speak ill of NGST. On the contrary we tried to be supportive and encouraging of the mission that was ultimately to become our successor. In Hubble press releases which I regularly reviewed, in public talks, and in official briefings, I took care always to emphasize the great science NGST would do, and how Hubble research was paving the way for important NGST research in future years.

Nevertheless, there was an uneasy tension about all of this. It was something like being in the position of training someone to take over your job, just before you're fired. I felt that I had to defend Hubble's continued existence by always emphasizing the NGST connection—even though Hubble was still fresh and already one of the greatest scientific endeavors of our time, and NGST was still a glimmer in astronomers' eyes.

Beginning in the mid-1990s, Cepi became a dogged advocate for SM5, an extra Hubble servicing mission beyond those already planned. As the years went by this became something of an obsession with him—so much so that we teased him about it, made jokes about it at Project staff meetings. It was good-natured ribbing, but with an undertone of seriousness. Deep inside we all hoped Cepi would succeed in influencing the outside world to support additional Hubble servicing, while at the same time we were rooting for the successful development of NGST. In virtually every encounter with astronomers, Cepi would bend their ears about the need to keep Hubble servicing going. "Don't you care about the future of your science?" "Why aren't you speaking up in defense of Hubble?" he would ask.

A few miles to the south at NASA Headquarters, Cepi's SM5 campaign was raising serious hackles, serious anger, so much so that Ed Weiler seemed to become obsessed with fighting back. It appeared Cepi had become the enemy, as far as Ed was concerned. Cepi could not legally lobby Congress, but the aerospace contractors who worked on Hubble servicing were not so restricted. Hearing that SM5 was being discussed in the halls of the Capitol infuriated Weiler all the more.

From time to time, Cepi would get "taken to the woodshed" by the Goddard Center Director, after the latter had gotten scolded by Weiler about Cepi's machinations. Cepi would come back contrite and looking sheepish, but that usually didn't last long. It was Cepi's purpose in life to keep Hubble healthy and at the cutting edge of technology for as long as he possibly could. Anytime a problem arose on Hubble, Cepi would dive into the challenges of how to repair it. He just wanted to get the job done. Naturally, he also wanted to keep the paradigm of orbital servicing alive and give it a future. He wanted to keep his servicing team together. But it was official NASA policy, enforced by Weiler, to keep the Hubble program within bounds.

Given the animosity that Weiler apparently felt toward the Project Manager for HST Flight Systems and Servicing at Goddard, one might wonder why Cepi was kept in that position. To me the answer was simple. All of the Hubble servicing missions were spectacularly successful. Cepi not only was the creative force behind shuttle-based servicing as a discipline, but he also had a remarkable skill for getting the very best out of all the people and organizations who were engaged in bringing each mission to fruition. Weiler evidently didn't want to mess with success.

Ironically, Weiler garnered extensive favorable national publicity during each servicing mission, while Cepi's name was almost never mentioned in the media. The person most responsible for NASA's success in this area went largely unsung and unappreciated in the publicity surrounding the servicing missions. However, in 2003 Cepi's peers did recognize him by inducting him into the National Inventors Hall of Fame in honor of his development of satellite servicing techniques.

In 1998 Ed Weiler was appointed Associate Administrator for Space Science at NASA HQ. Before accepting the position he asked my advice about whether or not it was the right thing for him to do. I'm afraid I wasn't very helpful, chiming in with some platitudes about how he should follow his passion and be certain that high-level management, with all of its pressures and headaches, was the career path he really wanted to go down. In that same conversation Ed warned me that, as Associate Administrator, he could not allow himself to be viewed as treating Hubble in any special way. He couldn't be perceived as playing favorites with his old mission. I worried at the moment about how this might work out in practice. But I was prepared to give him the benefit of the doubt and to be supportive.

To fill his old job as Head of Physics and Astronomy, Ed hired Anne Kinney. Anne was originally from Wisconsin. She did her undergraduate work at the University of Wisconsin at Madison and got her PhD in Physics and Astronomy from New York University. She was smart, witty, and a good public communicator. The latter quality led to her position as a regular "guest commentator" on NASA Space Science Updates, the occasional news briefings announcing new Hubble discoveries televised on NASA TV. Anne was also a fine scientist. Prior to moving to NASA HQ, she had worked for 14 years at the STScI as an Instrument Scientist, attending to the Faint Object Spectrograph and to the astronomers who observed with it. Later on she was the Chief Scientist in the Institute's Education and Public Outreach group.

Starting in 1999, Anne took on the responsibilities within NASA of managing a large portfolio of space astrophysics missions—Hubble, the Chandra X-ray telescope, the Spitzer infrared telescope, the Fermi gamma-ray telescope, as well as the nascent NGST. At the time that seemed to me to be an enormous jump in professional responsibilities from the jobs she had previously occupied. She had so little management experience. But I both respected and liked Anne and was glad to see her get the opportunity to advance.

When Dan Goldin first became NASA Administrator he vowed to reduce the size of the Agency. In particular he cut the number of employees at NASA Headquarters by about a factor of two. Whereas the Program Manager for any mission—the top executive who controlled a mission's performance requirements and funding—had formerly resided within Headquarters, Goldin moved that responsibility to the various NASA field centers—Goddard, Marshall, JPL, et al. The position of lead Project Manager on a mission at a field center morphed into the Program Manager position. At Headquarters, a new position was established, that of Program Executive. The Program Executive had a much-reduced role and level of responsibility relative to the prior headquarters Program Manager. Their role was primarily to serve as the technical and management liaison between Headquarters and the newly designated Program Manager for a mission at a field center. For her

Hubble Program Executive, Anne brought on board a former member of the Hubble Project at Marshall, Michael Moore.

Within the Hubble Project at Goddard, John Campbell and later Preston Burch took on the new title of "Hubble Program Manager." But Ed Weiler wanted to retain the title of Hubble Program Scientist at Headquarters. I was then the lead scientist within the overall Hubble Project at Goddard and the scientific advisor to the person at Goddard who was now called the Program Manager. But I was not the Hubble Program Scientist. These titles led to a certain amount of confusion in the world outside of NASA, especially in the news media. Ed and I agreed that my title would be Senior Project Scientist. I don't think the press ever understood the distinction or what our roles actually were.

Two days after the completion of STS-109/SM3b, NASA's internal quarrel about SM5 went public. The 2002 March 14 issue of Nature contained an article by columnist Tony Reichhardt entitled, "NASA urged to play waiting game on Hubble's retirement."[4] Reichhardt wrote in part:

"Even as NASA completes a successful service mission [SM3b] on the Hubble Space Telescope, a debate is simmering in the agency about how long to keep the instrument in action.

NASA's current plan is to return the telescope to Earth in 2010.... But some Hubble-project scientists and engineers at NASA's Goddard Space Flight Center have been arguing—albeit with little success so far—that the telescope should continue its observations ...

The dilemma of when to pull the plug on a successful mission is a familiar one at NASA. But the case of Hubble will be especially difficult. Not only is it the most powerful telescope in history, but also its observing grants are now established as a mainstay of support for US astronomers. And after last week's servicing mission ... the telescope has never been more capable.

But NASA needs to retire Hubble to pay for its successor, the Next Generation Space Telescope (NGST)."

During the mission I had given a telephone interview to Reichhardt in which I was frank about the issue and about my wish that Weiler would be "more open minded" about the possibility of an SM5, though I asked him to keep that off the record. He did include one quote from our interview, "Hubble senior project scientist Dave Leckrone of Goddard, who advocates extending the telescope's life, admits it will be an uphill struggle to convince NASA. 'The official position is that there will be no more servicing of Hubble after 2004,' he says."

The column continued:

"A committee representing Hubble users advised NASA last October to consider the extra servicing mission. Committee chairman George Miley of Leiden University in the Netherlands agrees that delaying the NGST would be 'bad for astronomy,' but adds that 'the effectiveness of an extra Hubble refurbishment mission should be

---

[4] 2002, Nature, Vol. 416, page 112.

seen in a wider context and considered within the framework of the NASA programme as a whole.'"

In other words Hubble was of extraordinary value to NASA as a whole, not just to the Office of Space Science. The funding for further servicing to extend its life should come from the overall agency budget. There should not be a life or death struggle between Hubble and NGST within the confines of Ed Weiler's budget—or at least, that's how I interpreted Miley's words.

Reichhardt's column concluded by noting the other side of the debate:

"But in a letter to Anne Kinney, director of NASA's 'Origins' programme, an advisory committee to that programme argued that **minimizing additional expenditures on [Hubble] is crucial to keep the development of NGST on track.**" [Bold lettering added here for emphasis.]

In my mind that statement from a segment of the astronomical community validated the worry I had all along. In the NGST world, Hubble was viewed as an impediment.

Accompanying the Nature article was a cartoon. It showed a large man with a big, toothy smile, identified as "NASA" by the pin on his lapel. He has his hand on the shoulder of a small girl, obviously unhappy judging by her frown. On her sweater is the label, "Hubble Hugger Club." She is eating what looks like a candy stick. The man is saying to her, "Now Honey, you throw that away and I'll give you a better one tomorrow!"

And so there it was. The dispute that had roiled in private for several years within the Hubble team was now out in the open for all to see. Weiler and Kinney were not happy. In particular they were unhappy with me for giving the interview to the Nature writer, although my quote in the article was a simple statement of NASA policy. Anne gave me an earnest scolding. I heard feedback from other Hubble people that Ed Weiler now thought Cepollina and Leckrone were in league in a campaign to promote SM5. That was not in fact the case. I had always pushed back, gently, on Cepi, trying to restrain him from excessive exuberance about SM5. It was a complex and divisive issue that involved a broad segment of the astronomical community. We had to serve that community. We also had to avoid undercutting NGST. Nevertheless, it appeared that I was now also a villain at NASA HQ.

As the feud regarding Cepi and SM5 continued over several years, the relationship between the Hubble Program Office at Goddard and Ed Weiler grew ever more fraught. Communications became strained, or at times non-existent. Ed developed a tendency to make decisions about the Hubble mission without consulting the Goddard office, the organization that would have to implement such decisions.

Perhaps the most problematic example of this was in 2002. At the time the Goddard Hubble Program Office was working toward a March 2004 SM4 launch date. We had just completed SM3b in March of 2002. So, we had only two years to get ready for the next mission, instead of the normal three or more. Moreover, budget reductions had pushed Cepi to reduce his servicing staff to about half its prior size. We had two-thirds the normal preparation time with half the normal staff—a major management and technical challenge, indeed.

At the time there was great pressure on the Office of Space Flight to move as quickly as possible to get the ISS constructed. That in turn meant there was major demand for shuttle launches to the ISS. Apparently, Ed Weiler had received a request to delay SM4 by about one year, to 2005 February, to free up a shuttle vehicle for an additional ISS flight. He agreed to do this. Normally one year of schedule relief would have come as welcome news to the Goddard team. It would give us a chance to slow down, take a deep breath and work on preparations for SM4 at a more deliberate pace.

The cost impact to the HST Program of slipping the SM4 schedule by one year would have been about $100 million. But in return for that extra funding Ed would have greater assurance of mission success. We would be buying down risk. That was a moot point, however. Ed directed Goddard to "manage manpower" to absorb the extra cost or that extra cost would come out of the NGST/JWST budget. In other words get by for two years on one year's funding. To the best of my knowledge, and in the recollections of Preston Burch and Cepi, Ed did not give us a chance to weigh in on the implications of his agreement before he made it.

When asked what "manage manpower" meant in Ed's view, he replied, "Work slower." This made so little sense to us that we had a hard time wrapping our brains around what he was actually asking us to do. One way to save the money to cover an extra year would be to delay the start of new contracts. But that opportunity had passed us by many months before. There were no new contracts coming along, and work under the current contracts was far along toward completion.

The only other way to spread one year's worth of funding over two years was to cut labor costs—either transition to part-time labor or lay people off entirely. But Cepi's team was the gold standard group of experts who knew how to do all of the jobs necessary to service Hubble and who had deep experience working with KSC, JSC, and the astronauts. Once they were laid off, it was highly unlikely that we would ever get them back. Replacing them later with less competent, less experienced engineers and technicians surely would spell an increased level of risk for SM4. In some circumstances it might even increase the risks to the shuttle crew, who relied on Cepi's team for training, tool development, EVA procedures, etc.

The Hubble administrative staff and contractors worked diligently for two months with the budget and manpower numbers, brainstorming how to accommodate Weiler's directive. How do you do the same work for $100 million less than what was needed, while still meeting all the milestone dates that had to be followed for a shuttle mission? We had to follow JSC's "Shuttle Template" that required us to produce specific products (mainly documents) by specific dates, leading up to launch. The schedule path through the template began almost two years before launch. This was the Human Space Flight Program's orderly way of keeping shuttle missions on schedule and with a high probability of success. A year's launch delay would obviously delay the starting date of the template, but the experts required to do that work still needed to remain on the Hubble payroll, doing long-term planning for the mission. The Project couldn't just let them go and expect to re-hire them a year later. The problem was intractable. Our folks could not find a workable solution. We were stuck.

One day in 2002 November, the Program team was asked to give a briefing on Hubble status at one of Ed Weiler's staff meetings at NASA HQ. After we had concluded our standard talks, Ed popped the question I, for one, had been dreading. Rather he phrased it in a declarative way. Paraphrasing, "I'm assuming you guys have figured out how to delay SM4 to 2005. And of course there will be no extra money. Just work slower." Preston Burch did not speak. In the moment, I assumed it would not have been a good career move for him to tell Weiler he couldn't do it. However, I had no such reservations. As calmly and professionally as I could, I spoke back to Ed. Again paraphrasing, "There is no way to achieve what you're asking for without laying off about half of the core team, the people who really know how to do servicing. If this were a production line, maybe we could just work slower, slow down the rate at which pencil sharpeners or whatever came off the assembly line. But we're not talking about that here. What you're requiring puts SM4 at significant risk of failure."

Ed shot back with remarks of his own, which I no longer can recall. But each time he made an assertion that I considered to be incorrect I attempted to rebut it.

This went on for several minutes. Eventually Ed stood up and said, "I like to fire shots across your bow to watch how you react." He then left the room. This was not the first time Ed and I had crossed swords. Ken Ledbetter said, "This has gone on too long." Anne Kinney stood up, looked at me and said, "That was bad. Arguing with him doesn't do any good."

Then Chris Scolese, Weiler's Deputy and a first-rate manager and engineer, asked me, "What do you want from us?" I responded, "We want our mission back!" It later turned out that, after the meeting, Scolese apparently had gone back to the Office of Space Flight and re-negotiated the agreement. Instead of a year's delay, we would only be delayed by six months. Chris found a compromise that greatly helped the Hubble Program. We could make a six months slip work with little or no extra funding.

In the months that followed I bore some short-term career ramifications for having argued with Ed Weiler. However, the situation was tolerable since I recognized that the result of that discussion was a more reasonable, less risky schedule for SM4. This episode rapidly faded into the rearview mirror a year later when NASA Administrator Sean O'Keefe canceled SM4.

The Congressional Appropriations bill covering the funding for NASA for fiscal year 2003 included the following language:

"The conferees direct NASA to carry out an in-depth study of an additional servicing mission (SM5) in the 2007 timeframe that would study operating HST until the Webb Telescope [JWST] becomes operational. The study should address the costs of an additional servicing mission and the potential scientific benefits."

Now Ed Weiler and Anne Kinney had no choice. The question of whether or not to support an SM5 was on the table. Congress required them to address it in a thoughtful and transparent way. The Office of Space Science chartered two separate committees. One of these was dubbed the "HST Post-SM4 Scientific Review Panel." David Black from the Lunar and Planetary Laboratory in Houston was its Chair. The second study group, called the "HST–JWST Transition Panel, had as its Chair

Hubble veteran John Bahcall. Distinguished astronomers whose judgment would be trusted by the community populated each panel. The two studies were carried out during the spring and summer of 2003.

Both panels recognized the extraordinary importance of Hubble to science. The Bahcall group said, "By any standards the HST has been a spectacular success—one of the most remarkable facilities in the entire history of science." Black's panel referred to it as, "… arguably the most effective and productive scientific facility ever launched."

Both panels were of the opinion that, if SM4 were fully successful and if an SM5 were implemented, then Hubble would continue to provide a vigorous program of cutting edge science as it had previously. The Black committee concluded, "… The Panel was unanimous in its view that HST would continue to provide the highest quality scientific return at and beyond the time of a proposed SM5 [2007–2008]." However, "The Panel was of differing views regarding the overall scientific value in the second decade of this century, when it must be compared with other missions that would stand on the shoulders of discoveries made by HST."

Both panels agreed that funding for an SM5 should not come at the expense of JWST development. It was desirable that there be a period of overlap between the two missions to allow synergy between their capabilities.

Bahcall's committee outlined three alternative scenarios for Hubble's end of life. Of these the group strongly preferred one option—two additional servicing missions, SM4 around 2005 and SM5 in 2010. The latter mission would allow a propulsion module to be attached to Hubble to assure that it could be safely de-orbited, but importantly also, it should include routine servicing to extend the observatory's life well into the decade of the 2010s. They believed Hubble should continue in operation until it could no longer produce outstanding science. It should not be de-orbited by an arbitrary date—2010—regardless of how well it was still performing. Both the Black and Bahcall panels recommended that NASA conduct a competition for a new scientific instrument to be installed on Hubble during SM5 to further enhance its potential for new discoveries.

Both Black and Bahcall gave scientific support to the concept of doing an SM5 without explicitly endorsing it. Both provided useful discussion, but hedged on a conclusive final recommendation.

Perhaps without realizing it at the time, both committees made prescient observations about what the future might hold for Hubble and JWST. First, there might not actually be an SM4. At the time of their deliberations the shuttle fleet was grounded because of the Columbia disaster. The future of the shuttle program, and thus the future of Hubble servicing was a big unknown. SM3b had been fully successful. But how long would Hubble last, if it turned out that, unexpectedly, SM3b in 2002 had been the final servicing mission?

Second, as the Bahcall committee pointed out, historically the average time for development of major space astronomical observatories had been 16 years from the time of NASA's Announcement of Opportunity for their scientific instruments to the time they were launched. And so the panel foresaw that JWST might not be launched until 2017 or later. A gap of a decade or more between the end of Hubble science and the launch of JWST was not out of the question. Moreover, termination of Hubble would mean loss of any major observing capabilities at ultraviolet and optical wavelengths for the foreseeable future. JWST could not observe in the ultraviolet at all, and it would have very limited capabilities for observing at optical wavelengths from about 600 nm longward.

On 2003 July 31 the Bahcall panel held a public forum in DC, inviting open commentary and discussion about their study and recommendations. At the end of the day it appeared that most of the 200 astronomers, engineers, and NASA employees present favored extending Hubble's lifetime (Guinnesy 2003), a view shared by Bahcall's six-member committee.

Anne Kinney pushed back, saying, "... continuing the HST servicing and operations beyond 2010 is not in the NASA budget. We believe the money is better invested in newer facilities." She asked the audience which NASA programs they would be willing to eliminate in exchange for extending Hubble's lifetime. There was no answer coming from the assembled group. Of course this was not an easy question. If that were indeed the tradeoff to be made, answering the question would require extended study and deliberation.

The question came up as to whether NASA could approach Congress for additional funding to continue Hubble servicing and operations beyond 2010. Anne said, "Congress can never provide enough money for everything that the science community would like to do.... And do you want Congress deciding the science program instead of the science community?" Her answer didn't seem to recognize any special status for Hubble, either scientifically or in public opinion.

If the community's desire to extend Hubble's lifetime were to be realized, while at the same time protecting JWST's funding, it would require additional money. That money would need to be earmarked both for Hubble operations beyond 2010 and for one additional Hubble servicing mission. It would have to be a NASA initiative, not just an Office of Space Science initiative, recognizing the importance of Hubble to the Agency as a whole. But the circumstances that prevailed at NASA in 2003 after the Columbia disaster would have made any campaign to obtain extra funding for Hubble moot. Things were about to get worse for Hubble—much worse.

The date was 2004 January 16. Preston Burch had spent the last two days at a management retreat led by his boss, Dolly Perkins, at the Wye River Conference Center on Maryland's eastern shore. Dolly was the Director of Flight Projects at Goddard.

That Friday morning Preston had driven directly from his home to Wye River, arriving very early. Dolly came up to him and asked, "Have you heard? SM4 has been canceled." Preston thought she was joking and kept waiting for the punch line. But Dolly insisted, "No, no, I'm serious. It's on the front page of the Washington Post this morning. You've got to get back to Goddard and get the troops together. There's some talk of [Administrator] O'Keefe coming out there today."

Preston was dumbfounded, having heard nothing about this directly from the Hubble office at NASA HQ. He made a dash for his car and headed back to Goddard, about a two hour drive away. There he confirmed what Dolly had said, that O'Keefe was indeed coming out to talk directly to Goddard management and to the Hubble project team. He quickly arranged for a conference room where the Administrator could meet with us.

Shortly after noon I was sitting in my office in Building 7. I was startled when Anne Kinney walked in. "I want you to know that I had nothing to do with this," she proclaimed, with a tone of urgency in her voice. "With what?" I asked. I had no idea what she was talking about. Anne shrugged and said, "You'll see." I stood up and walked to the door to greet her. Sure enough, down the hallway came Ed Weiler, Mike Moore, and a small parade of other HQ Space Science folks. Among them was John Grunsfeld.

That was a surprise. Grunsfeld was our most experienced and accomplished spacewalking Hubble repairman. He was fresh off two highly successful missions, SM3a and SM3b. And now here he was in a suit and tie. He had taken leave from the Astronaut Office at JSC to move to Washington and assume the role of NASA Chief Scientist. In that role he was, among other things, the chief scientific advisor to the NASA Administrator.

Leading this small parade was Sean O'Keefe. Grunsfeld later told us that he had been completely blindsided by the announcement O'Keefe was about to make. He had been given no opportunity to help shape O'Keefe's thinking about SM4.

What I most remembered about the meeting between O'Keefe and the Goddard Hubble team was my own feeling of stunned disbelief and ultimately sadness. Fortunately, Steve Beckwith, the Director of the STScI, took good notes and posted them on the web.[5] Between Steve's notes and my memories, here is what transpired:

The meeting lasted about 45 min with O'Keefe speaking impromptu, without notes. He clearly had given the matter a lot of thought. His bottom line was that future shuttle missions after return to flight would be devoted entirely to completion of construction of the space station. There would be no further shuttle flights devoted to servicing Hubble. He was totally committed to implementing all of the recommendations of the Columbia Accident Investigation Board (CAIB), led by retired Vice Admiral Harold Gehman. No single factor—safety, costs—drove his decision. Rather it was a combination of all the relevant factors. The decision had been O'Keefe's and his alone.

In order to assure that a servicing mission to Hubble was no more risky than a flight to the ISS, the shuttle program would have to develop methods for inspecting the orbiter's thermal protection system—the tiles and the reinforced carbon–carbon panels on the leading edges of the wings and the nose cap—for damage caused by debris flying off the external tank during launch. That had been the root physical cause of the Columbia failure. This inspection would have to be done after the shuttle had reached orbit, and there was currently no method for doing that from a stand-alone shuttle orbiter. For flights to the ISS, the crew within the space station could take photographs of the exterior of the shuttle, including its underside, as the shuttle pilot performed an end-over-end somersault maneuver in front of one of the

---

[5] http://spaceref.com on Friday, 2004 January 16.

ISS's windows. Experts searching for tile damage or wing leading edge damage could inspect these photos after they had been transmitted to the ground.

Next the shuttle team would have to invent methods for repairing damage to the thermal protection system in orbit. Repair materials could be stored on the ISS. Astronauts could do spacewalks to make the repairs as needed.

Finally, the ISS provided a lifeboat for the shuttle astronauts who were stuck in orbit awaiting a rescue of some kind—either another shuttle mission or by using the Russian Soyuz escape vehicle. No such safe haven would be available to Hubble servicing crews. A mission to rescue a stranded Hubble servicing team would require a second shuttle launch and probably would be very disruptive to other missions. Besides, there was no defined technique for transferring crewmembers from one shuttle orbiter to another.

One other telling factor mentioned by O'Keefe was that he had been informed that the Hubble spacecraft would probably only last another three years even after it had been fully serviced. This was a highly pessimistic number. It's not clear how O'Keefe came by this lifetime estimate. But it was a number that was widely tossed around in the media at that time. Hubble science operations had lasted six years—from 1993 to 1999—with gyros that were periodically failing. By 2004 Hubble operators had developed software that allowed scientific observing with only two gyros, or even only one. Our batteries had shown signs of degrading—losing charge capacity. But Hubble engineers had determined that the problem was induced mostly by the way they were testing the batteries. The act of measuring battery capacity was in fact degrading capacity. Our bad! When we started running capacity checks in another way, the rate of degradation was reduced considerably. Even Helen Wong's highly conservative statistical analysis from 2002 showed that there was still a 50/50 chance of Hubble being scientifically productive four years after successful completion of a servicing mission.

To O'Keefe, developing these new techniques to inspect and repair damaged exterior surfaces on an orbiter, or even to rescue a stranded Hubble servicing crew, in order to purchase an additional three years of Hubble operations was not worth the cost or effort. These were one-time mitigations that would never be used again.

After O'Keefe's briefing was over, a lot of glum people walked out of the room. Preston Burch called for an "all-hands" meeting of the Hubble team later that afternoon in the big Building 8 auditorium so that everyone could be informed of O'Keefe's decision. He asked me to say a few word of consolation or inspiration. That had been one of my special duties over the years, to use Hubble's remarkable scientific accomplishments to inspire and motivate the team about the importance of what they were doing. I guess I was the Program Chaplain. But what could I say to the several hundred people assembled that afternoon? Many of their jobs might soon be terminated, or at least re-directed to other work. Basically, I simply thanked them for their hard work and dedication over many years to keep Hubble operating as the premier scientific facility of our time. I believed that centuries into the future, people would still be talking about Hubble and how it turned our understanding of the universe on its head. I think I also mentioned that, someday, I'd like to write a book about their accomplishments.

Reaction to O'Keefe's announcement was swift. The following week, on January 21, Senator Mikulski of Maryland, the ranking minority member of the Subcommittee on VA/HUD and Independent Agencies, the Subcommittee that dealt with appropriating NASA's budget, wrote a letter to O'Keefe stating:

"I was shocked and surprised by your recent decision to terminate the next scheduled servicing mission of the Hubble Space Telescope (HST).... Given Hubble's extraordinary contributions to science, exploration, and discovery, I ask you to reconsider your decision and appoint an independent panel of outside experts to fully review and assess all of the issues surrounding another Hubble servicing mission." She concluded her letter with the statement, "Hubble has become the most successful NASA program since Apollo. It cannot be terminated prematurely with the stroke of a pen without a thorough and rigorous review while planning, preparation and training activities continue."

The press also chimed in. In an editorial dated 2004 February 29, entitled, "Premature death for the Hubble," the New York Times stated, "Cancelation of the servicing mission is being justified on safety grounds, but that is not the whole story. Indeed, it looks as if Hubble is being sacrificed primarily to make way for President Bush's grand new plans to send astronauts to the Moon and Mars in future years.... Our guess is that, with NASA on high alert after the Columbia tragedy, the next shuttle flights will be the safest ever. Astronauts are paid to take risks, and there would be no shortage of volunteers for a Hubble mission that seems no more risky than other flights and [would be] a lot more important scientifically.... The administration essentially argues that the scientific returns are not worth the risk and effort of a servicing flight. Our feeling is just the opposite. The gains from extending Hubble's life are real and achievable and should not be sacrificed for a distant exploration program that for now is mostly wishful thinking and can surely be delayed a bit."

USA Today had a similar take on what O'Keefe had done. An editorial in the 2004 January 19 edition put it this way. "Two days after President Bush announced a push to send a man to Mars, NASA doomed the Hubble Space Telescope by scrubbing a shuttle mission to upgrade the venerable instrument. The result is an inadvertent irony. In the name of sending more humans into space, NASA has pulled the plug on its strongest real-world argument for doing so."

The commentaries in the media were widespread and almost universally critical of what O'Keefe was doing. They tended to focus on the new Bush initiative to send people back to the Moon and to Mars, and the need to free up funds to begin that program as the real, unstated reason for O'Keefe's decision. Be that as it may, some of us in the Hubble Program at Goddard were a bit more forgiving. Our impression was that the NASA Administrator had been hit hard personally by the loss of human life in the Columbia disaster. It had happened on his watch. He was not from the world of high performance aircraft and space flight. He called himself a "bean counter." He probably was psychologically unprepared for the high risks that came with the jobs of test pilot and astronaut. He never wanted to be so close to that kind of tragedy again, especially when there was action he himself could take to prevent it. If he was risk-averse before, he was way beyond that after Columbia.

Initially O'Keefe resisted calls from Congress, the scientific community, the media and the general public to change his mind, or at least to seek the thoughts of others who were well qualified to consider the risks versus the rewards of one additional shuttle mission to Hubble. Eventually he relented. He requested that Vice Admiral Gehman, Chairman of the CAIB, weigh in.

In his response, the Vice Admiral noted that the CAIB report had required NASA to develop an autonomous capability to inspect and repair damaged thermal tiles without the ISS nearby. The need for such a stand-alone repair might be needed, for example, for a shuttle launched to rendezvous with the space station that encountered engine problems and didn't make it all the way up to the ISS. So, O'Keefe was wrong in his assertion that the need for a stand-alone inspection and repair capability would be unique to a Hubble servicing mission and would have no utility beyond a one-time mission to Hubble.

Gehman concluded that, "…. Only a deep and rich study of the entire gain/risk equation can answer the question of whether an extension of the life of the wonderful Hubble telescope is worth the risks involved…"

At that point, O'Keefe, to his credit, gave in and sent a request to the National Academies of Science and Engineering to conduct a thorough assessment of the risk and reward tradeoffs of another Hubble servicing mission. In response the National Research Council organized a study committee chaired by a widely respected member of the scientific and engineering communities, Louis Lanzerotti.

The Lanzerotti panel began work in April of 2004. Because of the urgency of the task with regard to NASA decision-making, they produced an interim report in mid-July of 2004 and a final report the following year.[6]

As usual, Cepi was forward looking, always trying to stay ahead of the game. A couple of months after the crash of Columbia he began to ponder how long it would be before the Shuttle Program returned to flight, or whether the shuttles would ever fly again. Was there any alternative to human servicing in orbit?

Cepi was aware of another possible option—using a robot to service Hubble. The Canadian company MacDonald, Dettwiler and Associates (MDA), the outfit that provided the Remote Manipulator System (RMS) arms that flew in the shuttle's payload bay and on the ISS, had developed a robot for the space station called the SPDM—special purpose digital manipulator. A version of the SPDM named Dextre was part of the Mobile Servicing System on the ISS (Figure 16.1). Used in conjunction with the RMS, it could do repairs that would otherwise require a spacewalking astronaut, but without any of the human risks. The question was, could Dextre be adapted to fly on an unmanned spacecraft and service Hubble by remote control from the ground?

Cepi sent one of his bright young engineers, Jill McGuire, on a field trip to MDA in Vancouver, British Columbia. Jill's plan was not just to discuss the possibilities, but also to collect a set of videos showing the different things Dextre could do that might work for Hubble servicing—open latches, disconnect or connect cables,

---

[6] 2005, "Assessment of Options for Extending the Life of the Hubble Space Telescope: Final Report" (Washington, DC: The National Academies Press).

Figure 16.1. Dextre, the two-armed tele-robot, shown on the outside of the International Space Station. It is operated by ISS crewmembers from inside the station, performing tasks that would otherwise be done by space-suited astronauts. (Credit: NASA, MacDonald, Dettwiler and Assoc.)

unlatch hooks, manipulate boxes, etc—small tasks that could be built into procedures to mechanically and electrically replace systems or instruments.

Jill brought back a mother lode of video evidence showing that such a job might be within Dextre's capabilities. Cepi put together a small team of his engineers to begin brainstorming ideas about how Hubble might be serviced by remote control with a spacecraft carrying Dextre docked to Hubble. Operators on the ground would control the moment-to-moment manipulations of the robot's two arms and hands. So, this would be servicing of Hubble by humans, but with the humans safely tucked away in a control room on the ground.

On 2004 February 21, less than a month after Administrator O'Keefe announced the cancelation of SM4, NASA issued an RFI—a Request for Information—to the community seeking good ideas about how to handle the end of the Hubble mission—how to extend Hubble's useful life and how to de-orbit it safely when the time came to do that. What were the methods and technologies at an advanced level of readiness that might be applied to the Hubble problem? One of the stated purposes of the RFI was "… to invite alternative mission concepts by which NASA may more fully accomplish its goal of maximizing HST science productivity, e.g., life extension approaches and techniques, with or without robotic servicing.…"

As he had so many times throughout his career, Cepi had already anticipated a need and moved ahead of everyone else to satisfy that need. Advanced Camera for Surveys Principal Investigator Holland Ford once called Cepi the "General Patton of HST." That was because when an order came from higher up the chain of command to capture the next village, it seemed that Patton would have already taken that village days before and would have moved beyond that to the next town or bridge.

Preston, Cepi, and Jill McGuire organized a "road show" and took it to Headquarters. It was a presentation about the work that had already been done during the preceding months on a concept for a tele-robotic servicing mission. Sean O'Keefe gladly embraced the idea. It gave him a face-saving way out of his quandary. He certainly didn't want to be the NASA official remembered in history for "killing" the Hubble Space Telescope. The robotics approach would provide him with a defense against the many critics who were excoriating him for his decision to cancel SM4. He gave Goddard the green light to proceed. He even managed to add funding to the Hubble budget to the tune of $85 million to start work on this alternative mission.

For a year from 2004 March to 2005 March, Cepi's tele-robotic team worked very hard, performing some remarkable engineering, developing the details of the servicing mission concept. They called it the HRSDM, which stood for Hubble Robotic Servicing and De-Orbit Mission. The team created hardware designs and operational procedures for how an unmanned spacecraft carrying Dextre, the Canadian shuttle remote arm, and the Hubble payload—batteries, gyros, two new scientific instruments, etc—could rendezvous and dock with the Hubble spacecraft. They demonstrated how Dextre could remove old components and install new ones, all controlled from the ground. They designed the many complicated tools that would allow Dextre to perform these tasks. They designed the robotic spacecraft with a propulsion module, the De-orbit Module that would remain attached to Hubble as the rest of the spacecraft, the Ejection Module, separated and pulled away at the conclusion of the servicing mission (Figure 16.2). Thus, NASA would have the means to safely de-orbit Hubble at the end of its lifetime.

It was estimated by Art Whipple, the Hubble Program Systems Engineer, that the robotic servicing of Hubble, in the end, could do about 75% of the servicing tasks that would have been accomplished by shuttle-based humans. It would take significantly longer than a shuttle-servicing mission—about 73 days versus 13–14 days total mission duration. This difference was due in part to the problem of "latency," that is the lag time between when a command was sent to Dextre from the ground and when controllers on the ground could see Dextre's response—about two seconds. Each move of the robot had to be observed and verified before the next move could be commanded—a very slow process. But with a robot doing the work, what was the hurry? If the mission could be designed to install the two new scientific instruments, COS and WFC3, to replace the batteries and to bring six healthy new gyros on board, that would constitute full success.

In March of 2005 the HRSDM was subjected to the rigorous Preliminary Design Review and Non-Advocate Review process normally required for every NASA mission. Teams of experts not associated with the Hubble Program scrutinized every detail of the design work that had gone into the HRSDM up to that time. Remarkably perhaps, given the complexity and trail-blazing nature of a robotic servicing mission, the HRSDM passed the reviews with flying colors. In the normal course of things, the team would be authorized to proceed onward to the critical design phase in which the final detailed design of every component and of the robotic

Figure 16.2. Artist's rendering of the Hubble Robotic Servicing and De-Orbit Mission. Dextre the robot is shown attached to the end of the Canadian shuttle remote arm. At the aft end of Hubble is shown the HRSDM spacecraft with an upper De-Orbit Module and a lower Ejection Module. At the bottom are solar arrays for electrical power and antennae for communications with the operators controlling the servicing process from the ground. (Credit: NASA, GSFC.)

system as a whole would be worked out. That effort would culminate in a Critical Design Review probably near the end of 2005.

Most of us associated with the Hubble Program at Goddard just moved along with the flow of this effort, hoping beyond hope that this radical new approach to in-orbit servicing might give us our new instruments and reinvigorate Hubble for the long run. One exception was Preston Burch. He had begun to have doubts. The estimated cost of the HRSDM was ballooning to something approaching $2 billion. And there were so many places where things could go wrong, so many opportunities for the mission to come to an unhappy outcome.

Regardless of whether we got permission to proceed with a robotic servicing mission to Hubble, the excellent engineering work Cepi's team had put into the HRSDM concept laid a foundation for potential future unmanned servicing missions, for example, to the JWST. It was designed to orbit the Sun one million miles away from Earth at the Earth–Sun L2 Lagrange point. Human servicing of JWST would be out of the question. But a robot might fill the bill someday. A future robotic mission to attach a de-orbit propulsion module to the back end of Hubble would likely be necessary, also. The time for something like an HRSDM would come eventually.

The Lanzerotti committee, the panel chartered to provide the "second opinion" Senator Mikulski and others in Congress had requested, was having a close look at both options—going forward with the HRSDM or overturning Sean O'Keefe's decision to cancel the shuttle-based human servicing originally intended for SM4. They favored the latter option, arguing that the technology for a robotic mission wasn't sufficiently advanced. They concluded:

"The need for timely servicing of Hubble imposes difficult requirements on the development of a robotic servicing mission. The very aggressive schedule, the complexity of the mission design, the current low level of technology maturity, and the inability of a robotics mission to respond to unforeseen failures that may well occur on Hubble between now and the mission make it highly unlikely that the science life of HST will be extended through robotic servicing.

A shuttle-servicing mission is the best option for extending the life of Hubble and preparing the observatory for eventual robotic de-orbit; such a mission is highly likely to succeed."

Lanzerotti's panel made the following three recommendations:
1. The committee reiterates the recommendation that NASA should commit to a servicing mission to the Hubble Space Telescope that accomplishes the objectives of the originally planned SM4 mission.
2. The committee recommends that NASA pursue a shuttle-servicing mission to HST that would accomplish the above stated goal. Strong consideration should be given to flying this mission as early as possible after return to flight.
3. A robotic mission approach should be pursued solely to de-orbit Hubble after the period of extended science operations enabled by a shuttle astronaut servicing mission, thus allowing time for the appropriate development of the necessary robotic technology.

Sean O'Keefe left NASA in December of 2004. He went on to become Chancellor of Louisiana State University.

In March of 2005 President George W. Bush nominated Dr Michael Griffin to be the new NASA Administrator. His confirmation by the Senate would come in April. Mike Griffin was a brilliant guy. He held seven academic degrees in Physics, various areas of Engineering, and Business Administration. He had occupied positions of leadership at a number of aerospace companies, and had been a professor at several universities. In the late 1980s he did a tour at NASA Headquarters as Chief Engineer and later as Associate Administrator for Exploration. On top of all of this, he was a pilot and certified flight instructor. Prior to being nominated as the new NASA Administrator, Mike had been named head of the Space Department at Johns Hopkins Applied Physics Laboratory (APL) in 2004.

One day in late March of 2005 a slightly built man, informally dressed and wearing a ball cap, walked into the Hubble Program office suite in Building 7 at Goddard. He looked vaguely familiar. It was Mike Griffin. It seemed likely that he would soon be confirmed as Sean O'Keefe's replacement at Headquarters and thus would become our new boss. That was a great relief to all of us. Here was someone who was intimately familiar with the kinds of jobs we did and who had the perfect technical background to be a good fit in running the Nation's space program.

Several of us sat around Preston Burch's conference table and had an easy chat for 45 min or so with this very accomplished man. He reminded us that he had once done some work on Hubble. My recollection was that he had been part of an effort in the 1980s at APL to come up with an alternate gyro design. He said to us straight away, "You guys need the shuttle! The technology is simply not ready to do a robotic mission."

That May, after Mike Griffin assumed the Administrator's chair at NASA HQ, he promulgated three conclusions about Hubble servicing:
1. Due to its complexity and risk, robotic servicing is "off the table."
2. NASA will prepare for a Shuttle servicing mission to HST while awaiting the results of Shuttle Return-to-Flight activities and lessons learned.
3. NASA will fully fund the HST servicing effort to the $291 million level provided by the Fiscal Year 05 Congressional appropriation.

At Goddard we were directed to phase out all work on HRSDM and to get back to work expeditiously on a Shuttle-based SM4. Joy spread throughout the Hubble work force. The extreme "low" we had all experienced on 2004 January 16 when Sean O'Keefe canceled SM4 was now fully offset by the exhilaration of knowing we could all resume doing what we were good at—teaming with JSC and KSC to prepare for another human expedition to our telescope in orbit.

Shuttle Return-to-Flight had a false start with STS-114 on 2005 July 26. Videos taken during launch showed debris shedding off the external tank during ascent—the same issue that had spelled doom for Columbia. As a result the next shuttle launch was delayed until 2006 July 4 to allow additional design modifications to be made. STS-121 on Discovery marked the resumption of regular shuttle missions, entirely for purposes of continuing to construct the ISS, to transfer crewmembers, and to resupply it with provisions.

Mike Griffin took Sean O'Keefe's worries about the safety of a shuttle mission to service Hubble very seriously. When he became Administrator he didn't cavalierly dismiss O'Keefe's concerns, nor immediately re-institute the plan for a shuttle-based SM4. Rather he commissioned a thorough review of the probability of loss of crew and vehicle during a shuttle flight to the ISS in comparison to the corresponding probability of loss during a flight to Hubble. Apparently he personally spent time at JSC going over the numbers, the probabilistic risk analyses carried out by the experts on such things. Mike was fully qualified to do that and to challenge their assumptions and calculations.

On 2006 October 26 Griffin called a meeting at NASA HQ of all the parties involved in planning a shuttle-based servicing mission. He reviewed the status of the risk assessments, the status of shuttle safety provisions instituted after return to flight, the readiness of JSC and KSC to support a mission, and the readiness of the Hubble Program to execute the mission. His conclusion about the risk of loss of crew and vehicle was formally about the same for both a flight to ISS and a flight to

Hubble—roughly 1 in 200. It helped the case for the Hubble mission that a second shuttle would be on the other launch pad at KSC, ready to launch a rescue mission, STS-400, within seven days of being called up.

Preston gave a presentation on the status of SM4 preparations. In particular he emphasized that a launch date no later than mid-2008 would be prudent to minimize the risk that major failures on the spacecraft would cause an interruption to science operations.

I gave a briefing entitled, "WHY SM4?" The message was clear:
1. We hadn't yet approached the limits of what Hubble could do,
2. At the conclusion of SM4 Hubble would be at its scientific apex,
3. As a "general purpose" observatory, Hubble provided diverse and powerful tools to attack problems at the frontiers of most areas of astronomy,
4. User demand for Hubble time continued unabated with a 5:1 oversubscription rate,
5. Scientific productivity, as measured by the number of peer-reviewed papers in professional journals, continued to grow, outpacing all other astronomical facilities,
6. Hubble continued to inspire the public; it was a national icon and source of pride.

On 2006 October 31 we all were called to assemble in the large Building 8 auditorium at Goddard. The place was packed with the Hubble team. Senator Mikulski was sitting in the front row. Ed Weiler sat next to her. Mike Griffin was called upon to say a few words. His words were galvanizing, "We are going to add a shuttle-servicing mission to the Hubble Space Telescope to the shuttle manifest to be flown before it [the shuttle] retires." The crowd rose and cheered. Hubble servicing was back.

Before concluding this chapter, I need to return to the subject of the tension that pertained between the Hubble and the NGST/JWST programs between 1996 and 2004. Are there any lessons to be learned from that situation? I believe there are.

The Hubble Space Telescope was the new crown jewel of astronomy, decades in the making. Being entrusted with stewardship of crown jewels brings with it a high level of responsibility and the need for a certain amount of humility. One's first duty is to take good care of what one has been given, to be grateful for it and to try to make the very best use of it, to exploit it the greatest extent possible. I believe the community got way out ahead of itself in its excessive exuberance for the NGST before the time was right for such a mission. In so doing, it came to view Hubble as an impediment. Instead of being developed as Hubble's successor, NGST became Hubble's near-term competitor. Had the plan for a truncated lifetime for Hubble actually been executed, the community would have faced a 10 or 12 year hiatus during which it would have had no major optical observatory in space—a very unfortunate outcome. But fate and tragedy—the Columbia accident—carried us in a different direction.

When Hubble was being constructed in the 1980s and we all were eagerly anticipating what doors of knowledge it would open, we often used the mantra,

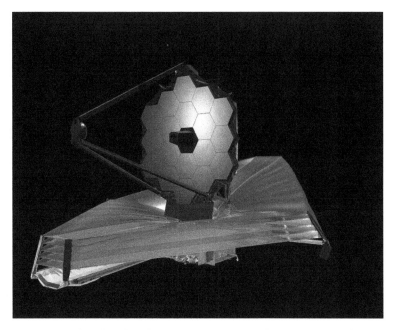

Figure 16.3. Artist's rendering of the fully deployed JWST. The area above the sunshield faces away from the Sun and remains perpetually cold. The structure hanging down below the sunshield is the spacecraft bus. The scientific instruments are in a large module behind the primary mirror. (Credit: NASA.)

"Conscious expectation of the unexpected." There were questions that we didn't yet know how to ask with answers that might upend our understanding of how the universe works. That same phrase might also be applied to the development of major space facilities. No matter how carefully one plans in advance, and how much cash one holds in reserve, one should always expect to be bitten by the unexpected. JWST is a classic example. It was originally expected to cost less than $1 billion and to be launched in 2007. At the time I write these words (Autumn of 2019) it is scheduled for launch sometime in 2021 (Figures 16.3–16.5). Its total life-cycle cost is now estimated at approximately $10 billion.

In Peter Stockman's 1997 compendium of all things NGST, referenced earlier in this chapter, the section on *Technology for the Next Generation Space Telescope* begins with the statement: "Advanced technologies are a crucial strategic element in NASA's mission plan for NGST. Based on the findings of three independent study teams, we have concluded that **no new invention is required** to carry out the NGST mission." Fast forward 21 years: on 2018 October 29 Tom Young, the chair of the Independent Review Board overseeing JWST development, said to a committee of the Space Studies Board, "I personally, have come to the conclusion that **JWST had too many inventions**, too much risk, and was a step too far (Foust 2018)." Clearly the

Figure 16.4. The fully deployed flight Optical Telescope Element of the James Webb Space Telescope being tested in the large clean room in Building 29 at the Goddard Space Flight Center in 2017. The primary mirror consists of 18 hexagonal mirror segments made of beryllium coated with gold. All together it measures 6.5 m (21.3 feet) across, giving it about 7 times the light collecting area of Hubble. The gold-coated secondary mirror, supported by three lightweight booms is round, 0.74 m (2.4 feet) in diameter. Two other mirrors are housed in the trapezoidal box at the center of the primary mirror. (Credit: NASA, GSFC, NGST Corp.)

fundamental assumptions underlying the NGST/JWST mission must have evolved over the intervening years.

By this I do not mean to criticize the NGST/JWST program. I firmly believe that JWST will be a superb new observatory, capable of expanding the frontiers of knowledge well beyond what Hubble achieved. I believe it will be the best it could be. And that probably is as it should be. I believe the people managing and directing the program were outstanding professionals who did their jobs well and acted in good faith. But to this day I still feel the sting of some senior members of the astronomical community applying intense pressure on us to ramp Hubble down as much as possible as quickly as possible in order to accelerate the development of JWST so that it could fly in 2007 or 2008 or 2009 or 2010…

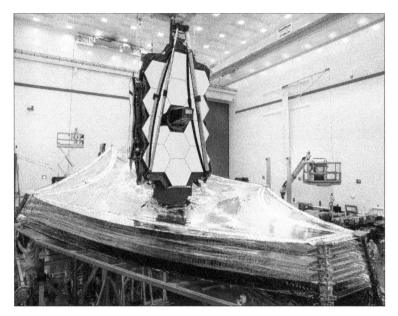

Figure 16.5. The completed flight JWST partially deployed at the Northrup Grumman plant in Redondo Beach, California. The tower housing the secondary mirror and two portions of the segmented primary mirror are not deployed in this view. For scale note the person standing at the lower left. (Credit: NASA, C. Gunn.)

From my perspective the lesson to be learned here is, *fully embrace the good thing that you already have in hand, and don't discard it before you're absolutely certain that the better thing to follow is real.*

References

Foust, J. 2003, Considering the Fate of Hubble, http://www.thespacereview.com, August 11
Foust, J. 2018, Independent board chair calls JWST a 'step too far', spacenews.com, November 1
Guinnesy, P. 2003, PhT, 57, 29
Illingworth, G. D. 2016, STScI Newsl, 33, 1

AAS | IOP Astronomy

Life With Hubble
An insider's view of the world's most famous telescope
David S Leckrone

Chapter 17

Replenishing the Toolbox: The Science Instruments of SM4

A carpenter cannot build a solid house or craft fine furniture without a complete set of the right tools—a hammer and chisel alone will not do. Similarly, an astronomer or astrophysicist can't develop a solid story about how things work in the cosmos without the right tools. A camera attached to a telescope taking a single image of a star cluster, or a galaxy, or a planet can provide information about the size, structure, and relative brightness of the object. But add a set of color filters to the camera and take images through each one, and you will get a treasure trove of information about the physical nature of the object—the age of the star cluster, or the state of evolution of different regions within the galaxy, for example. Use a spectrograph to spread the light out across a light-sensing detector and sort the light in order of the wavelengths of its intrinsic colors. Then you can go looking for patterns of dark or bright lines to find what chemical elements the object is made of, its temperature and density, its velocity of motion. Using these kinds of observations, astronomers can go searching for super-massive black holes at the centers of galaxies or investigate the atmospheres of extra-solar planets, for example.

As SM3b, the 2002 servicing mission, receded farther into the past, Hubble's scientific tool kit was becoming depleted. By 2008, the year NASA expected to launch SM4, only two of the scientific instruments remained usable—WFPC2 and NICMOS. The former was nearly 15 years old and technologically out of date; the latter, our only near-infrared instrument, was aging but was still functioning by the grace of advanced technology—the NICMOS cryocooler. The imaging spectrograph, STIS had failed completely. It had lost its Side 1 electronics to a short circuit in a power supply in May of 2001. Its redundant Side 2 electronics went down with a similar problem in August of 2004. A short circuit disabled a Side 1 power supply in the ACS—our best camera—in June of 2006. ACS continued operating on its Side 2 electronics. However, in January of 2007, a short circuit took out a power supply on Side 2. Two of the three camera channels—the High Resolution Channel and the

Wide Field Channel—were totally out of business. Only the ACS Far-Ultraviolet Channel—the least used of the three—remained in operation.

A decade before, NASA had gotten a head start in developing advanced new scientific instruments in anticipation of a final servicing mission in 2002. When STIS failed, the creative juices of Cepi's engineers started to flow once again. There was a great deal of optimism that a way could be found to replace the circuit board on which the blown-out power supply resided, either robotically or by a space-suited EVA astronaut. When ACS failed a couple of years later, there was additional motivation to be even more creative. If we could fully refill Hubble's scientific toolbox during SM4, then we would be leaving the observatory more capable than it had ever been before.

In November of 1996 NASA issued an Announcement of Opportunity (AO) to propose scientific investigations and a new scientific instrument for the 2002 Hubble servicing mission. Strange as it sounds, given everything that happened in subsequent years, in 1996 the 2002 mission was expected to be the fourth and final servicing mission to Hubble, SM4. By plan the Hubble Program was to end in 2005. Given the short time span between the 2002 servicing mission and the nominal end of the program, Ed Weiler wanted this AO to be open to niche instruments, anything the community wanted to propose as a last hurrah for Hubble. The selection of the new instrument was scheduled for the following summer.

In late July 1997, after the usual formal peer review of perhaps a half dozen submitted proposals, Associate Administrator for Space Science, Wes Huntress, authorized the selection of the Cosmic Origins Spectrograph (COS) proposed by Jim Green (Principal Investigator) and his team at the University of Colorado. This was indeed a niche instrument. It was optimized to observe the red-shifted lines of neutral hydrogen (and perhaps other elements) absorbed by the clouds of gas concentrated into filaments that spanned the vast, nearly empty regions between the galaxies—the "Cosmic Web." Green's team had designed COS to be as sensitive as possible to far-ultraviolet light with wavelengths between about 110 and 150 nm.

While the AO was still on the street, in the spring of 1997, Associate Administrator Huntress formally extended Hubble's mission termination date to 2010. So, Weiler's original rationale for a niche instrument no longer applied. I felt we needed a more versatile instrument that would have wider appeal to the astronomical community over the longer haul.

As proposed, COS was a relatively inexpensive instrument. Its funding profile put forward in the Colorado proposal would use up only about half of the budget Headquarters had allocated for new instrument development. Immediately after the peer review was over, when it seemed obvious to me that COS would be selected for SM4, I made a dash to Ed's office. I pointed out to him that, since the Hubble mission had recently been extended to 2010, our instrument needs had changed. I suggested that we use the balance of the allocated budget to do two things: (1) Protect Hubble's ultraviolet-visible (UVIS) imaging capability so that the observatory would not "go blind" during the next decade. We should build a new WFPC, perhaps a "clone" of WFPC2, much as Ed had argued for in the mid-1980s to back up Hubble's original WFPC; and (2) Expand COS's spectroscopic capabilities across a wider wavelength band, into the near ultraviolet, to make it more useful to a larger number of astronomers. Ed embraced these ideas.

Designing COS to cover near-ultraviolet wavelengths was relatively straightforward. Jim Green had teamed with Ball Aerospace, situated just a few miles away from his office in Boulder, to build and test his new instrument. Ball had already built three successful Hubble instruments—GHRS, STIS, and NICMOS—and was at work on a fourth, Holland Ford's Advanced Camera for Surveys (ACS). STIS contained two-dimensional imaging detectors sensitive to ultraviolet light called MAMAs, an acronym for "multi-anode micro-channel arrays." Gethyn Timothy of Stanford University originally invented them. Ball built a number of MAMAs, only four of which proved to be well-suited for space flight—two optimized for far-ultraviolet wavelengths and two optimized for the near ultraviolet. STIS carried two of these, one each for the far and near ultraviolet. That left two very good flight spare MAMAs on the ground at Ball. Holland Ford claimed one of these—the far-ultraviolet unit—to install in ACS. That left the remaining near ultraviolet MAMA for Ball to build into COS to give it the broader wavelength coverage we desired—out to about 320 nm.

Starting from scratch to build a new UVIS camera was another matter entirely. With a scheduled launch date for SM4 in 2002, there was simply no time to go through, once again, the many-months-long process of putting out a new AO to solicit teams to compete for the job of designing and building the new camera. Other than for the sake of process and appearances, it didn't seem necessary to do that. Goddard, the STScI, JPL, and Ball together had enormous expertize and resources to design, build, and test the new camera.

It would be developed "in-house" and provided to the community by NASA as a "facility instrument." No guaranteed Hubble observing time or other benefits would accrue to the scientists involved in its creation. After it was installed on the telescope during SM4 it would be a resource for all to use, based on the normal competition for observing time all Hubble observers had to go through. Associate Administrator Huntress approved this plan in a letter to the Goddard Center Director, dated 1998 March 11.

So Goddard and the STScI proceeded with all due speed to set up the engineering and scientific teams needed to create what would later be named Wide Field Camera 3 (WFC3). This included the selection of a voluntary, independent outside Science Oversight Committee (SOC), to establish the design parameters for WFC3 and to provide community-based oversight of its development. Bob O'Connell of the University of Virginia agreed to chair the SOC.

The work to design and develop a "baseline" UVIS WFC3 began in earnest in early 1998. Engineer Bryan Fafaul from Cepi's Flight Systems and Servicing Project was put in charge as overall manager of both WFC3 and COS. Project Scientist Ed Cheng (of NICMOS cryocooler fame; see Chapter 13) was designated as the lead Instrument Scientist and John MacKenty of the STScI became the Deputy Instrument Scientist for WFC3. I took on the Instrument Scientist duties myself for COS.

Within the Project Ed Cheng was campaigning for the addition of a near infrared channel to the WFC3 instrument. He reasoned that there would be a lot of unused space available in WFC3. With up-to-date (late 1990s) CCD light sensor technology, we could achieve a wide field of view with only one large detector chip (or more precisely two large chips butted together to make a 16 megapixel camera), plus one set of optics and one set of electronics to cover the UVIS spectral range, from 200 to 1000 nm. This was unlike WFPC2, which was limited by the CCD technology of the 1980s. It needed three smaller CCDs for its Wide Field Camera mode and one CCD for a higher resolution Planetary Camera mode (see Chapter 3). That added up to four separate cameras with four sets of optics, four sets of electronics, and a fully packed instrument. Moreover, the technology of infrared light sensors had moved way beyond those that had been used in NICMOS. An IR channel with a modern detector on WFC3 would give considerably better performance than its older predecessor and provide a valuable bridge to the future capabilities of JWST.

Meanwhile, the idea was making the rounds within the Hubble observer community that it might be possible to add a second channel to the WFC3 that would extend its wavelength coverage into the near infrared. Astronomer Charles Beichman of JPL pointed out that Hubble's heat output (the main background noise source for any infrared instrument in this un-cooled telescope) remained fairly low to just beyond a wavelength of 1700 nm. The thermal background noise then started rising precipitously toward longer infrared wavelengths. A near-infrared instrument limited to wavelengths shorter than about 1700 nm would not require a cryogen or any other complex cooling provisions such as the NICMOS Cooling System. It would only need a thermoelectric cooler (TEC) similar to the ones we were already using to cool our CCD cameras to moderately low temperatures (about −80°C in WFPC2).

A wavelength span from 200 to 1700 nm—the widest ever in a single Hubble instrument—would make WFC3 a truly "panchromatic" camera. The scientific rationale was similar to that for NGST. Extending Hubble images farther into the infrared, would allow it to penetrate deeper into the high-redshift universe, observing ever closer to the Big Bang, as well as enabling numerous other important science objectives.

The idea for an IR Channel to be added to WFC3 received an outpouring of support from the user community. Multiple senior review committees urged the Hubble Program to pursue it. The added funding needed could come in part from the pool of money earmarked for research grants to Hubble observers. In this way, the astronomers who were pleading for the new capability might bear a portion of the extra costs—they would have "skin in the game."

At first I resisted this groundswell. This was not what Ed Weiler and I had agreed to—a simple, inexpensive clone of WFPC2, to assure that UVIS imaging capabilities would be available all the way to 2010. But it didn't take long until the scientific arguments, the interesting infrared results already being produced by NICMOS (with older technology) and the high level of community pressure caused me to relent. I became a true believer in the merit of this idea.

Weiler informally approved of the IR initiative in August of 1999. He stipulated that it must be fully funded from within the Origins Theme (the part of Space Science at NASA HQ that included the budgets for Hubble, NGST, and several other programs), and that any re-phasing of funding between Hubble and other Origins Projects must be accomplished to the mutual satisfaction of all the Project Managers involved.

When Ed Weiler became Associate Administrator for Space Science in 1998, Harley Thronson replaced him in an acting capacity as Origins Theme Director while a permanent replacement was being recruited (that would be Anne Kinney, who came on board in October of 1999). Harley was successful in negotiating a re-distribution of funding between two projects, the Space Interferometry Mission (SIM) and Hubble, which in fact solved budgetary problems for both of them. On 1999 August 12 Harley authorized the Hubble Program at Goddard to proceed with the addition of the IR Channel to WFC3.

As a year or two went by, Weiler and Kinney seemed to become more ambivalent about the IR Channel. Adding the second channel had essentially doubled the instrument's projected cost. The cost of WFC3 (and everything else associated with Hubble servicing) was also growing because of the steady slippage of the projected launch date for SM4 from 2002 to 2003 to 2004 and ultimately 2005, a result of the vagaries of scheduling shuttle launches during the era of intensive International Space Station construction. On several occasions they appeared to be looking for a pretext for removing the IR channel as gracefully as they could, while not bluntly directing the Hubble Program at Goddard to take it out. But they were aware of the very strong support it had in the community. After a massive lobbying effort by the American Astronomical Society, by individual astronomers and by contractors, all requesting restoration of Space Science funding that had been cut from NASA's budget, a Congressional Appropriations Conference Committee had even included language supporting the IR Channel in its Conference report in October of 1999. I assumed Ed and Anne didn't want to be seen as the bad guys responsible for killing such a popular potential new Hubble capability.

I was told by a reliable source that when a representative from Ed's office approached Senator Mikulski's staff at one point to broach the subject of de-scoping WFC3 by removing the IR Channel, the person was told that the Senator didn't want that to happen, or words to that effect. Apparently the STScI and its parent corporation, AURA, had been tipped off that this move was afoot and headed it off by getting to Mikulski's staff beforehand. The Director of the STScI at that time, Steve Beckwith, was an ardent champion of the IR channel. He had expended a great deal of his own "political capital" to assure that the IR channel remained in WFC3. After a while, with the final two-channel design of WFC3 approaching its completion and construction beginning, it would probably have been more expensive to take the channel out than to leave it in anyway.

In 2002 Project Scientist Ed Cheng, who had been Cepi's go-to guy whenever a difficult technical problem with the science instruments had come up, left the Civil Service to form his own private consulting firm, Conceptual Analytics. But he still came when Cepi called, now as a contractor instead of a government employee. Ed continued under contract to help monitor the development of the new detectors for the WFC3 IR channel being manufactured at Rockwell Scientific (later to become Teledyne Imaging Sensors) in Camarillo, California.

Astrophysicist Randy Kimble replaced Ed as Project Scientist for Flight Systems and Servicing, "just at the moment," as Ed said, "when the job got really hard." Randy also took over as the WFC3 Instrument Scientist. Randy and John MacKenty shepherded the instrument successfully through its difficult testing and calibration program. They were responsible for assuring that WFC3 actually worked.

The instrument ended up going through three major thermal-vacuum (TV) tests. A full-up end-to-end TV test is essential for a newly built science instrument intended to fly on a satellite. The TV test is where an instrument either functions properly and proves its mettle in meeting its scientific performance specs—sensitivity, background noise levels, image quality, etc—or it doesn't. Usually it doesn't, the first time it's tested under the harsh environmental conditions it will encounter in space. It takes a technically astute scientist, leading a capable team, to find all the "bugs" and to figure out how to fix them. That was Randy Kimble's special talent. (After SM4 he was asked to do the same kind of work on the JWST observatory, as Integration and Test Project Scientist.)

The first TV test of WFC3 took place in late 2004, after Sean O'Keefe had canceled SM4. Headquarters had the forlorn hope that WFC3 and COS could find rides on other satellite missions. It was a "forlorn" hope because the two instruments had been built specifically to work with the Hubble telescope optics and with the unique interfaces—power, commanding, data flow, pointing control, etc—provided by the Hubble spacecraft. No other mission was likely to possess those same interface characteristics. I thought the two new instruments were destined to end up in a museum as a curiosity of America's space program. Nevertheless, we were ordered to give them a thorough set of tests, document what they could do and what their problems were, and put them into mothballs for some undefined future use.

It was under those circumstances that Randy, John MacKenty and a group of scientists from Goddard and from the STScI, put WFC3 through its paces inside the giant Space Environment Simulator (SES) at Goddard in TV test #1 (Figures 17.1 and 17.2). A COS team from the University of Colorado, headed by astronomer Cynthia Froning, went through a similar process at Goddard with their instrument.

Going into the test, it was known that the IR detector—an early-generation sample of the particular recipe that had to be developed for WFC3—had disappointing performance, and this was confirmed by the end-to-end testing. The CCD detector covering the ultraviolet-optical wavelengths worked well. However some of the UVIS filters had terrible problems with ghost images produced by internal light reflections. The IR grisms (gratings scored on the side of prisms that produced very low resolution spectra) were mounted with a rotation error of nearly 90°. In addition several heat pipes required for cooling the detector housings under-performed significantly and needed to be improved. Finding basic problems such as these was exactly why instrument-level testing was essential.

Figure 17.1. Space Environment Simulator at Goddard Space Flight Center in which both WFC3 and COS received thermal-vacuum testing to make sure they would function properly under the conditions to be found in Earth orbit. The SES is 40 feet tall and 27 feet wide. (Credit: NASA, GSFC.)

TV test #2 was executed in 2007 after SM4 had been re-instated. Most things in the WFC3 instrument now worked well. The original IR detector had been replaced with a later-generation device with significantly improved performance. However, the replacement had an inordinately large number of "hot" (high noise level) pixels. In flight it would have been a major operational nuisance to produce "clean" images with the replacement detector. In addition the UVIS detector's thermoelectric cooler had cracked during earlier testing, requiring a late, delicate, but ultimately successful repair. There were also some problems with the fine control of the temperature of the IR detector that required electronics adjustments after the test. But in general the earlier bugs had all been worked out and the performance of the instrument was judged to be acceptable.

One of the maxims by which I have always lived is that "even the darkest cloud has a silver lining." That was demonstrated once again in the case of the WFC3's IR Channel. The HgCdTe (mercury–cadmium–telluride) infrared detectors originally

Figure 17.2. WFC3 being set up for a thermal-vacuum test inside the Space Environment Simulator at Goddard. To the right, wrapped in insulation, is a device that simulated the aberrated images produced by the Hubble telescope. (Credit: NASA, GSFC, the WFC3 Team.)

manufactured by Rockwell were not good performers. However, if we had launched WFC3 in 2004 or 2005, we would have had to make do with them.

With Hubble Program support at Goddard, Ed Cheng had started a new laboratory, the Detector Characterization Laboratory (DCL) on our campus in Greenbelt, initially to support the work for WFC3. Scientists in the DCL were able to perform very sophisticated tests and characterization of light-sensing detectors, including those coming from Rockwell. The DCL went on to support numerous flight projects, large and small, in addition to Hubble. These included the JWST, the European Euclid mission and the Wide Field Infrared Survey Telescope (WFIRST).

Over the years leading up to SM4, Rockwell had used DCL test data to better understand and improve their own IR detector fabrication processes. In 2006 the Rockwell team asked to give us a briefing at Goddard on the improvements they had made in their devices, based on the DCL data. This was to be a "closed" briefing because sensitive, proprietary information would be discussed. At the time I found it ironic and sad (if not a little funny) that, because of U.S. Government regulations, Bob Hill, the scientist who had led the detector test program at Goddard that generated all these new data on the Rockwell detectors, was forbidden from attending the briefing. You see, he was a foreign national—a Canadian. "You just can't trust those Canadians!" I said facetiously at the time.

Rockwell had concluded that they would now be able to produce infrared detectors for WFC3 that were far superior to what they had done previously. They offered to produce a new batch at no cost to the Hubble Program for Bob Hill and his team to test. If we liked them, we could purchase at cost a new batch of flight candidate devices that might provide a better choice for the IR Channel. We referred to this late initiative as "Cepi's Last Stand."

Many years before, Cepi had insisted that the Hubble instruments be designed so that their detectors could easily be removed and re-installed. (He had visions of them being replaced by space-suited astronauts during servicing missions.) So, once the new batch of improved detectors became available, it was a relatively straightforward matter to identify and select the best one, based on DCL testing. That device was superior to anything that had been available to us before (and, by the way, was purchased for a relatively small sum). The team packaged it into a flight camera head and mounted it into WFC3 just in time to be tested in TV test #3 in the spring of 2008, shortly before it was to be sent to the Cape for launch. That TV test confirmed how wonderful the performance if the IR Channel, with its up-to-the-minute detector technology, truly was. Its scientific capabilities became extraordinary essentially overnight. Having extra time to bring the IR detector technology to maturity and to get the very best detector available installed into WFC3's IR Channel before it was launched was a small silver lining within the tragically dark cloud that had hung over us after Columbia was lost.

Only once before, in the history of human space flight, had a space-suited astronaut opened up a scientific instrument in orbit and repaired it. That was Pinky Nelson who replaced the interface control board in the Ultraviolet Spectrometer/Polarimeter on the Solar Maximum Repair Mission in 1984 (see Chapter 2). Without special tools he had great difficulty doing the job.

Fast forward to August of 2004. Sean O'Keefe had canceled SM4 the prior January. And now Hubble's primary spectrograph, STIS, had blown a low voltage power supply in its redundant Side 2 electronics. The instrument was dead. Cepi's team was in the midst of figuring out how to service Hubble from the ground with a robot in orbit. Naturally, they started to think about how they might bring STIS back to life with a robotic repair.

Of the two failed redundant sides of the STIS electronics, the one that had failed earlier, Side 1 in 2001, would provide the easiest accessibility through Hubble's opened +V2 aft shroud doors, and so would be the easiest to repair (Figure 17.3). Getting access to the single faulty power supply circuit board involved removing the cover from the Side 1 main electronics box (Figure 17.4). The cover was conveniently located on the outside of the instrument as part of its enclosure.

First the robot would use special tools to remove the EVA handle that spanned the MEB-1 cover and was in the way. Then it would attach an automated tool (AT) to the cover. The job of the AT was to unscrew the 111 screws that held the MEB cover in place and to capture them so that they wouldn't go floating away and cause mischief inside Hubble's aft shroud or in the shuttle's Payload Bay. Why so many screws one might ask? The screws served not only to hold the cover in place and to maintain mechanical integrity under the stresses of launch, but also to help transfer heat away from the 13 circuit boards inside the MEB.

The AT consisted of a rectangular frame that could move a selection of screwdriver heads to any point within a two-dimensional surface, engage a screw head, and rotate it until the screw came free. A computer program controlled the

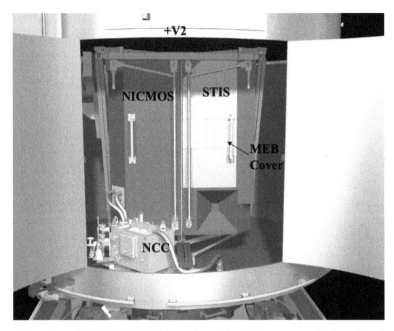

Figure 17.3. Orientation of STIS inside Hubble's aft shroud. By 2004 power supplies in both main electronics boxes (1 and 2) had failed. MEB-1 was the only one that was accessible. STIS resides next to the NICMOS instrument, and the box in front of NIMCOS is its cryocooler. (Credit: NASA.)

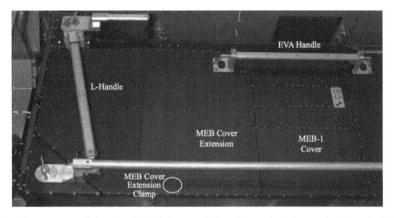

Figure 17.4. Close-up view of the side of STIS that would be addressed during repair. Note the EVA handle that had to be removed by unfastening four screws that held it in place, and the 111 screws holding down the MEB-1 cover that had to be unfastened, captured, and stored. (Credit: NASA, Ball Aerospace.)

process of aligning the screwdrivers to each of the 111 screws. After they were all unfastened the robot would remove the AT, still tightly coupled to the MEB cover with the loose screws all safely captured between the two. The robot would use other special tools to remove and replace the faulty circuit board (Figure 17.5). It would then replace the MEB cover with one of alternative design that did not require all

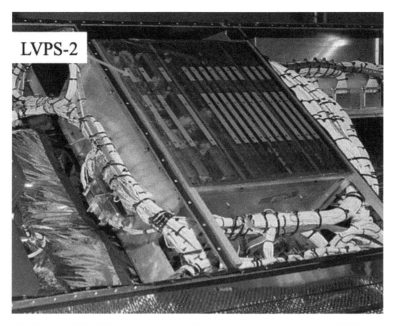

Figure 17.5. The interior of the Side 1 main electronics box that powered and controlled STIS up until its failure in 2001. This image was taken at Ball Aerospace before STIS was launched. The low voltage power supply that failed is labeled. Astronauts would gain access to the interior of this box, remove the failed LVPS-2, and replace it during SM4. (Credit: NASA, Ball Aerospace.)

those screws to do its job. This would all be controlled or monitored from the ground.

In May of 2005, shortly after he took over as NASA Administrator, Mike Griffin had taken the robotic servicing mission to Hubble off the table and directed us to continue preparing for another shuttle mission, though that had not yet been fully vetted and approved. So, the Goddard engineers started thinking about repairing STIS in the usual way, with space-suited astronauts doing the job. But they remained enamored of the automated tool. The thought was that the AT could do its job of removing the 111 screws from the MEB-1 cover between EVAs, while the astronauts were sleeping. That would save precious EVA time.

The team still wanted to have control of the tool from the ground even though the work would be done on the shuttle. They needed to be able to send commands to it from the ground in the same way we normally sent commands to the Hubble instruments. They needed video coverage that could be monitored on the ground. This would entail some kind of camera system that could peer inside the tool's enclosure.

An alternative approach, that would allow the EVA astronauts to do the job of removing the MEB cover, was invented by an engineer named Jason Budinoff and his colleagues at Swales Corp. It involved an ingenious tool that would, like the AT, keep the 111 loose screws corralled. They called it a fastener capture plate (FCP). The idea was to create a plate with 111 small holes drilled precisely at the locations

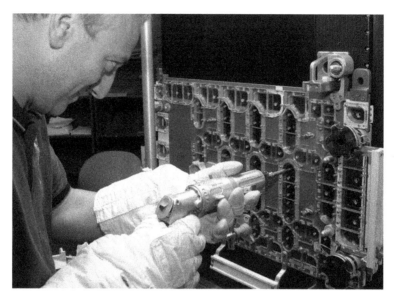

Figure 17.6. Astronaut Mike Massimino practicing removing 111 screws from the MEB-1 cover, using the mini power tool inserted through the 111 holes that had been precisely positioned in the fastener capture plate. (Credit: NASA, GSFC.)

of the 111 screw heads. The holes were big enough to allow a screwdriver head to pass through, but were too small to allow the loose screws to escape. The FCP would be sealed around the MEB cover and both would be removed together as a sandwich, with the screws floating like bees around a hive in between the two layers.

One day in October of 2005 we were sitting in a review of the STIS repair planning. Cepi loved the scheme with the telerobotic AT used to remove the MEB cover. It was "sexy" to an inveterate inventor like him. But the task had become very complex. It was not clear that it really would enable a more efficient use of EVA time. And the cost of the system had grown to $15–20 million. My much-respected Deputy Senior Project Scientist, Mal Niedner, had seen enough. After all, the job was just to get STIS repaired, reliably and by the simplest means possible—nothing more. This robotic contraption was obviously getting out of hand.

So, Mal wrote Cepi an email with the radical suggestion that we just trash the AT idea and allow the astronauts to do the job in the simple, old-fashioned way. They could use a power tool to unscrew the EVA handle, remove it, and then unscrew the 111 fasteners through the holes in the FCP (Figure 17.6). In either scenario, an astronaut would have to use a specially designed card extraction tool to remove and replace the faulty circuit board from the MEB. He or she would close out the job by affixing a new MEB cover where the old one had rested, but this time by simply rotating and locking two handles, rather than screwing in 111 fasteners.

Cepi responded to Mal skeptically, but he said he would have his engineers look into it. A week or two after Mal sent his email, the AT was abandoned, and the idea of simply using the fastener capture plate plus an EVA astronaut wielding a power tool became the baseline approach for repairing STIS. This change paid dividends later when, in 2007, a short circuit in the Advanced Camera for Surveys took out two of its three channels. There were only 32 fasteners to unscrew to open up the ACS electronics box, but the same technique used for the STIS repair was also applicable to replacing circuit boards in ACS. The ACS repair would involve its own tailor-made fastener capture plate, with an astronaut operating a power tool to unscrew and capture the fasteners, and then using a special card extraction tool to pull out four circuit boards.

Ed Cheng was on one of his frequent trips to California to spend some time at Rockwell, consulting with the engineers who were fabricating the detector chips for the WFC3 IR channel. It was February of 2007. Early one morning the phone rang in Ed's hotel room. It was Cepi. He said, "I need you back at Goddard for a meeting tomorrow. All the astronauts are coming. We're going to talk about ACS. It's broken." "But, Cepi, I'm in California," Ed protested. Cepi was unmoved. "Be back in the morning," he said.

So, Ed got on his laptop, pulled up the airline's web site, and got a ticket for a flight out of Los Angeles that night—on the "redeye." The next morning he showed up at the meeting, bleary-eyed but present as commanded. Astronaut John Grunsfeld was surprised. "Oh, you got here!" John exclaimed. "How could I not get here?" Ed replied. "I got THE call from Cepi to come back."

They sat around the table in Cepi's conference room with engineers from the Project, brainstorming about what to do. Any scheme for repairing ACS was already heavily constrained. To make the launch in October of 2008, they would need to have any hardware required for the repair ready to go into its testing program by June of 2008. That gave them about 16 months to design and build a fix. Then there was the issue of EVA time. The mission was already fully subscribed. To make room for an ACS repair, something else would probably have to be taken off the SM4 manifest (that turned out to be the Aft Shroud Cooling System, which was built but never flown). Specialized tools would have to be created. The astronauts would have to train for a repair both in one-*g* at Goddard and in the Neutral Buoyancy Lab at JSC. There was much to do and not a lot of time to do it.

But first they had to diagnose what had failed in the ACS. Was it something that they could repair? The short circuit had been a big one. It had blown a fuse in the spacecraft's power distribution unit. It probably produced smoke, as a pressure sensor in the instrument had detected gas. It also may have produced collateral damage in the surrounding circuitry.

The failure was probably in the low voltage power supply (LVPS) on Side 2 of the ACS electronics. The LVPS provided electricity to the circuit boards that controlled the CCD detectors in both the Wide Field Channel (WFC) and the High Resolution Channel (HRC). Those circuit boards resided in two separate CCD electronics boxes (CEBs), one devoted to each channel. The Side 1 LVPS had failed previously and neither the WFC nor the HRC CEBs were operable on Side 1. The remainder of the ACSs electronics in the main electronics box (MEB) still seemed to be working. So, filter wheels could be rotated and mechanisms could be moved in the instrument.

The third channel, the Far-Ultraviolet Channel, was still working with its own power supply on Side 1. But the two primary cameras were not functioning.

Lead electrical engineer Ed Cheung (a different person than Ed Cheng; we always had fun with their similar names) could narrow down where in the power supply circuitry the short might have happened. But he could not identify the specific location. (An Anomaly Review Board came to the same conclusion later on.) In STIS the location of the short was isolated to a specific circuit board, so the astronauts only needed to remove and replace that one board during an EVA. It was different for ACS. The experts couldn't tell the astronauts specifically which component(s) needed to be changed out. It was impossible to dig into the interior of the instrument to diagnose the problem. It was about 350 miles above the surface of the Earth. There was considerable uncertainty about how a repair could even be attempted.

After the meeting Ed Cheng spent some time contemplating this quandary. In a hallway conversation Hubble Program Manager Preston Burch suggested to Ed, "Just do with ACS what you did with the NICMOS cooler." Ed replied, "Preston, you have no idea what that was like!" Ed might have asked, "How many miracles do you want me to perform?" But he didn't. He simply said, "We'll do our best. No guarantees."

In fact creating the NICMOS Cooling System and having it work perfectly as it did, was really tough. An EVA repair of ACS would likely be even tougher to pull off, not least because time to get everything together before the mission was very tight. It would take a small team of very capable people working with an intense sense of urgency. That's the kind of team Ed and Cepi had experience putting together. But even if they could meet the tight schedule, the EVA itself would likely be difficult—time-consuming and physically challenging.

So the question on the table was, how do you repair the ACS without getting inside the instrument and without making the task impossibly difficult for an EVA astronaut? If they couldn't get into the ACS to repair what had failed, Ed reasoned, why not simply bypass all the failed stuff and create a new power supply that could be attached and hooked up on the exterior of the instrument? And why not create entirely new circuitry to control the CCDs? Essentially, they could build a new camera electronics system on the backbone of the old one, and not have to fool with the guts of the ACS.

In principle this same approach might be used to repair both the Wide Field Channel and the High Resolution Channel. However, it wasn't clear that they would have enough time to build and test two sets of the required hardware and still make the launch date, not to mention enough EVA time for the astronauts to handle two separate repairs. So, Ed recommended that they concentrate on the channel recognized to be of higher scientific priority, the WFC, and his scientific colleagues concurred. The WFC, which had about ten times greater efficiency than WFPC2 in surveying large fields on the sky, had been used for about 70% of all the observing done with ACS prior to its electrical failures.

An alternate, simpler approach was taken as a possible path to restoring the HRC by providing power to the HRC electronics through the new WFC electronics and power supply. The idea was that the HRC electronics were likely fine, but the power supply had failed. If it failed in a way that was an open circuit instead of a short circuit, then this approach would work. As we found out when we got on orbit, the failure was in fact a dead short and so restoring the HRC in this manner was ultimately not successful.

The electronic circuits that controlled the WFC detector were contained on four circuit boards housed in its CEB. But the task was not as simple as just pulling out those four boards and replacing them with identical new ones. The old boards were densely packed with discrete components and were really too complex simply to replicate in the time available. Moreover, their architecture was obsolete.

Another complication was that the new boards needed to be able to communicate with each other much as the original boards had. For a long time it was assumed that the original wiring that had interconnected the four old circuit boards, would serve the same function with the four new boards in place. But then Ed Cheng had a dream—a nightmare. In his dream he imagined that this old wiring within the CEB was faulty. What made this a nightmare was that the wiring couldn't be tested from the ground. There was no way to know in advance that the wiring was bad. What if they proceeded with Ed's scheme for repairing ACS and the fix didn't work because of faulty old wiring they were never able to test beforehand? This situation was not acceptable. Another means of communication among circuit boards had to be found.

The solution to the problem originated with a venerable career electrical engineer at Goddard named Ike Orlowski. He designed a "backplane," a compact and flexible harness that connected the new boards together and allowed them to talk to each other. A backplane is something like a motherboard in a computer. It is a circuit board or other device into which other circuit boards can be plugged, allowing them to work together as a coordinated system. So, the eventual architecture ended up having two backplanes, the new one for servicing the four new boards and their communications needs, and the old one that talked to the other detectors and electronics within the ACS.

For an astronaut to reliably align and insert four new CCD control boards and the backplane into the CEB, it would be necessary to mount the entire set of boards rigidly into some kind of cartridge. A space-suited astronaut could then, after having removed the old circuit boards, push the cartridge and the new boards as a unit into the CEB with relative ease.

One problem with this approach was that the old connectors within the CEB, the ones into which the old circuit boards had been locked, were known to be "finicky." They could not be relied on to give a good electrical contact to the new circuit boards as the cartridge was installed. Greg Waligroski, the lead mechanical person on the ACS repair project at Ball Aerospace, came up with an elegant solution to the problem, and performed numerous engineering tests to show that the solution worked. But much time had passed before it became clear that mechanically replacing the old CCD control circuit boards was going to be practicable.

There was another issue. The cartridge would be too large to fit into the available space in the CEB if the four new circuit boards were the same size as the ones they were replacing. The boards had to be made smaller.

Once again new technology came to Hubble's rescue. The same people who were producing the light-sensing detector chip for the new IR channel in WFC3 also were working on a new kind of device to operate detectors in the focal planes of telescopes, particularly in telescopes in which the detectors had to be cooled to very low temperatures, like JWST. Engineer Markus Loose led the group at Teledyne. The new device was called an ASIC, which stood for application specific integrated circuit. The key words here were "application specific." The ASIC was a small chip that could be adapted to operate a variety of detector types (Figure 17.7). In the present case, it could be configured to operate the CCD in the Advanced Camera's Wide Field Channel. The ASIC required little power and generated little heat. It could be mounted close to the detector. In old-fashioned designs, a large box of warm electronics operated a cold detector at the focal plane of a telescope. By necessity there had to be a long cable separating the detector from the electronics box. That set-up always produced a lot of background noise in the data. This problem was solved by the ASIC.

So, the ACS repair (ACS-R) hardware consisted of a new power supply to be mounted on an EVA handle on the exterior of the instrument; a cleverly designed cartridge containing four circuit boards and a backplane; and two cables for routing Hubble internal power to the new power supply and from there into the CEB (Figure 17.8). One of the four circuit boards would contain an ASIC, programmed

Figure 17.7. An example of an "Application Specific Integrated Circuit" similar to the one produced by Teledyne for use in the repair of ACS. The chip in the center is about 0.9 × 0.6 inches in size. It is programmable and is fully capable of controlling the CCD detector in the Wide Field Channel, as well as processing its data output. (Credit: Teledyne Imaging Sensors.)

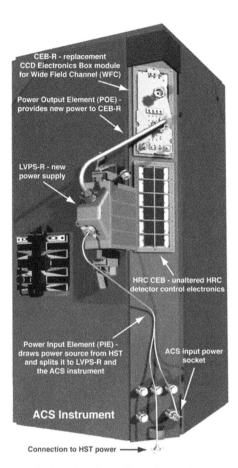

Figure 17.8. Artist's rendering of the layout of the ACS repair. The four new circuit boards, including the ASIC, would be contained in the CCD electronics box module for the Wide Field Channel (WFC) at the top. The new low voltage power supply would be attached to the EVA handle near the center. No attempt would be made proactively to do a similar repair on the High Resolution Channel (HRC) whose electronics box is shown below that of the WFC, although there was hope that the HRC might also come back on when power was applied to the repaired WFC. (Credit: NASA, GSFC, Ball Aerospace.)

to control the WFC detector, process all the data it produced and pass the output along as digital bits to be transmitted to the ground. The other three circuit boards would do supporting functions for the ASIC such as talking to the rest of the ACS electronics and making it look like the original electronics were still there, that nothing had changed.

New tools would need to be developed by the Project: A grid cutter tool for cutting away the metal grid that covered the CEB; a fastener capture plate, aligned to the 32 screws that fastened down the cover of the CEB, similar in principle to the larger FCP used in the STIS repair; and a card extractor tool to allow the EVA astronaut to pull out the four old circuit boards from the CEB without worries about contacting the boards' sharp edges and potentially cutting a spacesuit glove.

Design work and proof-of-concept testing of the ACS-R began in earnest in the spring and summer of 2007. Because Ball Aerospace was the original prime contractor who designed and built the ACS, it was obvious that it needed to be the outfit responsible for designing and building the ACS-R. Environmental testing of the hardware was scheduled for July and August of 2008, enabling shipment to KSC by late August to support a launch of SM4 in early October. That meant the team had about one year to complete the entire job. They were racing the clock to get things completed in time. A headlong sprint was required. Keeping this intense activity on track required the dedication of many senior and junior engineers, under the leadership of an outstanding Goddard systems engineer, Kevin Boyce.

By early June of 2008 the ACS-R flight hardware had been delivered to Goddard and an intensive testing and verification program to prove that the system was flight-worthy was about to begin. Back in Boulder the Ball Systems Lead for ACS-R, Becky Emerle, was diligently reviewing and closing out all the paperwork associated with the project, when she made a distressing discovery. In the rush to get things built, they had temporarily put some power supply modules into the new LVPS-R that were of questionable reliability—that would not be acceptable for flight. Interpoint, the manufacturer of the high-quality flight modules had delivered them to Ball very late. The ACS-R team had in many instances substituted non-flight parts just to keep the hardware construction moving. They thought they had eventually replaced all of those lower-quality modules with the flight parts. But Becky's fastidious attention to the details in the paperwork demonstrated that was not the case. Some of the non-flight boards had slipped through.

So, at Goddard, the ACS-R team pulled the boards out of the power supply units and sent them back to Ball to be retrofitted with the correct, flight-quality modules. To keep the ACS-R effort moving on a tight schedule, engineering models of the CEB-R and the LVPS-R were initially used for some of the early preliminary testing. The flight-quality units caught up in short order. And Becky Emerle entered the pantheon of Hubble heroes.

AAS | IOP Astronomy

Life With Hubble
An insider's view of the world's most famous telescope
David S Leckrone

Chapter 18

A New Beginning: The Last Servicing Mission

2009 May 11 was a warm, breezy, partly sunny day at KSC. A high bank of thin cirrus served as a backdrop for the rafts of soft white cumulus clouds drifting below. It was a good day to launch a space shuttle. Three miles off in the distance, Atlantis, stacked with its External Tank and Solid Rocket Boosters stood at the ready on Pad 39A. Farther away still, our rescue craft, Endeavour, stood in place on Pad 39B (Figure 18.1). If the Hubble servicing mission, STS-125, ran into trouble, Endeavour could be launched as STS-400, to provide a lifeboat for the Atlantis crew and bring them home safely. While the odds of actually needing a rescue mission were very low, having that capability lowered the risk sufficiently to allow STS-125 to take place as the only mission not to go to the International Space Station after the Columbia accident. In short order, after it had become clear that it would not be needed for a rescue mission, Endeavour would be released to resume its flow toward a launch to the ISS as STS-127.

The Press Site was crowded with reporters, some of them with microphones and cameramen in tow. As usual, I was fielding interviews with a lot of help from my Deputy, Mal Niedner, and a colleague from Goddard, Jennifer Wiseman. Jennifer had been for a time the Hubble Program Scientist at NASA Headquarters and in 2010 would become my successor as Hubble Senior Project Scientist at Goddard. Ed Weiler was there with his impressive new wife, an entomologist from the University of Maryland. He joked that his new job was to be her field assistant. Ed had been re-appointed Associate Administrator of the Science Mission Directorate at NASA Headquarters by Mike Griffin in May of 2008.

The Hubble Program had come down a rough and twisting road to get to this day—to the long-postponed Servicing Mission 4 (see Chapter 16). Actually SM4 would be the fifth servicing mission to be flown to Hubble, thanks to the split of SM3 into two separate flights—SM3a and SM3b. More than seven years had passed since the previous servicing mission to the telescope, SM3b on STS-109 in March of 2002. The observatory was not in very good condition.

Figure 18.1. Shuttle Atlantis poised on Pad 39A at Kennedy Space Center for launch as STS-125, Hubble servicing mission SM4. In the distance on Pad 39B is Endeavour, ready to be launched as STS-400, a rescue mission for the crew of Atlantis if required. (Credit: NASA.)

We had two spectacular new instruments ready to be installed on Hubble—the Cosmic Origins Spectrograph (COS), and the Wide Field Camera 3 (WFC3). The COS would give astronomers a new tool that was very sensitive to light deep into far-ultraviolet wavelengths. The WFC3 could take images in near-ultraviolet and visible light with high sensitivity and excellent resolution over a wide field of view. Its real claim to fame, however, resided in the new Infrared Channel, that would extend Hubble's ability to image light 70% farther into the infrared than WFPC2 or ACS. In the near-infrared, WFC3 had much higher sensitivity, a wider field of view and much better angular resolution than NICMOS, although NICMOS reached to even longer infrared wavelengths (out to 2400 nm, versus the 1700 nm limit for WFC3). With this new IR Channel, there was no need for any special cooling provisions beyond the kind of thermoelectric coolers we had already been using with our visible-light CCD cameras in WFPC2 and ACS.

Three of Hubble's six gyroscopes installed during SM3a in 1999 had failed. With the advent of the 486 spacecraft computer, also installed in 1999, and with new software developed in the STOCC at Goddard and at the STScI, Hubble could carry out its observing programs with reasonable efficiency using only two gyros in the pointing control loop. Since 2005 the spacecraft had operated in this "Two Gyro Mode," while the single remaining gyro was held in reserve. The plan was to replace all six gyroscopes during SM4.

Hubble's six nickel–hydrogen batteries had been on the spacecraft for 19 years. They had never been replaced. They were slowly losing their capacity to hold electrical charge. It was high time they were retired, replaced, and brought back to Earth where they would continue to be used for research in a laboratory environment.

Two of the three Fine Guidance Sensors (FGSs), essential for holding Hubble precisely pointed and tightly locked on its targets, had been replaced in prior servicing missions. FGS number 1 (FGS1) was replaced with the only flight spare FGS we had, designated FGS1R, during STS-82/SM2 in 1997 (the "R" designated "replacement"). FGS1R had been updated with an adjustable internal optical element that corrected its performance for spherical aberration and restored the FGS to its original pre-launch performance specifications. The old FGS1 was returned to Earth, fully refurbished and later flown to replace the original FGS2, becoming FGS2R, during STS-103/SM3a in 1999. Our hope was to re-furbish the old FGS2 and use it as FGS3R to replace the original FGS3 during SM4. However, those plans went awry when FGS2R began to experience numerous failures in acquiring the guide stars needed to align Hubble to observers' targets. It seems a simple LED light source in FGS2R was failing. So, the FGS unit in position 2 had to be replaced a second time with the FGS included in the payload of the current mission. As a result, the original 19 year-old FGS3 would never be replaced. FGS3R was re-designated FGS2R2. FGS1R and FGS2R2 would be used as the primary units on the Hubble Telescope with old FGS3 demoted to a backup status.

On 2008 September 27 one of the two redundant sides of the electronics in the Scientific Instrument Command and Data Handling (SIC&DH) computer—Side A—failed. Only Side B remained useable. As its name implied, this computer was essential to the operation of all the scientific instruments on board the observatory. Without it, Hubble science would cease. Administrator Mike Griffin promptly made the decision that he didn't want to go through with all the challenges of SM4 only to have this critical computer totally fail, shortly after the shuttle had left the scene for the final time. He ordered that the launch scheduled for October of 2008 be scrubbed. Cepi's team at Goddard had to pull out the spare SIC&DH computer from storage, bring it up to operation, find test equipment to check out this device that had been on the shelf for 18 years, thoroughly test it, and prepare it for flight. As a result it became necessary to delay the launch of SM4 for seven months.

The most adventurous task of all assigned to the spacewalking astronauts during the mission was the in-orbit repair of both STIS and ACS. Both of these were extremely capable instruments heavily in demand by astronomers using Hubble. Both of them failed because of electrical short circuits in low voltage power supplies. Their primary components—optics, filters, and light-sensing detectors—were unaffected.

For the first time on Hubble, a daring rescue would be attempted wherein the EVA astronauts would open up the two instruments and use specialized tools to extract and replace circuit boards. Only once before, during the Solar Maximum Repair Mission in 1984, had an instrument been opened and a circuit board replaced by a space-suited astronaut. The manufacturer, Ball Aerospace, had not designed STIS and ACS to be opened and repaired or modified in orbit. It was a complex job for both instruments, and yet another example of the resourcefulness and creativity of Hubble's engineering team.

There was concern at NASA Headquarters and elsewhere that we were piling too much complicated work to be accomplished in the limited time available on the astronauts. The ACS failure occurred after the astronaut crew had been assigned to the mission. They already had a full schedule—five EVAs worth of tasks to perform. Then the SIC&DH failure came along. That made the workload heavier still. One planned addition to improve Hubble's performance in the long term—an Aft Shroud Cooling System (ASCS)—was removed from the manifest. Its installation, which included mounting another external radiator similar to that installed for the NICMOS Cooling System on the outside of the spacecraft, would have required a major expenditure of EVA time. The ASCS was built but was never flown.

Alan Stern, the Associate Administrator for Space Science at the time when ACS had failed and an ACS repair was being proposed was, at first, reluctant to approve adding the repair to the already long list of EVA tasks. But the spacewalkers were eager to give it a try—John Grunsfeld for the ACS repair and Mike Massimino for STIS. After much debate and creative mission planning, the ACS repair was added to the mission timeline.

There was no room on the manifest for a propulsion module that could be used someday to propel Hubble safely out of orbit. That would have to await a (presumably) robotic mission sometime in the future. However, on this mission the astronauts would be attaching a ring-shaped device called the soft capture mechanism (SCM) onto the aft bulkhead of the spacecraft to give engineers a leg-up in planning Hubble's eventual removal from orbit. The SCM contained targets and structures that would make it easier for a future robotic spacecraft, containing a de-orbit propulsion system, to softly dock and latch onto Hubble's tail end.

Finally, we hoped to complete the installation of the remaining NOBLS (New Outer Blanket Layers). These were specially coated sheets of stainless steel foil, cut to size for each Equipment Bay and supported by steel frames. They were to be mounted over the tattered blankets of multi-layer insulation (MLI) that were discovered by EVA astronauts as early as SM2 in 1997. It was essential to protect the various equipment bays from overheating when exposed to sunlight or from getting too cold during orbital night. The ability of the original MLI blankets to do that had been seriously compromised. The NOBLS would take care of the problem for as long as Hubble was in orbit.

As the launch countdown proceeded beyond the built-in hold at $T - 9$ min, a hundred or so people gathered along the shore of the Turning Pond. We were simultaneously nervous and excited, anticipating the spectacle we were about to witness and the adventure that was about to begin. There were numerous cameras waiting to record the launch, most notably a camera on a crane labeled, "IMAX 3D." In fact Atlantis was carrying an IMAX camera and about 8 min worth of IMAX film in its payload. The crew had been trained to shoot scenes during the mission that would later be used in a new IMAX film, "Hubble 3D." Pilot Greg Johnson, better known as Ray J, was the principal IMAX cameraman on the flight with assistance from Commander Scott (Scooter) Altman.

The countdown proceeded without interruption: "$T - 10, 9, 8, 7, 6$, Main Engine start, 4, 3, 2, 1 and liftoff of Space Shuttle Atlantis for the final visit to enhance the

Figure 18.2. Launch of STS-125 on 2009 May 11. (© Michael Soluri/all rights reserved.)

vision of Hubble into the deepest grandeur of our universe," (Figure 18.2). As with all the previous shuttle launches I had witnessed, there were huge billows of steam, then flame, followed about 15 s later by the loud roar of the Solid Rocket Boosters and the staccato crackling produced by acoustic shock waves in the rocket exhaust—all of this beating like a drum against my torso. Soon, we lost view of Atlantis as it passed above a bank of clouds. Ed Weiler and I engaged in an awkward hug with mutual pats on the back in a modest ritual of congratulations.

To us on the ground this seemed to be a perfectly normal shuttle launch, beautiful and exhilarating. However, in the crew cabin of Atlantis, things had become a bit more sporting. Eight milliseconds (0.008 s) after liftoff, the Master Alarm went off in the cockpit with its irritating repetitive loud bleating sound.

Commander Altman had heard this sound many times before in the flight simulator at JSC as he practiced responding to whatever contingencies the Simulation Supervisor threw at him. However, he had never heard a Master Alarm go off while he was "going uphill." He had never climbed into an orbiter in an orange suit and not ended up in orbit in three previous flights. At that moment the vehicle was experiencing incredible vibrations. That was normal. But what was not normal were those powerful vibrations being accompanied by the soundtrack of the Master Alarm.

Scooter reached up and punched the alarm button off with his gloved left hand. He then looked down, scanning his display. It was showing FCS Channel 1 with a down arrow next to it. The FCS abbreviation referred to the Flight Control System, the system in the orbiter that controlled the active flight surfaces—the rudder/speed brake, elevons (a combination of elevator and aileron), and the body flap. It was critical to the controlled maneuvering of the orbiter during launch and re-entry.

The FCS down arrow told Scooter that one channel of the Flight Control System was out. There was a more thorough FCS display to look at. To get to it, Scooter punched "SPEC 53 PRO" into his keyboard. The full FCS display came up showing all down arrows. One channel had completely lost power.

At 8 s after launch the orbiter would begin a roll maneuver. Scooter was trained not to miss that. He counted "6...7...8" s. He watched an indicator needle displace, showing that the vehicle was rolling properly. He made his first call to the ground, "Houston, Atlantis, roll program with the FCS channel." That told the Shuttle Ascent Flight Director, Norm Knight, in Mission Control that Scooter was doing what he was supposed to do and also that he was aware of the FCS problem. He'd seen the error message. Was there anything the controllers at JSC could tell him that he could do about it? Knight quickly consulted with his team and directed the CAPCOM (Capsule Communicator), Greg "Box" Johnson, sitting next to him to relay the message, "Bypass across the board, Scooter, no action." In other words, the failed channel was one of four redundant units. The mission could proceed with the remaining three that were working properly.

At Launch + 35 s and again at Launch + 46 s, two more alarms sounded. This sequence—one alarm after another after another—was a little unnerving to Ray J, the pilot sitting next to Scooter, who was on his first shuttle flight. The gaseous hydrogen (GH_2) pressure gauge for the left Space Shuttle Main Engine was behaving erratically, setting off the alarm. In pre-flight simulations, Scooter's experience was that these pressure sensors (transducers) could misbehave for any one of several reasons. But in this case, with an apparent gradual decrease in the GH_2 pressure, the symptoms suggested the possibility that one of the shuttle's main engines was indeed failing. He and Ray J discussed what they were seeing. Ray J said, "Okay, I see this dP/dt" (the rate of change of pressure over time). Scooter noted, "We don't have any indications of a data path problem. If we lost an engine here, it would mean a Return To Launch Site (RTLS) abort."

One of Altman's jobs as a mission Commander was always to be aware of the boundaries ahead of him. At what point could they do a Transatlantic Landing (TAL) abort? When would a "Once Around Abort" or an "Abort to Orbit" become possible? He also had to make sure his crew knew what their situation was. Everyone needed to be on the same page and prepared to act, based on the "play" that Scooter called. He didn't really think they were going to have to do an RTLS abort. But he started talking about it with the rest of his crew, so everyone would be mentally prepared if it became necessary in the next minutes to go in that direction. Scooter had been through thousands of simulations of such mission abort scenarios and so was comfortable thinking it through. That wasn't true for the members of the crew who hadn't gone through those simulations. Down on the mid-deck, they were thinking, "Holy Cow, we might be going back!" Drew Feustel later told Scooter drolly, "It ruined my enjoyment of the launch experience."

In the 30 year flight history of the Space Shuttle Program no crew had ever needed to execute an RTLS abort. It was widely considered to be the most dangerous of all ascent abort modes. Former astronaut Rick Hauck once told a group of us on a VIP tour at KSC that an RTLS abort gave you just enough time to "kiss your fanny goodbye." After Solid Rocket Booster separation at about 2 min after launch, the orbiter would continue to fly down range to burn off fuel. It would be going too fast to allow a normal bank and turn back toward the Cape. But it, along with the External Tank, could be flipped over. In that configuration, the shuttle would be thrusting backward at Mach 5, shedding velocity, with its exhaust plume enveloping the vehicle. Eventually the Orbiter's down-range velocity would be reduced to zero by this backward thrusting maneuver. Then it would reverse course and begin flying back toward the launch site. Just before the shuttle's main engines were cut off, the Orbiter would be pitched nose down, so that when the External Tank was jettisoned, the drag of the atmosphere would pull it downward, safely away from the orbiter. The possibility of a catastrophic collision between the External Tank and the Orbiter vehicle was considered the greatest danger of an RTLS abort. After External Tank separation, the orbiter would behave as a glider, returning to land at KSC about 25 min after it had lifted off.

All of this was going through Scooter's mind when the call came up from Houston that the problem they were seeing was due to a failed pressure transducer. Flight Director Knight's team had much better insight than the crew on Atlantis into how the three main engines were performing. The problem did not affect the operations of the shuttle engines, and they should ignore it. The Ascent Team at JSC had been efficient and incisive, diagnosing the alarms and rapidly getting calls back up to Atlantis. It had just been a "routine" day at the office for the crew of STS-125.

After the thunderous excitement of launch, the remainder of the first two flight days of a shuttle mission was usually of only passing interest to the press and the public. In the "old days," the flight crew would have some time during those two days to adapt to the alien sensations of weightlessness. Otherwise the time would be spent routinely in opening the Payload Bay doors, activating the remote arm, and preparing spacesuits and tools for the spacewalks to come. However, ever since the Columbia disaster this routine had been upended into a more intense, time-pressured drill.

As soon as the Shuttle Main Engines cut off, at about 8½ min after launch, the Mission Specialists in the crew had to hustle out of their seats, turn around, and begin to do a lot of space-sickness-inducing moves to begin quickly taking pictures of the External Tank as it was jettisoned and fell away toward a fiery re-entry. The point was to photograph spots where the external foam insulation might have come loose, potentially hitting and damaging the Thermal Protection System of the orbiter. They hurried to re-configure the orbiter, get as much of their equipment out as they could, break out the remote arm and use it to perform a video survey of the Payload Bay and the exterior of the Crew Cabin—all in a hectic five hours before going to bed that first day.

Flight Day 2 used to be relatively relaxed for the crew, a time for them to figure out what this being-in-orbit thing was all about and how it affected the systems of their own bodies. Now it was devoted to the newly developed procedure required by the CAIB report (Chapter 16) for self-inspection of the orbiter, looking for serious damage that may have been inflicted on the vehicle during launch. For this, they attached a rigid boom to the end of the remote arm, doubling its length from 50 to 100 feet. The rigid boom was called the Orbiter Boom Sensor System (OBSS). At the far end of the OBSS was a package of cameras and lasers. With this extra-long arm Megan McArthur, the arm operator, was able to move the cameras and lasers to view the entire Thermal Protection System, searching for damage. Megan had a PhD in Oceanography from the University of California at San Diego. Now here she was at work in the vast ocean of space. Scooter was the prime backup for Megan and relieved her for a part of the inspection. Other members of the crew watched this process carefully to help Megan and Scooter avoid hitting some part of the orbiter, potentially causing damage in the process of looking for damage. The survey images acquired by the crew revealed only minimal damage to the thermal protection system. A final survey was planned near the end of the mission, to search for any new damage from space debris and micrometeoroids incurred while they were in orbit.

Flight Day 3 was the day of rendezvous and docking with Hubble. I was sitting next to Preston Burch in the blue flight control room or BFCR in Building 30 at JSC. We called it the "Blue Ficker." It had replaced the old customer support rooms (CSRs) of missions past, and was a much better facility.

Preston was steaming. He had just learned that the STOCC team at Goddard had neglected to switch the Hubble Data Management Unit from the 1 Mbps science-data format to the 32 kb rate required for downlink through the Shuttle communications system. Preston growled, "Here we are on the very first day that we're supposed to do something and we messed it up." So the rendezvous process was stopped temporarily until the correct communications link from Hubble through Atlantis to the STOCC could be established.

Because of the delay in establishing communications, the preplanned final roll maneuver of Hubble to position it correctly to be grappled by the shuttle's remote arm was not performed. Scooter was expecting Hubble to be in the correct attitude when he brought Atlantis close enough for Megan to grapple it. That was the plan. That would be simple. Then the call came up from the ground, "Uhhh, ya know Hubble's not in attitude. We think it would be easier if you just flew the rendezvous out instead of us." As you would expect, Scooter replied, "Roger, Houston, that's no problem." But, he had concerns. This kind of rendezvous was more challenging than if Hubble had been in the correct orientation. Scooter recalled practicing through a similar scenario during a Joint Integrated Simulation in Houston. The SIM had not gone well. He was determined that the outcome would be better here in the real world.

Scooter started the rendezvous thruster burns in his front seat position, but as they got closer to Hubble he gave up that seat to Ray J and moved to the back so that he could complete the process looking out of the overhead window. He flew the orbiter up underneath Hubble to a distance of about 150 feet, and waited there until they were stable. He then moved closer, leaving Megan with only a short distance to reach the grappling fixture on the telescope with the remote arm.

Everyone in the crew was excited, pumped up. You can't fix the telescope if you can't bring it on board Atlantis. That had gone beautifully. As Megan was moving Hubble down into the latches on the Flight Support System, it seemed like an IMAX moment. Scooter had been glad when they had agreed to record some IMAX footage during the flight. He was a great fan of the IMAX movies. But when the subject first came up he had assumed it would be a very simple process—push a button at the moment called for in the Flight Plan and take a few minutes of footage—turn it on, turn it off. What could be simpler?

Actually it was not so simple. They had to choose the lens, set the exposure time, focus the camera and choose the scene. They had a list of scenes that Toni Myers, the Producer and Director, wanted them to try to get. But the final choices were theirs to make. They had only 8 min worth of film on board and could only shoot 30 s at a time.

Ray J and Scooter hustled to try to get an IMAX shot of the aft end of the Hubble spacecraft coming to rest on the Flight Support System. But the motion downward of this massive body being carried by the Canadian arm was so slow that it turned out to be barely perceptible in the footage they shot. Later, to register some actual motion in a scene they decided to shoot one of the doors in Hubble's aft shroud being opened during an EVA. "Okay, are you ready?" Scooter asked Ray J. And then, in the blink of an eye, the door popped open. It was already open as they activated the camera. The shot was over practically as soon as it had begun. So, they were going to have to find something else to shoot the next time. After all, it was supposed to be a **motion** picture.

It was Thursday, 2009 May 14, Flight Day 4 for STS-125. I was eager to get to the Blue Ficker. This promised to be a very important day for Hubble and for science. The first spacewalk of SM4 was scheduled for that morning. The first big task on this first EVA was to remove WFPC2 from its radial instrument bay on Hubble and to replace it with the magnificent new panchromatic WFC3. WFPC2 had operated dependably and well for over 15 years. Now it was time for it to retire from the field. In the list of mission priorities, the WFC3 came in at number 2, behind only the six replacement gyroscopes. Successful installation of the new camera was the most important scientific objective of the mission.

John Grunsfeld and Drew Feustel egressed from the Airlock at 1 min before 8 am Houston time (CDT), a bit later than planned. After their routine setup procedures in the Payload Bay, Grunsfeld ("John") helped Feustel ("Drew") mount the Manipulator Foot Restraint (MFR) at the end of the remote arm. John would be the free floater during the EVA. Mike Massimino ("Mass") and Michael Good ("Bueno") would be their IVA (intra-vehicular activity) assistants, walking the EVA guys step-by-step through the detailed written Flight Plan. Megan McArthur ("Megan") would be in the driver's seat transporting Drew around on the remote arm. (The names shown in parentheses were used as the call names for the crewmembers, so that there was no ambiguity, for example, between Mike Massimino and Mike Good.)

Megan lifted Drew to the work site, the location where the WFPC2's radiator stuck out from the side of the Hubble spacecraft. To remove WFPC2 from its bay Drew first had to install a metal frame onto the back of the instrument's radiator. That provided a way to grab hold of the instrument and move it without actually touching it. He then used a power tool with a long shaft to reach through holes in the WFPC2 radiator to unfasten an electrical grounding strap and a "blind mate connecter." The latter's function was to complete power and data circuits into the instrument. The next step was to release the mechanism that mechanically attached WFPC2 to Hubble's internal Focal Plane Structure—the "A-Latch." The reverse steps would be followed later to insert, fasten and hookup WFC3 after WFPC2 had been removed and temporarily stowed on the sill of the Payload Bay.

To open the A-Latch and release WFPC2, Drew had to use a simple manual ratchet wrench—similar to the ones I have in my toolbox at home—to rotate a very long shaft that was threaded like a bolt into the A-Latch deep inside the spacecraft near the front end of the instrument. The ratchet wrench could be set to rotate the bolt either clockwise (to tighten) or counter clockwise (to loosen). The main difference between my tools at home and the Hubble servicing tool was that extender rods could be attached to the Hubble wrench, allowing it to reach the bolt head.

To provide the prescribed torque, or rotational force, to the bolt head, the plan was for Drew to attach a Mechanical Torque Limiter (MTL) to the wrench. The MTL had five different torque settings. If those torque levels turned out not to be adequate, Drew could switch to a Contingency MTL (CMTL) to exert higher levels of torque. When the amount of torque pre-set on the MTL or CMTL was reached, the torque limiter would slip, not allowing any stronger force to be applied to the head of the bolt.

When he installed WFPC2 during SM1 in 1993, Astronaut Jeff Hoffman had tightened the A-Latch of WFPC2 to a torque of about 35 ft-lbs—or so he thought. Drew set the MTL to 38 ft-lbs as he attempted to rotate the A-Latch bolt. It didn't budge before the MTL slipped. He tried two more times. The MTL slipped both times. Drew said into his microphone, "MTL slips at 38."

He pulled the wrench off the bolt head, double checked all the settings, and re-inserted it to make sure it was properly seated. He tried to loosen the bolt twice more. The MTL slipped each time. Clearly he needed more torque.

The design limit for the A-Latch bolt was 57.1 ft-lbs. A torque higher than that would likely cause the bolt to break. But Drew had a lot of margin, allowing him to go to torque levels higher than 38 ft-lbs. without doing any damage.

Meanwhile, John had gone to one of the protective enclosures to retrieve the CMTL. He carried it to Drew. Drew set the torque limit to 45 ft-lbs. He attempted to turn the A-Latch bolt. The CMTL slipped. He tried two more times with "no joy." "I put in three attempts, got three CMTL slips, see no motion on the A-Latch bolt," Drew reported.

I was sitting at my console in the BFCR next to Preston Burch. We could see much of what was happening on the big video display at the front of the room. I had also been following the conversation on "Air to Ground." At the point where Drew said that he still had "no joy," I started to feel alarmed. I needed to get closer to where the action was taking place—with Cepi's engineers one floor above. I ripped my headphone jacks out of their sockets, dashed to the elevator, and strode into Cepi's area a minute later. I asked the first guy I came to, Clay Fulcher, "So what's going on?" "They can't get the A-Latch to open. So, at this point we can't get WFPC2 out," he responded.

John suggested to the ground that they try another A-Latch tool, a shorter extender rod. Drew installed it on the CMTL, verified it was properly attached, verified that the torque limit was set to 45 ft-lbs, and that the ratchet was seated on the bolt head. He made four more attempts to release the A-Latch. The CMTL slipped each time. The CAPCOM suggested that they first try tightening the bolt a little bit, just to break it free, and then go back to the counter clockwise direction to see if it had loosened up. That didn't work either.

Finally, I heard Drew ask, "How far can we go with this? What is the implication if I over-torque and break the bolt?" Houston replied, "There actually is no issue with having the latch taken all the way to the failure point. The instrument [WFPC2] will still function. If we get to that point, we just reconnect the ground strap and blind mate connector. We just leave it as it is."

When I heard this, I really began to panic. It was unthinkable to me that we would leave Hubble for the final time with the aged WFPC2 still in place and bring WFC3 back to the ground—uninstalled, unused for science. At that point, I broke a rule. We had all agreed months before the mission on a set of contingency plans. In the present situation, our plan said keep trying as best you can up to the point that the bolt breaks. If it breaks, it breaks! I had agreed to that plan. And our protocol was that we would never try to change a contingency plan in real time, during the mission. "Follow the plan," was our mantra.

I pleaded that they ask Drew to stop trying and let us think about this overnight. If we failed to install WFC3, then according to the success criteria to which we had also agreed and which HQ had set in stone, this mission would be, by definition, a failure, no matter what else we accomplished. But I got strong pushback from the rest of the team, including the managers in the BFCR. If we delayed overnight, that would completely disrupt the timeline for the rest of the mission. And it was unlikely that we would be able to find some kind of engineering solution to free up the stuck bolt anyway.

Then I personally think that Hubble's "Guardian Angel" came to rescue us. Days later, after all the in-orbit work was finished, I was in a building at KSC mingling with the astronauts' families as we waited for the opportunity to watch Atlantis land. I struck up a conversation with Drew Feustel's Dad. He told me that Drew's hobby was restoring automobiles. At any given time, Drew had three or four autos in his garage in various states of tear down and re-assembly. Drew was good with his hands, and highly experienced at using wrenches, for example, to unscrew recalcitrant lug nuts holding a wheel in place. Of course I did not know this while we were having the emergency days earlier during EVA 1. But I did know that Drew was a Purdue graduate (as was I), and so a good person to deal with the problem.

Drew was instructed by the ground to get rid of the MTL and the CMTL, and simply try to unfasten the stuck bolt by hand, by feel. And that's what he did. He re-inserted the ratchet onto the head of the bolt and slowly, carefully began turning. "I think I got it," he exclaimed. "It is loose. It's turning easily now, and I don't think it's broken." John asked him how the torque he had applied by hand compared with the 45 ft-lbs he had gotten to with the CMTL. Drew replied, "It was close, it's hard to say. I tried to give it a little extra push right at the end. I definitely got my ½ turn."

Figure 18.3. The author, unaware he was being photographed, at the instant he heard astronaut Drew Feustal call down that the WFPC2 A-Latch bolt was turning freely. Cepi is the person with his back turned just to my left. (© Michael Soluri/all rights reserved.)

From that point on, fully unfastening the A-Latch and removing WFPC2 proved to be straightforward.

When I heard Drew announce, "it is loose, it's turning freely and it is not broken," my response was instantaneous (Figure 18.3). I didn't realize I was being photographed in what I later thought was an embarrassing posture. There was an instant release of an enormous amount of tension, coupled with a huge wave of relief, and a small surge of anger—at no one in particular. I suppose it was anger at the fates for putting us through this agony unnecessarily. I felt I had lost years off my life in just a few minutes of overwhelming worry. But now the wonderful new camera we had struggled so hard to bring into existence could find its rightful home, in the radial instrument bay of the Hubble Space Telescope.

Later the hard-nosed, totally rational Hubble engineers chided me. If the CMTL could have been raised a little bit above its 45 ft-lb limit, the bolt would have come loose. It was simple mechanics. Perhaps it's the Romantic in me, but I persist to this day in believing it took the finely tuned, highly experienced touch of a real human being to save the day. That's why I prefer humans to robots in dealing with unexpected problems in space.

Drew and John completed their work for the day by replacing the partially failed old SIC&DH computer with the spare unit—the task that had caused the mission to be delayed for months. They also unfastened the soft capture mechanism from the FSS platform, to which it had been attached at launch, and re-latched it to the bottom of Hubble. All went smoothly.

After the high drama of EVA 1, I was expecting the next spacewalk, EVA 2, to be a calmer experience. Mass and Bueno first had the job of replacing the three rate sensing units (RSUs), each of which contained two gyroscopes (Figure 18.4). We had changed out RSUs on two previous servicing missions—SM1 and SM3a. So, the procedures were well established and there was a deep experience base to handle any problems that might arise—or so we thought.

Figure 18.4. Astronaut Richard Good transporting a remote sensing unit from its storage container to be mounted inside the aft shroud of the spacecraft. (Credit: NASA.)

The three RSUs were mounted behind the three large Fixed Head Star Trackers (FHSTs) that peered out into space through openings in two large Axial Bay doors. The major difficulty was getting access to them without bumping into the precisely pointed FHSTs. It was the job of the free floater astronaut, in this case Mass, to wiggle his way carefully up inside Hubble's aft shroud behind the FHSTs. Bueno, riding the RMS arm, would work from outside the open Aft Shroud doors (Figure 18.5).

One new wrinkle on this mission was the use of a new tool called the RSU changeout tool or RCT. John Grunsfeld discovered it in a hardware store. More precisely, John saw a commercial gadget called a PIKSTIK in a hardware store that allowed a person to grab objects that were hard to reach. He reasoned that the EVA astronauts could really use a tool like that on Hubble servicing missions to reach and grasp the RSUs (Figure 18.6).

The RCT was a long narrow metal bar-like structure, more robust and versatile than the commercial version. It would give Bueno extended reach into the aft

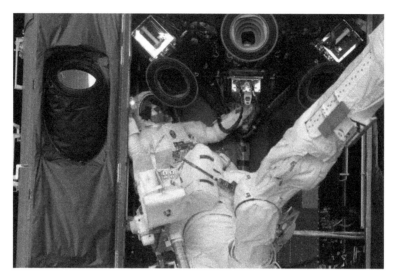

Figure 18.5. EVA astronauts Mike Massimino in the background behind a Fixed Head Star Tracker and Richard Good in the foreground, attached to the RMS arm, shown changing out the RSU in the center #1 position. (Credit: NASA.)

Figure 18.6. On the left is a commercial product called a PIKSTIK that allowed a person to grasp hard-to-reach objects. On the right is the RSU changeout tool (RCT), modeled on the PIKSTIK, that extended an astronaut's reach into the area behind the Fixed Head Star Trackers. The jaws grasped the handle of an RSU and firmly locked onto it. (Credit: NASA, GSFC, J. Cassidy.)

shroud. When he pulled the trigger on one end of the tool, the jaws on the other end would close and lock over the handle of the RSU, allowing an old RSU to be removed or a new one to be installed.

Mass was held in place inside the cavity by a portable foot restraint (PFR). He would assist in aligning the new RSU onto two guide pins on a base plate, which was already attached to the spacecraft structure, assuring that the new unit was firmly seated. Bueno would then use a power screwdriver with a long shaft (the pistol grip tool or PGT) to bolt the new RSU down to a pre-determined torque setting. Mass would conclude by connecting the two cables for power and communications that he had previously disconnected from the old RSU.

The first RSU to be removed and replaced was RSU#2, the one in the upper right in Figure 18.5. The task went smoothly, entirely according to script.

Then came RSU#3 (upper left). The first hint of trouble came after the cables were disconnected from the old unit and its three bolts were released. Mass reported to John, who was working the IVA position inside the cabin, that "This one was a little 'stickier' coming off than the last one." There were two metal guides sticking up from the bottom of each base plate, one on the lower right and the other on the lower left. In addition to being lined up with the guide pins and the boltholes, the lower end of the RSU had to rest between those two guides. It was in trying to pull the old RSU#3 away from the guides that Mass felt the resistance or "stickiness" he had referred to.

Using the RCT, Bueno began to install the replacement RSU. He reported that this unit, dubbed Serial Number 1007, was difficult to seat on its base plate and within the two base plate guides. He tried screwing in one of the bolts and reported, "RSU is not seated." Mass inspected the RSU and base plate for obstructions that would prevent it from seating properly. He verified, "Nothing is sticking out from the plate or the bottom of the RSU." They made another attempt. Bueno reported, "It's still not seated. It feels like it's hitting something. I can't bring it down any further." They try fastening the bolts in a different order. Bueno asks Megan to move him a few inches here and there to see if that would give him a better alignment for inserting S/N 1007. After multiple attempts there was simply no joy.

Bueno removed S/N 1007 completely for inspection. He reported to John, "This one is a lot harder [to remove] than the other one." He comments again, "This one is different than the last; it does not seat flush and feels as though it is 'rocking.'" The two spacewalkers tried several more times, testing the result each time by attempting to drive the bolts and checking to see if they were engaged with the threads in the boltholes. Nothing worked.

From the ground the CAPCOM instructed John that if their next attempt failed, they should retrieve the RSU originally intended for position #1 (S/N 1006) from its protective enclosure and mount it in position #3 in place of S/N 1007. That move produced a good outcome. S/N 1006 seated properly on the base plate at position #3, and all three of its bolts engaged and tightened, as they should. John reported, "Houston, we're ecstatically able to report that RSU-3R connectors are mated, S/N one-zero-zero-six." Bueno discussed with John that it was probable that S/N 1007 was the problem, and John concurred.

John queried the ground as to which RSU they should attempt to install in position #1, S/N 1007 or the spare unit that was always carried on a servicing mission, S/N 1005. CAPCOM responded, ".... ground would prefer to attempt installation of S/N 1007 first, as this is the up-graded unit (the spare is not) and a higher priority, in hopes that the difficulty experienced with installing it in the RSU #3 position was associated with the unique tolerance stack-up of S/N 1007 with the RSU #3 position." That was the rub. S/N 1007 was the newest RSU. Both of its gyroscopes had all the improvements intended to reduce the risk that the hair-thin flex leads that carried the electricity needed to run the gyroscopes would break—silver coating on the flex leads and a less corrosive fluid in which the flex leads were suspended (see Chapter 14). So, they wanted to give S/N 1007 every chance.

In the end S/N 1007 could not be successfully mounted in position #1 either. Mass and Bueno installed the spare RSU instead. Little harm came from that. It left Hubble with six good gyroscopes, sufficient hopefully to last for many years.

Back on the ground after the mission, Goddard engineers inspected the problematic S/N 1007. The culprit was identified as excessively thick layers of epoxy and silver tape that had been applied over the heads of some of the screws that held the RSU box together. The dimensions of the box exceeded their specifications by a small amount—small, but sufficient to prevent the box from fitting where it was supposed to fit. It was an unfortunate human error.

Before Mass and Bueno closed the axial bay where they had been installing RSUs, they installed one end of a power cable that would be needed for the attempted repair of the ACS during EVA 3. That was a "get-ahead" task to give John and Drew a head start on that challenging task the next day.

After they were finished with the RSU changeout, the two spacewalkers replaced one of Hubble's two battery packs. Needless to say, the battery replacement was planned with care to avoid accidentally giving the spacewalkers a nasty shock and preventing sparks and electrical arcs that might damage spacecraft components. While Hubble was latched onto the FSS in the Payload Bay, it received its electrical power from the shuttle through an umbilical cable. So, the solar arrays were not needed during the mission. They were disconnected by opening circuit breakers in Hubble's power control unit. The old batteries were drained of most of their charge. There was a subtle reason for that. It prevented the batteries from conveying electric current from the rest of Hubble's circuitry to the batteries as the astronauts were demating the battery connectors. After disconnecting the battery cables, the astronauts covered their electrodes with protective caps. Otherwise, the replacement of a battery module was a simple matter of unbolting the old one and bolting in its replacement.

After EVA 2 was over, I sat on the panel that gave that day's Mission Status Briefing to the media. During the briefing, referring to the fact that EVAs 1 and 2 had encountered unexpected difficulties, I said only half joking, "I have a prediction. We've always said EVA 3 was going to be the most difficult and the most challenging, but I predict it's going to go more smoothly than any other EVA on this mission. I just think that some version of Murphy's law is going to lead us in that direction." I had in mind the fact that it was John Grunsfeld who was tasked with doing the ACS repair job the next day. I recalled how efficient John had been on STS-109, changing out Hubble's power control unit. I fully expected his work on repairing the ACS to be to the same standard.

In the EVA timeline planned for the mission, the ACS repair task was divided into two parts. Part 1 was planned for EVA 3 and Part 2 for EVA 5. Mission planners created that division out of an abundance of caution, because of the threat that the repair would run long. If it became necessary to finish the repair during

EVA 5, something else would have to be taken out of the EVA timing, most likely the installation of the replacement FGS.

I was prophetic. EVA 3 proceeded without a significant hitch. Drew, on the RMS arm, and John free floating, removed the now obsolete COSTAR from its axial instrument bay, stored it temporarily on the sill of the payload bay, retrieved the new Cosmic Origins Spectrograph instrument from its protective enclosure and successfully installed it in the bay previously occupied by COSTAR without a problem. COSTAR had done remarkable service on behalf of science. But it was time to return it to Earth to take its place of honor in the Smithsonian's National Air and Space Museum in Washington. As the two astronauts completed the COS installation, John radioed to the ground, "Thanks to all the folks at the University of Colorado and Ball Aerospace for getting COS ready after all these years." The instrument had been installed seven years after its originally planned date.

Now came the job about which there had been the greatest anxiety within the Hubble Program prior to the mission—the attempt to repair the ACS. While John did the step-by-step surgery on the ACS, Drew rode the RMS between the work site and the protective enclosures carrying tools and components back and forth. The job was intricate. It required approximately 13 different tools—some specialized to perform only one small step in the procedure.

John's first task was to cut away a metal grid on top of the Wide Field Channel's CCD electronics box (CEB) with a special cutting tool. The second arm from the bottom on the right side didn't appear to be cut all the way through. So John flexed the cutting tool back and forth by hand until the grid broke off.

Next he loosened, but did not unscrew, each of 32 fasteners holding down the cover of the CEB. Then he placed the specially designed fastener capture plate over the screws, attaching it to the CEB cover with latches and "sticky pads." With so many fasteners to be removed, John had to work quickly. He later said that he used a "Zen" approach—not thinking about the fasteners he had already removed or the fasteners that still remained to be unscrewed, but only about the one fastener that had to be removed now. He kept his mind "in the moment." All together the 32 fasteners holding the CEB cover were removed in a little over 10 min—about 19 s per fastener on average.

John transferred the Fastener Capture Plate, still attached to the CEB cover with 32 screws floating around between them, to Drew, who bagged them and carried them to a protective enclosure. John reported to the ground that the Fastener Capture Plate had been removed from Hubble's aft shroud. A challenging step of the ACS repair job was now complete.

The next step was to remove the four original circuit cards that had controlled the ACS's Wide Field Channel CCD detector. The cartridge holding the four newly designed smaller cards, one of which contained the new high-tech ASIC chip, would replace those (see Chapter 17). John began to release each of the eight card lock fasteners (two per card). He paused for a moment to crank up his spacesuit cooling flow—a lot of work had made him too warm.

He used another special tool to pull on the upper and lower edges of each of the four original cards, assuring that they were disconnected. One-by-one he proceeded to extract the cards. A strut in the telescope structure partially interfered with the removal of the final card. This had been anticipated. John was able to bend the card away from the strut as planned. He stored all four cards in a caddy to be transported by Drew back to a protective enclosure for the trip home.

From the ground CAPCOM advised John and Drew that they are running about "40 min ahead of schedule." He gave them a "GO" to complete the entire ACS repair job today. Now that they were ahead on the timeline, it would not be necessary to do Part 2 of the repair during EVA 5. That was great news! We now had assurance that the entire ACS repair would be accomplished and that time would be available to install FGS2R2 two days hence.

John inspected the now empty interior of the CEB. All the internal connectors were "pristine," and there were no broken pins, he reported. He then pushed the replacement CEB cartridge containing the four new circuit boards into place. Each card was then driven farther into its internal connector by a built-in mechanism to assure good electrical contact.

Next Drew handed the new low voltage power supply to John. He latched it to a handhold on the exterior of ACS. All that remained was to hook up the power cable that Mass and Bueno had installed as a get-ahead task the previous day, and another cable that attached the new low voltage power supply to the newly installed CCD electronics box.

Bueno, who was working as the intra-vehicular activity crewman during the ACS repair, reported to the ground that the ACS repair task was complete and that the ground was "GO" to perform the ACS aliveness test. As John and Drew were completing all their payload bay cleanup chores, CAPCOM announced that there was a "Good AT on ACS!" The aliveness test was followed by a functional test that checked out a number of the critical capabilities of the repaired ACS. The newly restored Wide Field Channel worked splendidly.

As this news was communicated to all of us in the BFCR, I noticed a few tears rolling down Ed Cheng's cheeks. Designing, building, and testing the ACS repair kit in time for SM4 had been a difficult and technically challenging job done on a tight schedule. And now the ACS Wide Field Channel was working well again. Ed's concept for the repair was proven valid. It had worked on the actual instrument on Hubble. He was greatly relieved, and felt exceptionally proud of the team that produced this minor miracle.

Through some "backdoor" wiring that connected the Wide Field Channel to the High Resolution Channel, it had been hoped that the latter might also come back to life when the former was powered on. That didn't happen. Apparently the powerful electrical short had not produced an open circuit, but was instead a dead short. Electrical current was moving with zero resistance through a path where it was not intended to go. The new power supplies provided by the repair could not overcome a dead short. But recovery of the HRC had only been a possibility, a hope, not a promise.

Later John Grunsfeld noted that EVA 3 had gone by so smoothly and efficiently that there had been time remaining during which he might have also installed a separate repair kit for the HRC. The EVA had consumed 6 hr, 36 min—much less

Figure 18.7. The EVA handle on the side of STIS, spanning the main electronics board cover. Before the cover could be removed, the handle had to be removed. Note the handle's rail in the center, the two end caps to the left and right containing two retaining screws each, and the two stanchions that supported the two end caps to which the handle was screwed. (Credit: NASA, GSFC, Ball Aerospace, J. Cassidy.)

time than the other four spacewalks on SM4. But that was not in the cards. We were thrilled to have the ACS Wide Field Channel back and working better than it ever had before.

EVA 4 on Flight Day 7 was perhaps the most bizarre EVA in the history of Hubble servicing. That day's job was to repair the Space Telescope Imaging Spectrograph by replacing one circuit board containing a low voltage power supply that had failed due to an electrical short in 2001. Bueno rode the RMS arm, while Mass was the free-floating spacewalker during this EVA. After collecting together the assortment of tools required for the STIS repair and stowing them on their workstations for easy access, Bueno and Mass proceeded to open the two large doors on the +V2 side of Hubble's aft shroud to get access to STIS.

Their first step was to remove the EVA handle that was attached to the side of STIS and spanned the main electronics box #1 (Figure 18.7). This should have been relatively simple to do. The handle was attached to two mounts (stanchions) by four screws that passed through two handle end caps. Cepi's tool experts had designed handle removal tools (HRTs) that Mass would attach over the end caps, one at the top of the handle, the other at the bottom (Figure 18.8). He would use the pistol grip tool (PGT) as a power screwdriver or nutdriver, inserting its hexagonal bit into two holes in each HRT to engage the hexagonal receptacle on the head of each screw, then turning each one counter clockwise to unfasten it. The point of the HRTs was to capture and retain the loose screws and the end caps, so that they would not float away as the handle was removed.

As Mass was operating the PGT to remove the handle, Bueno was close by on the RMS arm assisting in the transfer of tools and other hardware back and forth. Mass started with the HRT at the top of the handle, unfastening the screw on the right side. He rotated it 9 turns. He then moved to the left side of the top HRT and rotated that screw 13 turns, completely releasing it. "I don't think I got the top right," Mass said, and returned to that fastener to give it a total of 13 turns. "Top looks good,

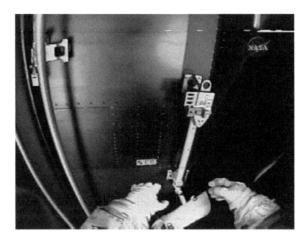

Figure 18.8. Picture taken by Mass' helmet camera showing the EVA handle spanning over STIS' main electronics box #1 cover. To remove the cover he first had to unscrew and remove the handle. At the top of the handle one of the handle removal tools had been attached. It served to capture the unfastened screws and an end cap so they wouldn't drift away. (Credit: NASA.)

moving to the bottom," he reported. He inserted the bit into the bottom HRT on the right side and triggered the PGT to release the fastener. He stopped and started the PGT several times, for a total of approximately 7 turns. He backed away from the bottom right screw and moved over to the bottom left. He applied 13 turns to the bottom left screw and it came free easily. He then returned to the bottom right side commenting, "… having trouble with this one on the lower right." He applied 1 turn, stopped, 1 more turn, stopped, and then 2 more turns and stopped. Mass removed the bit from the screw head, re-inserted it, and applied 12 additional turns. He then pulled his left foot out of the portable foot restraint holding him in place so that he could move to get a better view of the bottom HRT. He commented, "… the bit looks worn."

Mass made several more attempts to unscrew the lower right fastener with various combinations of speed and other settings of the PGT. Nothing worked. That one screw remained stubbornly fixed in place. He removed the bottom HRT and had a closer look at the screw head. He commented, "… the bolt is pretty beat up, the one on the right." The receptacle for the drill bit in the top of the screw was supposed to be hexagonal in shape. He commented, "… it no longer looks like a hex shape."

Evidently, during those first few rotations of the PGT bit in the lower right hole of the HRT, Mass had accidentally reamed out the screw head, so that the hexagonal bit no longer could engage and turn the screw. Watching this on the video screen in the BFCR, it looked to me like Mass was inadvertently squeezing the trigger of the PGT just a click to soon, before the bit was fully engaged into the screw head. I thought at the time that that would be an easy mistake to make, as there was no tactile information coming through that bulky spacesuit glove to his hand. Accidentally touching the trigger and giving it a slight squeeze would be easy to do. And a nutdriver bit that was already rotating when he tried to insert it into the hex head would ream out the hexagonal receptacle.

So now what should Mass do? The first three screws had unbolted without a problem. But the EVA handle was still firmly held in place by one screw that could no longer be turned by the power screwdriver. On the ground one might simply drill through the screw head and pop it off its shaft. But during an EVA that would be a time-consuming and very messy operation. We couldn't have tiny metal shards floating around in the aft shroud of Hubble. A lot of valuable EVA time was clicking off the clock as the crew in orbit and a large team on the ground pondered what seemed like an intractable problem.

As he was sitting at his console in the BFCR, Jim Corbo, the Hubble Systems Manager, began brainstorming about the situation Mass was facing. "If I encountered a problem like this in my garage at home, what would I do?" he asked himself. Then he answered his own question. "I'd just take a brute force approach. Pull on the handle with enough force to break it off."

Over a communications loop, Corbo contacted James Cooper in the STOCC at Goddard. James was the Mechanical Systems Manager for SM4. James in turn called Jeff Rodin, leader of the Mechanical Response Team, on a speakerphone in Building 29 where Cepi's office suite, the large clean room, and other facilities were located. During the mission, we had teams of experts in various technical disciplines off in the background ready to troubleshoot in their particular areas if problems arose. The Mechanical Response Team was one of them.

"Jeff, you've been watching this, right?" Cooper asked. "Yes, of course," Jeff replied. Cooper continued, "Corbo has suggested we look into pulling the EVA handle off by hand. The question is, how much force would that require? Is there some way you could quickly determine that?" Hearing this conversation, another member of the response team, Bill Mitchell, volunteered that he had a mockup of the STIS panel with two attached handles in the clean room. "We could take out the three screws from one of those and then do a pull test on it with the fourth screw in place," he said.

Jeff Rodin and his colleague, Ken Dickinson, quickly came up with a plan for how to rig up the test. But they needed a couple of tools. Jeff literally went running through the empty halls of the building; it was Sunday and nobody was around. He ran to an area where he expected to find some technicians. He came across Gene Mcalicher, who was there working on another project. From Rodin's body language, Gene picked up on the urgency of the situation. "What do you need?" he asked. "I need a torque wrench and a digital fish scale," Jeff replied. So, Gene went off to look for those items and Jeff proceeded to Room 190 where Ken Dickinson was already waiting.

A few minutes later Bill Mitchell, still dressed in his clean room bunny suit, came tearing into the room carrying the EVA handle mockup. They got the test rig all set up, and texted some pictures of the rig to James Cooper. Cooper gave them the green light to go ahead with the test. Gene stepped up onto the table, attached the digital fish scale to the handle and began pulling upward with considerable force. The handle broke free—violently. The liberated handle became a projectile flying high above their heads. They reflexively ducked, not knowing where the broken metal piece would come down. Gene had measured the force required to break the handle away from the screw—60 lbs.

This improvised test was captured in a video that Jeff Rodin made with his cell phone. He forwarded it to James Cooper who sent it on to Jim Corbo in Houston. Only a few people at JSC had the opportunity to see it. Preston Burch and Cepi were among those few.

Word of this test result quickly made its way to the shuttle team. Preston and Cepi got a message to Lead Flight Director Tony Ceccacci about it and suggested that the Atlantis crew be given that option to consider. The idea was communicated to Scooter who, as Mission Commander, had the final say about whether or not Mass should give it a try. Scooter discussed with Mass whether he was OK with attempting to pull off the handle. Mass was a large, strong man. He should easily be able to apply 60 lbs of linear force pulling outward on the top of the handle.

Mass agreed to do it. He wrapped three pieces of Kapton tape over the bottom of the handle, covering the end cap and the screw holes. The idea was to prevent any particles that might be created by the forceful ripping of the handle from flying into the aft shroud or the Payload Bay. He then took hold of the handle, moved it back and forth a few times trying to put some fatigue into the metal to weaken it, and then he gave a forceful pull. In an instant the handle was free and firmly in Mass' grasp. "Disposal bag please," he calmly requested.

The video of the test at Goddard later became widely available for viewing on the web and elsewhere. It tended to provoke nervous laughter, and gasps, because of the violence of the rupture of the handle from its mooring. Scooter later told me that, had he seen the video in advance, he might not have allowed Mass to pull the handle off. He had visions of the handle being violently propelled into Mass' visor and doing serious damage. But all ended happily. Mass reported that he saw no debris created during the process. The Kapton tape had contained it. The entire episode had consumed about 1.5 hours of unproductive EVA time. But now he and Bueno could proceed with the STIS repair.[1]

Their next step was to attach the fastener capture plate to the MEB cover (Figure 18.9). The FCP and MEB cover formed a sandwich with a gap in between. The small holes in the FCP, precisely aligned over each of the 111 screws holding the MEB cover in place, allowed the bit of Mass' power screwdriver (in this case a tool smaller than the PGT called the mini power tool or MPT) to penetrate down to the head of each screw. The unfastened screws were too large, however, to fit through the holes, so they simply floated between the FCP and the MEB cover. The complete sandwich—FCP, MEB cover, and 111 floating captive screws in between—would be stashed away in a container for return to the ground.

Except for the physical challenge of unfastening 111 screws in a relatively short period of time, the entire task was straightforward. After all the stresses Mass had faced with the reamed-out screw head, the unconventional removal of the EVA

---

[1] This episode, as related by the people who experienced it, can be seen online in a Video, "Hubble Memorable Moments: Brute Force," at https://www.youtube.com/playlist?list=PLiuUQ9asub3Ta8mqP5LNiOhOygRzue8kN.

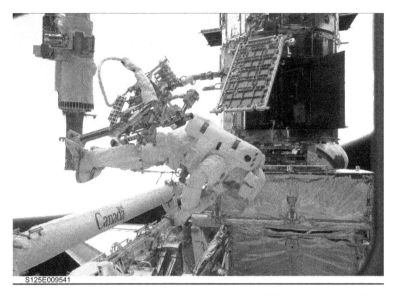

Figure 18.9. Astronaut Michael Good (Bueno) preparing for the STIS repair during EVA 4. The blue rectangle attached to a tether is the fastener capture plate used to facilitate the removal and capture of 111 screws holding down the main electronics box side panel. The panel had to be removed to allow access to the power supply circuit board that needed to be replaced. (Credit: NASA.)

handle, and the 1.5 hours of lost EVA time, he hoped that the remainder of the EVA would go more smoothly. But Mass' headaches were not over. After the FCP was locked down onto the MEB cover, he began to go through the motions of inserting his power tool into the first hole to unfasten the first screw. But when he pulled the trigger, nothing happened. The MPT didn't turn. The battery was dead. Mass let out a long sigh and said something like, "Well, shoot!" To us on the ground it seemed obvious that a string of much stronger expletives must be coursing through Mass' mind.

CAPCOM passed along instructions that Mass should go retrieve the spare mini power tool with its battery pack from the airlock. Mass moved to the airlock, ingressed, stowed the MPT he had been using and retrieved the spare. After passing it out to Bueno, Mass connected his suit to an umbilical that would replenish the oxygen supply in his backpack. That process took about five min during which, we guessed, Mass must have regrouped and reset himself mentally—undoubtedly necessary given the day's setbacks up to that point. Meanwhile Bueno had tested the new MPT and said, "this one works." Mass reported that he believed he was fully charged with O_2 and "was 100%."

The two astronauts made their way back to the work site in the aft shroud. To us listening to their voices on "air to ground," Mass sounded energized and ready to go back to work removing the MEB cover on STIS. There were a couple of small glitches as Mass proceeded to unfasten the 111 screws. He commented that one of the screws was "stubborn." After a while the screwdriver bit began sticking to the heads of the removed screws. He had to bump them up against the FCP plastic

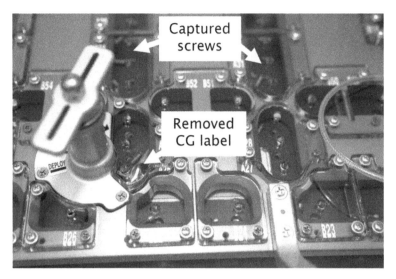

Figure 18.10. The fastener capture plate used to remove the 111 screws from the MEB cover during the repair of STIS. This photo was taken after the mission. Note the loose screws lying on their sides retained between the FCP on top and the MEB cover on the bottom. (Credit: NASA, GSFC, J. Cassidy.)

window to knock them loose. Otherwise, the process of removing all the fasteners went smoothly. It took Mass 33 min, 25 s to remove all of the fasteners, including time to change the screwdriver bits. He did a visual inspection of all the fastener locations under the FCP window and verified they all had been successfully removed (Figure 18.10).

At that point, Mass was able to maneuver the combined FCP/MEB cover around a couple of obstacles and away from STIS, as Bueno reached underneath with a wire cutter and severed a pair of wires that had originally been attached to the cover. The two men stowed the FCP/MEB cover stack in an enclosure on one of the equipment carriers in the Payload Bay.

Now the real repair could be done. Mass and Bueno fetched and carried two card transport enclosures to the work site. One of the enclosures was empty, awaiting the stowage of the old power supply card. The other contained the new replacement card. Mass used a special tool to unfasten the locks that held the old card in place. It was the top card in the stack of 13 contained in the MEB, so it was easy to find and remove the correct card. Mass attached the card insertion/extraction tool onto the failed power supply card, removed it from the MEB, and stowed it in its transport enclosure.

He then examined the inside of the MEB chassis, and reported, "It looks pretty good." The last time it was open was in 1996 in a clean room at Ball Aerospace. Now, almost 13 years later, its interior was exposed again, this time to the vacuum of space. The people working on it were space-suited astronauts, not technicians in clean room bunny suits. Amazing! No one would have imagined such a thing back in 1996.

Mass attached the insertion/extraction tool to the replacement card and pulled it out of its transport enclosure. He slipped it into the empty slot in the MEB, pushed it inward to assure it was properly situated, and then re-fastened the card locks.

After stowing the two card transport enclosures, the two spacewalkers retrieved and installed the replacement MEB cover. No longer would it be held in place by 111 small screws. Mass had only to push two levers down and lock them into place with four locking pins. Intra-vehicular (IV) astronaut Drew Feustel announced, "Houston, the MEB Cover is installed."

The two spacewalkers spent some time tidying up the STIS work site. Mass did a routine glove inspection. He reported, "The right glove seems a little dirty but other than that it looks good. The left glove looks like we might have picked up a little more of a tear here, on the left ... and actually I see a little tear of the cloth." John Grunsfeld, now the IV astronaut, confirmed the location, in the "palm region."

This was potentially hazardous for Mass. A relatively deep tear in one of his gloves risked loss of internal pressure in his spacesuit. The oxygen he depended on could possibly leak out into space. The CAPCOM called up to get a better description of the glove tear. A short time later Commander Altman informed Mass and Bueno that the EVA day would end after they had closed the two aft shroud doors.

Not only had the EVA run long, but also they couldn't risk the hazard of further damage to Mass' glove. That meant that the final job originally planned for this EVA, the installation of a New Outer Blanket Layer to enhance the thermal insulation covering the door of Electronics Bay 8, would have to be postponed. But the primary task for the day, repairing the dormant STIS instrument, had been accomplished. The CAPCOM informed Atlantis' crew that the STIS aliveness test was successful.

On the ground we all cheered at that news. STIS had been dead for five years. And now it had been brought back to life.

The word "routine" seems inappropriate to describe the intricate work required of EVA astronauts on any Hubble servicing mission spacewalk. In terms of the tasks to be performed, EVA 5 of SM4 was less challenging than the preceding and really stressful four. The challenging part came from the call from the ground that this final EVA would not be allowed to go long, regardless of what had been completed. As a result, the Atlantis crew decided to get up before the usual wake up call and to get an early start on preparing for the EVA. In fact they got up so early that the night shift CAPCOM, Shannon Lucid, was still on console when Scooter called down that the crew was ready to continue with the EVA Checklist! Only by getting a head start did they have enough time to complete all the scheduled tasks for the mission.

John, riding the RMS arm, and Drew, the free-floating member of the pair, had three assigned tasks. They had to remove and replace the second pack of nickel–hydrogen batteries, install the replacement FGS #2, and complete installation of the last two New Outer Blanket Layers to bolster the thermal insulation of the spacecraft's equipment bays.

They completed the battery pack installation without difficulty. That had also been the case for Mass and Bueno during EVA 2. The change out of FGS2R, however, presented a now familiar problem. During EVA 1, to remove WFPC2 from its radial bay, Drew and John had to unfasten a long bolt that locked the instrument into its A-Latch. That task had proven very difficult (and a little bit scary, see Figure 18.3). In the case of WFPC2 the bolt had been tightened much tighter than had been expected. It took Drew's mechanic's touch to loosen it by hand with a socket wrench.

Now they encountered the same problem with the A-Latch bolt locking FGS2R in place. When they used the Mechanical Torque Limiter (MTL), the A-Latch bolt refused to break free. They switched to the Contingency Mechanical Torque Limiter (CMTL). John gave it a try. "No joy," he reported. Then Houston directed him to remove the CMTL and try the ratchet wrench without any torque limiter attached. The direction was to "apply the torque evenly without any impulse." That is, don't give the wrench a sudden jerk, but try to apply force by hand smoothly and evenly. John did this and the A-Latch broke free. He turned the bolt ½ turn.

The two astronauts installed the handhold frame to the back end of the old FGS2R, the same frame that they had used to maneuver WFPC2 and WFC3 in and out of the spacecraft. John then used the pistol grip tool, the large power screwdriver, to rotate the A-Latch bolt 22 turns and disengage it from the latch. The FGS could then freely be removed from its radial bay.

The replacement FGS2R2 went in without a hitch. The team on the ground and spacewalking in the Payload Bay had learned their lesson from EVA 1. They had become experts in unfastening recalcitrant, overly tightened bolts that were impeding their ability to replace old instruments with new, much improved models.

As they had during EVA 3, John and Drew had worked very efficiently so that at this point they were well ahead of schedule. They proceeded to install the remaining New Outer Blanket Layers (NOBLs). There were seven NOBLs in all. They were almost always last in line to be installed on any servicing mission. There had been time to install three of them on STS-103/SM3a and one on STS-109/SM3b. That left three that might be installed on the current mission if EVA time permitted. EVA 4 had been concluded abruptly, due to concerns about a tear in one of Mass' gloves. The NOBL for Equipment Bay 8 had been scheduled as the last task on EVA 4, but its installation was left undone.

John and Drew began by installing the Bay 5 NOBL. They followed that with the Bay 8 NOBL and still had time remaining to do Bay 7. Going into the mission the expectation had been that there would likely only be time to install two of the NOBLs. But the spacewalkers on EVA 5 had completed all three.

The job was done. Everything that could be done to service Hubble on this mission had been done. In giving public talks later, I always said that SM4 was 110% successful, despite there having been so many impediments and near disasters—the frozen A-Latch bolt holding WFPC2 in place on EVA 1, the balky RSU that refused to fit into the spacecraft in EVA 2, the stripped screw head in the EVA handle on STIS in EVA 4, and another overly tightened A-Latch bolt in FGS2R in EVA 5. I am often tempted to attribute the success of SM4, despite all odds against it, to Hubble's "Guardian Angel." But perhaps it had more to do with the training, skill, creativity, tenacity, and courage of the entire SM4 team, in orbit and on the ground, and also to Cepi's formula for mission success. "Test, test, and re-test. Train, train, and re-train."

The last human being to lay a hand on the Hubble Space Telescope was John Grunsfeld at the conclusion of EVA 5—fitting I think, given that John had spent more time in orbit with Hubble than any other astronaut. Hubble was un-berthed and released from the RMS arm by Megan McArthur at about 7 am Houston time on 2009 May 19 (Flight Day 9). The Hubble team at JSC was gathered in the BFCR, looking at the large video display at the front of the room, watching the separation between Hubble and Atlantis grow progressively larger, until the telescope faded into the distance. I was filled with pride, dazzled by the beauty of our incredible telescope, and wistful that this was, in all likelihood, the last time humans would ever set eyes on it.

Atlantis was scheduled to land back at KSC on Friday, May 22. I had never watched a shuttle land, and so I made my way back to KSC from JSC to see Atlantis touch down. Before Atlantis was launched, the crew had been told that there was no way to keep it in orbit for more than one wave-off day—one extra day in orbit because of bad weather conditions at the Cape. However, because the advance weather forecasts for the planned landing day and for several days thereafter were not good, the ground called up changes to the orbiter's nominal "Group C" power down to reduce their draw on the available power reserves. They wanted to reduce the usage of the electrical power provided by their fuel cells to the lowest level that was still safe in order to buy another day or two of time in orbit, hoping that the weather at the Cape would improve. Fortunately, when the time came on landing day, everything powered back up successfully. But it seems likely that no shuttle had ever powered down as far as Atlantis did during the concluding days of STS-125.

Several landing opportunities were waived off because of out-of-limits weather conditions on Friday, Saturday, and early Sunday. All of us who had gathered at KSC were disappointed when the Entry Flight Director, Norm Knight (the same person who had been the Ascent Flight Director almost 2 weeks before), at last made the decision that conditions were too threatening at KSC and the orbiter would have to land at Edwards Air Force Base in California. We were disappointed, but thankful that the crew would be brought back to earth safely. Atlantis touched down on Runway 22 at Edwards at 8:39 am Pacific Time on Sunday, May 24.

The following September 9 we released the first EROs—Early Release Observations—obtained with our two shiny new instruments, WFC3 and COS. Two of the EROs are reproduced here. Figure 18.11 demonstrates the power of wide wavelength coverage enabled by WFC3—our panchromatic camera. The pillar of dust and gas is a part of a large nebula in the southern constellation Carina (which is the Latin word for the keel of a ship). The pillar is illuminated by hot massive stars outside the field of view at the top of the picture. It is being eroded away by the intense radiation emitted by those stars. Small fingers sticking out of the pillar are denser areas that are eroding less quickly. In some cases at their tips we can see the light from newly born stars. In visible light the dust in this pillar makes it opaque. We really can't see what's going on inside. But when we observe the pillar in infrared light, the interior becomes clear. The photons of infrared light are too long in

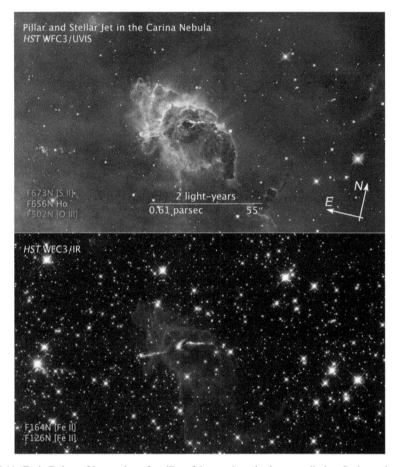

Figure 18.11. Early Release Observation of a pillar of dust and gas in the constellation Carina, taken with the two channels of the new WFC3. At the top is a composite of images taken through three different filters in visible light. At the bottom is a composite taken through two filters at near-infrared wavelengths. Each filter is designated in the lower left by the central wavelength in nm of the band of light it transmits. The "N" next to the central wavelength stands for "narrow band." Next to that is the chemical element or ion that emits light in that filter band. (Credit: NASA, the WFC3 Team, STScI 2009-25.)

wavelength to be scattered by the fine particles of dust, and so they penetrate the dust, allowing us to look inside the nebula. The most prominent thing we see in the interior is a bright, newly born star emitting jets of hot gas. Conditions inside the pillar are well suited for the formation of new stars—it's a stellar nursery.

Figure 18.12 is a spectrum taken with COS of a distant quasar, named with its catalog designation PKS 0405-123. One of the major science objectives for the COS instrument is to study the filaments of gas that span the vast distances between galaxies. These filaments are condensed into their form by the gravitational pull of concentrations of cold dark matter. They define the skeletal framework, the scaffolding, upon which almost all of the ordinary matter in the universe has concentrated to form stars, galaxies, and clusters of galaxies. Because of its web-like

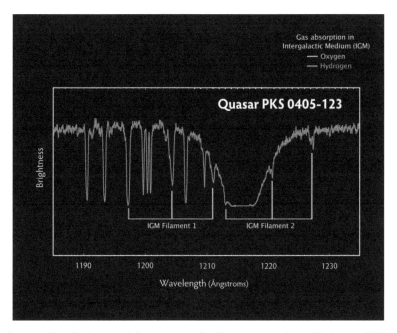

Figure 18.12. A portion of a far-ultraviolet spectrum of a distant quasar taken with the new COS instrument. The measured brightness of the light is plotted on the vertical axis and the wavelength of light is plotted on the horizontal axis. The wavelengths are given in units of angstroms, or one tenth of a nanometer. The neutral hydrogen gas in our own galaxy makes the broad absorption trough in the spectrum. Some of the narrow lines also originate in our own galaxy. But a few of them are produced by hydrogen and oxygen concentrated in filaments of the Cosmic Web, as labeled. (Credit: NASA, the COS Team, STScI 2009-25.)

structure, this framework is called the Cosmic Web. The strategy is to use the enormously bright quasars as intergalactic flashlights, illuminating the intergalactic gas between the quasar and us. Light from the quasar is absorbed by the intervening gas at specific wavelengths associated with atoms and ions of particular chemical elements. These elements are constituents of the concentrated gas in the filaments of the Cosmic Web. By studying filaments at different distances, astronomers can assess how the Cosmic Web has evolved over the lifetime of the universe.

AAS | IOP Astronomy

Life With Hubble
An insider's view of the world's most famous telescope
David S Leckrone

Chapter 19

Worlds Without End: Exploring Exoplanets

Do exoplanets—planets orbiting stars other than the Sun—exist? Is there life elsewhere in the universe? In the early 1980s when Hubble was being built, we had no clue. But professional astronomers and ordinary citizens alike had developed an intense interest in the subject. At a meeting of the Hubble Science Working Group during that period, the subject came up in conversation—would Hubble be able to detect a planet orbiting another star? I timidly spoke up, saying I was a "Trekie" and was very interested in the subject. I was very surprised when some of the more esteemed, senior scientists in the room—John Bahcall, Bob O'Dell, Ivan King, and others—chimed in with, "Oh, I'm a Trekie too." Apparently the famous but short-lived TV series of the 1960s had cleared the way so that it was now socially acceptable among professionals to ask these questions seriously.

As to the question at hand, someone went off and did the calculations. For Hubble to directly image an exoplanet with WFPC or the Faint Object Camera, Mother Nature would have to be very kind to us. A large planet would have to be orbiting a bright, relatively nearby star. Such a planet would likely appear no brighter in reflected light than one ten-billionth (10^{-10}) the brightness of the parent star. It was like looking for a firefly next to the lantern of a lighthouse. The parent star itself had to be very bright for us to see the planet at the limit of the telescope's sensitivity. In addition the planet would need to be separated from the star by about 0.5 s of arc or more as seen from Earth. Otherwise its image would be swamped by light in the outer periphery of the point source image produced by the star. It might be worth having a look, but the odds of directly imaging an exoplanet seemed small and greatly dependent on good fortune.

What we didn't realize in that discussion was that a direct image of one exoplanet, while satisfying a deeply seated curiosity, would be of limited interest scientifically. After Hubble was launched, and its spherical aberration corrected, astronomers discovered that there were much richer sources of knowledge about alien planets to be explored.

The first revelation came from Bob O'Dell's images of the great nebula in Orion's belt (O'Dell & Wen 1994). This spectacular view of a star-forming region only 1400 lt-yr away from Earth (Figure 19.1) was a composite of images taken through four filters with the WFPC2. One striking feature was the large number of compact structures that stood out in sharp relief from the background nebulosity. Some looked like bright nodules, others were small dark disks. These were what O'Dell termed "proplyds," short for protoplanetary disks. He first discovered them in spherically aberrated images taken in 1992 with the original WFPC. But now, with the aberration fully corrected, they were much more clearly defined.

Zooming in on the image (Figure 19.2), one could see several examples of the proplyds. Buried within the bright nodules were dark flattened disks of dust and gas encircling newly forming stars. The nodules were apparently made of gas blown out of the cloud of material from which the new star was condensing by intense radiation and winds from the cluster of hot, massive stars embedded in the Orion Nebula, the Trapezium. The isolated dark disk to the right in the figure, surrounding a young star, was apparently in a shaded region less affected by the presence of the hot stars.

O'Dell and his collaborators continued to survey the Orion Nebula over the next few years (McCaughrean & O'Dell 1996). They identified at least 152 objects that showed evidence for circumstellar disks surrounding young stars (Figure 19.3). The disks ranged in size from about 0.5 to 10 times the diameter of the Kuiper Belt, the circumstellar disk outside the orbit of Neptune that is a remnant of the formation of our own solar system. The stars centered within the disks had masses ranging from 0.3 to 1.5 times the mass of our Sun, and ages from about 1 up to about 3 million

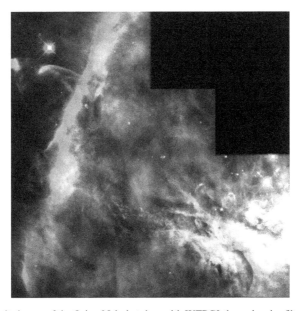

Figure 19.1. Composite image of the Orion Nebula taken with WFPC2 through color filters, transmitting light emitted by three chemical elements—red for nitrogen, green for hydrogen, and blue for oxygen—and a fourth filter that transmitted a broad band of visible light. (Credit: NASA, C. R. O'Dell, Z. Wen, STScI 1994-24.)

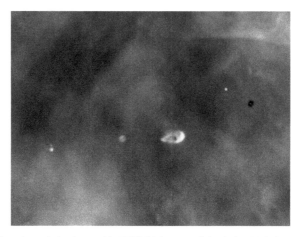

Figure 19.2. Examples of proplyds observed in WFPC2 images of the Orion Nebula. The bright nodules are clouds of gas being evaporated away from the material out of which a new star has formed by the intense radiation and winds of hot, massive stars in the center of the nebula. The dark object to the right is a circular disk of dust and gas, tilted to our line of sight, centered on a newly forming star. It is apparently within a shaded region protecting it from the hot stars. (Credit: NASA, C. R. O'Dell, Z. Wen, STScI 1994-24.)

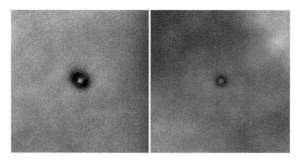

Figure 19.3. Examples of flattened disks of dust and gas surrounding young stars in the Orion Nebula. (Credit: NASA, ESA, M. McCaughrean, C. R. O'Dell, STScI 1995-45.)

years. McCaughrean and O'Dell concluded that the disks "are strong candidates for progenitors of planetary systems similar to our own." Our solar system most likely looked like this (Figures 19.3 and 19.4) at the time of its condensation out of a dense cloud of molecular gas and dust some 4.6 billion years ago.

During the mid-1990s additional observations of numerous circumstellar disks were made with WFPC2. These provided evidence in support of theories of how planetary systems likely form (Figure 19.5).

A large concentrated clump of gas and dust—one of thousands—embedded in a very dense and very cold part of a giant molecular cloud begins to collapse under its own gravity. The giant cloud is made mostly of molecular hydrogen, helium, and traces of heavier material. Material within the clump becomes concentrated and grows warmer as it grows denser—gravitational potential energy is converted into heat. As the clump collapses it becomes symmetrical in shape. Whatever small

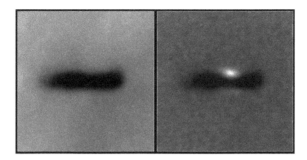

Figure 19.4. An edge-on protoplanetary disk seen in the Orion nebula. View at the left is in full color showing disk silhouetted against the bright nebula. View at the right has all the nebular light filtered out, revealing the central structure in the disk more clearly. The disk is doughnut shaped. The central star is not seen, but the scattering of light from the central star by surrounding dust and gas produces the bright areas above and below the disk. (Credit: NASA, ESA, M. McCaughrean, C. R. O'Dell, STScI 1995-45.)

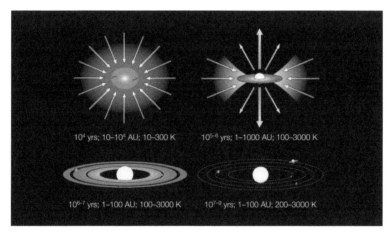

Figure 19.5. The major phases of planetary system formation with a time scale, size scale and temperature range of the system. Upper left: Deeply embedded protostar. Upper right: Circumstellar disk and jet. Lower left: Agglomeration and planetesimals. Lower right: Mature planetary system. (Credit: M. McCaughrean.)

amount of angular momentum it had initially must be conserved as it contracts. Like a spinning ice skater pulls in his or her arms to spin faster, the collapsing cloud spins faster as it grows smaller. The spinning cloud begins to flatten into a disk shape with the spin axis of the disk perpendicular to its surface.

Material from the inner part of the spinning disk accretes onto the concentration at the center of the system, causing it to grow ever more massive and dense. That central region is becoming a protostar. Some of the material moves out of the disk along magnetic lines of force and accretes onto the poles of the protostar. There it is heated and ejected sporadically as blobs of material along the star's spin axis, accelerated along open magnetic field lines, creating bipolar jets. The jets carry away angular momentum from the inner disk and from the protostar, reducing their rotational velocity. The inner part of the spinning disk flattens more rapidly than its

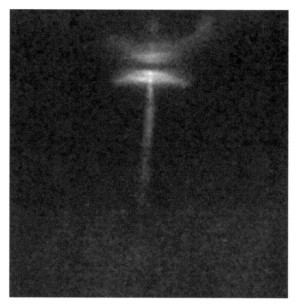

Figure 19.6. A WFPC2 image of HH30, a young protostar and protoplanetary system in the process of being formed. A dusty disk, seen here edge-on, surrounds and obscures the protostar. Above and below the center of the disk are two jets of material sputtering off the star, accelerated by magnetic fields along its rotation axis. Above and below the plane of the disk, material illuminated by the protostar is still in the process of collapsing down onto the disk, with the central region already more concentrated than the outer periphery. (Credit: NASA, C. Burrows, the WFPC2 Team, STScI 1995-24.)

outer periphery. Many aspects of this process can be seen in WFPC2 images of an object called Herbig-Haro 30, or HH30 for short (Figure 19.6).

Eventually the accretion of material from the disk onto the protostar ceases, as the rotation of the disk diminishes and the material in the inner disk is more or less depleted. The jets die away. Dust grains that formed in the protoplanetary disk begin to collide. Sometimes the grains stick together creating conglomerates. Sometimes the conglomerates are broken apart by impacts with each other. But eventually, after this process has gone on for a long time, the conglomerates build up in size to create a population of small planetesimals, which in turn collide, stick together and become the seeds from which rocky planets may eventually grow. If enough gas—primarily hydrogen and helium with traces of carbon monoxide, methane, and other molecules—is still around, the gravity of a rocky planet may attract enough of it to form a primitive atmosphere.

The protoplanetary system evolves into a mature planetary system of rocky planets, gas giants, and leftover material. The residual material remains distributed in what's left of the original disk, similar to our own Kuiper Belt. Of course, there is much more detailed physics going on in reality than I've portrayed here. But in broad outline, this is the general idea.

If the protostar achieves a mass greater than about 8% of the mass of the Sun, it will be capable of igniting nuclear reactions at its center, converting hydrogen into helium and releasing enough energy to support its own weight; eventually it stops contracting and settles into a stable configuration on what is called the hydrogen-burning main sequence. Anything less than 8% of one solar mass becomes a brown dwarf, or "failed" star, incapable of sustaining a continuously running nuclear furnace in its core.

None of Hubble's observations in the 1990s directly revealed an image of a planet. However, the observatory did open an entirely new field of observational astronomy—the detailed processes and environments involved in the formation of planetary systems observed at high resolution.

When STIS and NICMOS were installed on Hubble in 1997, they brought new tools that advanced this fledgling field of study even further—precision coronagraphs. An even more advanced coronagraph was included in the ACS, installed in 2002. These features within the new instruments gave observers the ability to cover the images of bright stars with physical masks of some sort—disks or bars—allowing long exposures without the risk of burning out the picture. In this way astronomers could bring out the faint details of the environs surrounding the stars. They could block out the "lighthouse lamp" to see the adjacent "firefly."

Good examples of what NICMOS and STIS could do with their coronagraphs are shown in Figures 19.7 and 19.8. The two stars shown here are older than the protostars O'Dell discovered in the Orion Nebula. Those were around one million years old, give or take, and were still in the process of collapsing onto the hydrogen-burning main sequence. These stars, HD 141569, and HR 4796, are probably approaching 10 million or so years of age. The disk around the former and the ring around the latter are most likely made up of residual debris left over from the

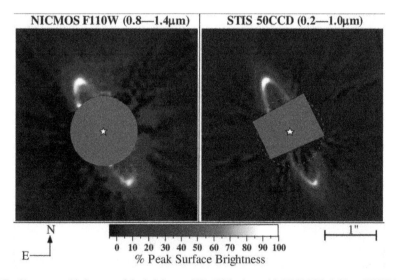

Figure 19.7. Chronagraphic images of the bright star HR 4796 taken with NICMOS (left) and STIS (right). By suppressing the bright light from the central star, the faint ring of dust around it can be imaged. The narrow ring around HR 4796 could not persist for long without the gravitational influence of one or more planets. (Credit: G. Schneider.)

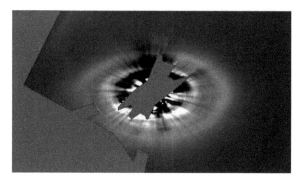

Figure 19.8. Chronagraphic image of the star HD 149768 taken with STIS. The dust disk surrounding the star is made of the debris left over from protoplanet formation. There are at least two gaps in the disk that could be evidence of the presence of protoplanets. (Credit: G. Schneider.)

formation of their protoplanetary systems. There are at least two gaps in the middle of the disk around HD 141569. The gravity of one or more large planets may be responsible for creating and sustaining these gaps. The ring around HR 4796 wouldn't last long unless it was being shepherded by one or more planets. Otherwise it would rapidly spread out and vanish from sight. Even without providing a photograph of planets, STIS and NICMOS provided indirect evidence for their existence.

The ACS coronagraph did provide an image of what may be a planet in orbit around the very bright southern hemisphere star, alpha Pisces Austrinus (the brightest star in the Southern Fish constellation) also known as Fomalhaut. The star is relatively close to us, a mere 25 lt-yr away. Ground-based sub-millimeter observations, as well as infrared observations from the Spitzer Space Telescope, had suggested that there was some kind of dusty structure around Fomalhaut, which was not centered on the star.

Astronomer Paul Kalas from U.C. Berkeley and colleagues acquired observations of Fomalhaut in 2004 using the ACS' coronagraphic mode in the instrument's High Resolution Channel. Hubble's sharp imagery verified the existence of a dusty ring about 21.5 billion miles across (Figure 19.9). The center of the ring was indeed offset from the star by 1.4 billion miles or about 15 times the distance between the Earth and the Sun (15 au). For comparison that is about half the distance across our own solar system. The observers concluded that the gravity of one or more unseen planets had perturbed the ring. Otherwise it should be centered on the star.

Kalas and his team noted a few intriguing bright spots in the 2004 image that could be candidates for a planet. In 2006 they repeated the ACS observation. One of those spots had moved a little (Figure 19.10). Fitting a tentative orbit to the two positions, he concluded that the planet orbited Fomalhaut every 872 yr. It could have a mass no greater than three times the mass of Jupiter. Otherwise it would have distorted the sharp inner edge of the ring (Kalas et al. 2008).

Ironically, on the same day that Kalas and NASA announced the discovery of Fomalhaut b to the media, 2008 November 13, a team of ground-based

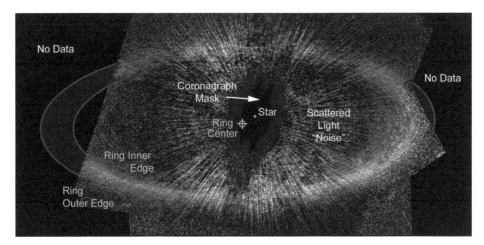

Figure 19.9. Hubble observation of the debris ring around Fomalhaut, acquired in 2004 with the coronagraph in the High Resolution Channel of the Advanced Camera for Surveys. This image verifies that the ring is not centered on the star. Its center is offset from by about 1.4 billion miles, suggesting that one or more planets have gravitationally tugged the ring off-center. (Credit: NASA, P. Kalas, et al., STScI 2008-39.)

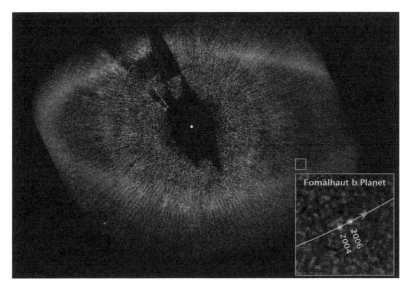

Figure 19.10. A second coronagraphic image of Fomalhaut taken in 2006 showed that one of the bright spots seen in the 2004 image had moved. It might be the planet that had de-centered the ring relative to the star. (Credit: NASA, P. Kalas, et al., STScI 2008-39.)

astronomers, led by Christian Marois of the Herzberg Institute of Astrophysics in Canada, announced the discovery by direct imaging of a system of three planets (a fourth was discovered later) around the bright star, HR 8799, 129 lt-yr from Earth in the constellation Pegasus (Marois 2008). They had made the discovery with ground-based telescopes in Hawaii, using adaptive optics (optics that can be changed in

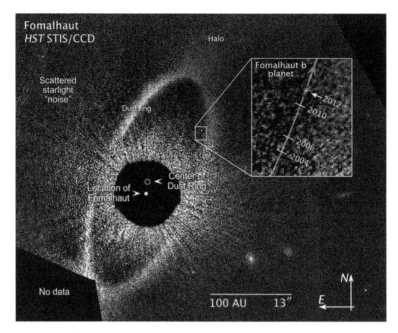

Figure 19.11. The four observed positions of the apparent exoplanet Fomalhaut b in 2004, 2006, 2010, and 2012. (Credit: NASA, P. Kalas, et al., STScI 2013-01.)

shape or orientation very rapidly to compensate for the rapid variations of the Earth's atmosphere) in the infrared. These three planets were each about seven times the mass of Jupiter.

The ACS failed in 2007, and its repair in 2009 was not successful in bringing the High Resolution Channel with its coronagraphic mode back to life. But the coronagraphic mode still worked on the repaired STIS. So, undaunted, Paul Kalas and colleagues repeated the coronagraphic imaging of Fomalhaut with STIS in 2010 and 2012. The bright spot interpreted as a planet in 2006 was still there and still moving on a regular course (Figure 19.11).

What was very surprising, however, was the discovery that the planet was moving on a highly elliptical orbit, coming as close to the star as 4.6 billion miles and reaching a distance of 27 billion miles at the outermost point in its orbit. It apparently orbited Fomalhaut once every 2000 yr (Figure 19.12). Kalas and his team speculated that an as yet undetected second planet had gravitationally interacted with Fomalhaut b, ejecting it from its original orbit closer to the star. That kind of gravitational interaction is probably common as planetary systems evolve.

So, Hubble may in fact have actually visually seen an exoplanet orbiting another star, as the Science Working Group had pondered almost three decades earlier. But this was a strange planet. It was brighter than it should have been for a planet with three times the mass of Jupiter, suggesting it might have a system of rings around it reflecting additional light from the parent star. And it was in a weird orbit.

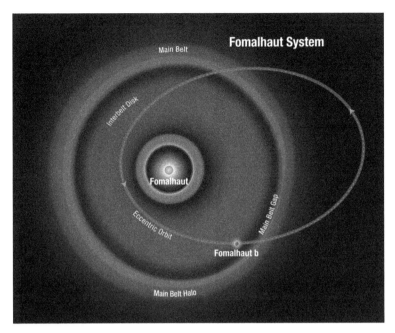

Figure 19.12. Artist's concept of the architecture of the Fomalhaut system, showing the highly elliptical orbit of Fomalhaut b, derived from four separate coronagraphic observations. (Credit: NASA, ESA, P. Kalas, A. Feild, STScI 2013-01.)

In any event, at the same time, ground-based astronomers had demonstrated the capability to do the same. No matter. Hubble would go on to show unique capabilities to learn about the properties of exoplanets that could not be done nearly as well, or not done at all, from the ground.

In the April 20, 2020 issue of the Proceedings of the National Academy of Sciences (Gáspár & Rieke 2020), astronomers András Gáspár and George Rieke of the University of Arizona discussed their re-examination of all the Hubble observations of Fomalhaut b made in 2004, 2006, 2010, 2012, 2013, and 2014. They found that the object grew larger and fainter with time, disappearing entirely in 2014. Also, although it was seen clearly in visible light, it emitted no radiation in the infrared—none of the heat that a planet would be expected to emit. They concluded that Fomalhaut b is not a planet at all, but rather "it is probably an extraordinary super-catastrophic planetesimal collision observed in an exoplanetary system!" The collision produced an expanding dust cloud containing copious amounts of dust. Collisions like this must have been common in the early epochs of our own solar system.

Back on Earth, astronomers at mountaintop observatories were perfecting a technique using Doppler shifts of starlight to indirectly detect the presence of planets around other stars. A planet orbiting a star exerts a gravitational force on the star. The star and the planet share a common center of mass, the location of which depends on the masses of the two objects. Since the star's mass dominates that of even the largest planet, the center of mass of the two will fall close to the center of the star. The star and the planet both orbit around their mutual center of mass. In other words the planet is gravitationally tugging the star around a little bit.

The component of the star's motion along our line of sight is called its radial motion, and the corresponding component of velocity of this motion is termed its radial velocity. When one or more planets cause the parent star to move around, astronomers using very sensitive spectrographs can measure the small variations in the star's radial velocity using the Doppler shift of absorption lines in its spectrum. Its light is shifted a tiny bit toward the red as the planet pulls the star away from us and a tiny bit toward the blue as it pulls the star in our direction. When the motion of the planet, and of the star, is perpendicular to our line of sight, the radial velocity becomes zero. So, measuring the small changes in the star's radial velocity with time and plotting it on a graph yields an oscillating curve. If a single planet moving in a circular orbit induces the motion, the curve looks like a sine wave. It will have a more complicated appearance if elongated orbits or multiple planets are involved.

The first exoplanet orbiting a main-sequence star discovered with this technique was 51 Pegasi b (Mayor & Queloz 1995). Its parent star, 51 Pegasi, is about 50 lt-yr from Earth. The planet's mass is about half that of Jupiter. Its distance from its star is much less than the distance of Mercury from the Sun. Its "year" is only 4.2 days in length. Astronomers estimate the temperature in its atmosphere to be in the neighborhood of 1000°C. Exoplanets with these characteristics—gas giants orbiting very close to their stars—are called "hot Jupiters." The radial velocity variations of 51 Pegasi induced by the gravitational tug of 51 Pegasi b, measured by Mayor and Queloz, are illustrated in Figure 19.13. The two astronomers were awarded the Nobel Prize in Physics in 2019 for this discovery.

Following the discovery in the mid-1990s of 51 Peg b and a few other exoplanets by the radial velocity method, teams of astronomers both in the US and Europe, began organizing surveys of large numbers of stars, looking for the small, tell-tale Doppler shifts in their spectra that might indicate the presence of planets. Hot Jupiters orbiting close to their star were the easiest exoplanets to detect with the radial velocity technique. The Doppler shifts they produce were relatively large and easy to see. And their short periods allowed them to be tracked around their orbits many times in just a few weeks.

Another kind of ground-based survey focused on monitoring the brightness of tens of thousands of stars over many months, looking for small fluctuations that might indicate a planetary eclipse. When an exoplanet's orbit is inclined so that the plane of the orbit is parallel, or nearly so, to our line of sight, the planet will pass once per orbit between its host star and us, blocking out a small fraction of the star's light. The measured brightness of the star may dip by only a few percent at the middle of such an eclipse. So, the sequence of measurements, the photometry, requires good precision (repeatability)—a small fraction of one percent—if the eclipse is to be detected from the ground. The primary hurdle for ground-based searches is the reduced precision of brightness measurements due to the rapid fluctuations produced by the turbulence in the Earth's atmosphere (that's why stars appear to "twinkle"). That problem is eliminated if the observations are done in the vacuum of space.

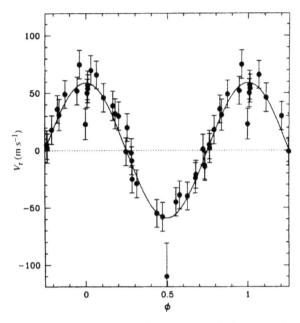

Figure 19.13. The small, periodic radial velocity variations induced in the star 51 Pegasi by its companion planet, 51 Pegasi b, measured by Mayor and Queloz at the Haute-Provence Observatory in France and announced in 1995. Note the scale of the vertical axis gives the measured radial velocity in meters per second. The horizontal scale is in fractions of the orbital period, 4.231 days. (Credit: Reprinted by permission from Springer Nature: Mayor & Queloz 1995.)

In 1999 David Charbonneau and his colleagues made the first ground-based observations of an exoplanet transiting across the disk of its parent star, a star similar to the Sun named HD 209458 (Charbonneau et al. 2000). Astronomers who had been monitoring the star for radial velocity variations had tipped them off about the possible presence of an exoplanet in this system.

An artist's concept of what HD 209458b might look like transiting across the disk of its parent star is shown in Figure 19.14. The opaque disk of the planet reduces the star's brightness as measured at the Earth by an amount that depends on the projected area of the planet's disk as a fraction of the projected area of the star's disk—perhaps a few percent. This transit is seen as a dip in the apparent brightness of the star that repeats on a regular basis—every few days for a planet orbiting close to its host star. The combination of the radial velocity variations of the star, the orbital period of the planet, the inclination of the orbit relative to our line of sight, the timing of various phases of the transit, and the intrinsic variations in brightness of the star itself across its disk (its limb darkening), allows astronomers to calculate the radius and mass of the exoplanet. Those two values determine the planet's average density, placing strong constraints on its internal structure.

The gaseous atmosphere of the planet transmits some of the starlight. Astronomers can compare the spectrum of the star observed during the transit to a baseline spectrum of the star taken outside of the transit. They can then measure

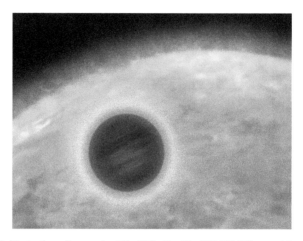

Figure 19.14. Artist's illustration of a gas-giant "hot" Jupiter like HD 209458b transiting across the face of its parent star. The opaque disk of the planet blocks a few percent of the star's light. The planet's atmosphere transmits starlight, some of which may be absorbed by gas atoms or molecules that are constituents of the atmosphere. (Credit: NASA, ESA, G. Bacon, STScI 2007-07.)

the wavelengths at which light from the star has been absorbed by the gases in the planet's atmosphere and how much light has been absorbed. These observations provide information about the chemical composition of the atmosphere, and about some of its physical properties such as whether it is cloudy or clear—pretty neat for an object you can't even see directly.

The *piece de resistance* for Charbonneau, his colleague Tim Brown, and their collaborators was their observation of transits of HD 209458b with Hubble (Figure 19.15; Brown et al. 2001; Charbonneau et al. 2002). They observed four transits of the planet, taking spectra with STIS every 80 s. The spectra spanned the wavelength interval 581.3–638.2 nm. There were two reasons to obtain spectra instead of direct brightness measurements. Since the spectra were spread out over many pixels in the STIS CCD detector, no single pixel would be overexposed. By adding up the measured signals in every pixel over the entire wavelength span, they could achieve a very large number of counts, and thus, a very high precision in the measured brightness—over a factor of 12 better precision than they had achieved in their prior ground-based observations. Second, by comparing the spectra taken within the transit and outside of it, they could determine if HD 209458b actually had an atmosphere and perhaps glean some information about its chemical composition.

From the transit observations, and prior observations of the radial velocity variations of HD 209458, Brown, Charbonneau, and their team derived a radius for the planet about 35% larger than that of Jupiter, while its mass was only 69% that of Jupiter. Its average density was remarkably low, only about 0.35 g cm^{-3}—about one third the density of water. Would this planet float?

Charbonneau and his colleagues used 417 of the STIS spectra, now fully spread out in wavelength, to look for absorption by sodium atoms in the atmosphere of HD 209458b (Figure 19.16). They wanted to search for sodium because its presence had

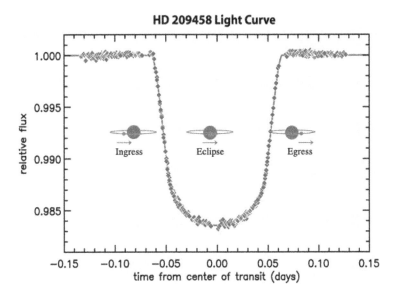

Figure 19.15. Observations of the brightness of HD 209458 obtained with the STIS instrument on Hubble. Each data point represents a STIS spectrum in which the data values in all the pixels have been summed together to compute a single brightness measurement. The central depth of the eclipse by the exoplanet HD 209458b is only 1.64%. The precision (repeatability) of the individual data points is one one-hundredth of one percent, thanks to the very stable image of the star obtained by Hubble above the atmosphere. (Credit: NASA, T. Brown, R. Gilliland, D. Charbonneau, STScI 2001-38.)

been predicted by several recent theoretical studies of hot gas giant planetary atmospheres. Indeed, during the transit the strong pair of spectral lines produced by sodium atoms at wavelengths of 589.0 and 589.6 nm did grow deeper, while the surrounding parts of the star's spectrum remained unchanged. After careful analysis of all the possible sources of error, the astronomers concluded that HD 209458b had an atmosphere (that discovery was an important first), and that sodium was present in that atmosphere. But there wasn't as much extra absorption by sodium in their observations as had been predicted, assuming the abundance of sodium in the planet's atmosphere was the same as in the Sun. The depth of the sodium lines should have increased by a factor of three more than was actually observed. So, this is a good example of how science works. The search for an answer to a question, leads to further questions. Are some of the sodium atoms tied up in molecules like sodium sulfide, sodium chloride, or sodium hydroxide? Are there high clouds in the planet's atmosphere that are partially obscuring the view? Perhaps HD 209458b was originally formed with less sodium (or perhaps lower metal abundances in general) than in the Sun. These are obvious topics to pursue in further research.

The major point made by the story of HD 209458b, is that Hubble and other space observatories like the infrared Spitzer Space Telescope and (the future) JWST are able to explore the atmospheres of planets orbiting other stars. This would have been unimaginable to us in the Hubble Science Working Group back in the early

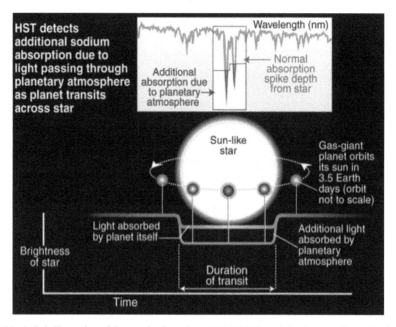

Figure 19.16. Artist's illustration of the transit of exoplanet HD 209458b and the changes observed in its spectrum due to absorption by sodium atoms in the planet's atmosphere. The small segment of the star's spectrum, covering 582–598 nm, shown in the inset at the top, is a plot of the actual observations made with STIS inside and outside of the transit. The two deepest absorption lines fall at the expected wavelengths of the strongest spectral lines produced by neutral sodium atoms, 589.0 and 589.6 nm. (Credit: NASA, D. Charbonneau, et al., STScI 2001-38.)

1980s when we were wondering if Hubble would ever be able to observe other worlds outside our solar system. Simply amazing!

The near-infrared wavelength range (1000–2500 nm) covered by Hubble's NICMOS and WFC3 instruments was another very promising spectral region in which to explore the atmospheres of exoplanets, looking for evidence about their chemical compositions and their physical properties. Important molecules like water, carbon monoxide, carbon dioxide, ammonia, and methane absorb light in broad wavelength swaths within that spectral range. Unlike the sharp, closely spaced lines produced by atomic sodium observed by Charbonneau and his team in HD 209458b, these near-IR molecular bands are broad. They do not need high spectral resolution to be detected. The low spectral resolution of the NICMOS and WFC3 IR grisms (gratings scored on the sides of prisms) is well suited to this kind of observation. Otherwise the observing technique is the same—observe the spectrum of a star during the transit of a planet across its face and compare that with the spectrum of the star acquired outside of the transit. In all cases data of very high precision must be acquired.

In 2007 Mark Swain from JPL and his collaborators used one of the NICMOS grisms in the wavelength range 1400–2500 nm to observe a planet transiting the star HD 189733 (Swain et al. 2008). This star is a bit cooler than the Sun. Their

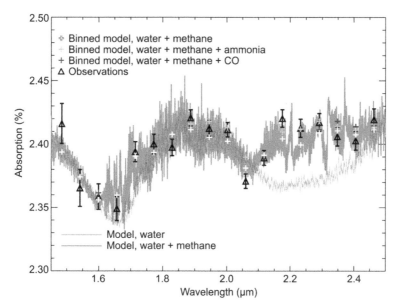

Figure 19.17. Near-infrared transmission spectrum of the atmosphere of the exoplanet HD 189733b, acquired with a grism on Hubble's NICMOS instrument while the planet was transiting its parent star. The observed low-resolution spectral data points are shown as triangles, with their estimated error range marked by vertical lines. The spectra predicted by theoretical models of the planet's atmosphere containing various constituents—water vapor, methane gas, ammonia, and carbon monoxide—are shown as colored plots and symbols. The symbols represent theoretical spectra binned in wavelength to the resolution of the observations. (Credit: Reprinted by permission from Springer Nature: Swain et al. 2008.)

observations are shown in Figure 19.17, together with spectra predicted by several theoretical models.

Both water vapor and methane gas are clearly present in the planet's atmosphere. Adding in ammonia or carbon monoxide improves the model fit to the observations slightly, but that's insufficient for a definitive identification of either of those molecules.

The big story here was the discovery of methane. It was the first identification of an organic molecule in the atmosphere of a planet outside the solar system. Obviously the "Holy Grail" of this kind of research would be the discovery of biomarker molecules that provide strong evidence for the existence of life in some form on an exoplanet—water, methane, oxygen, ozone, carbon dioxide, nitrous oxide, etc. Discovery of an abundance of such molecules all together in the atmosphere of an exoplanet where the temperatures were moderate and water could exist in liquid form would strongly suggest the presence of living organisms.

So far, most of the exoplanets that have been studied in this way have been hot gas giants with atmospheric temperatures in the neighborhood of 1000°C, give or take—an environment likely inhospitable to life. On the other hand, demonstrating that Hubble and other orbiting observatories provide the wherewithal to identify water, methane, carbon dioxide, etc on an extra-solar planet is an important step.

The studies described above took advantage of Hubble's ability to monitor a star known to possess an exoplanet moving in an approximately edge-on orbit as seen from Earth. As the planet transited, crossing in front of its host star, precise measurements with Hubble's instruments were used to acquire the spectra of starlight passing through its atmosphere. This was called the "transit method." The data acquired in this way were called "transmission spectra."

There was an alternative technique that also provided important information about exoplanet atmospheres. As the exoplanet moved around its orbit approaching the point where it was about to pass behind or to be eclipsed by its parent star, the dayside of the planet, the side facing the star, would briefly come into view (Figure 19.18). Starlight reflected off of the daylight face of the planet or light emitted by the atoms and molecules in the atmosphere on the dayside contributed to the spectra being taken with the Hubble instruments. When the planet passed behind the star a short time later, it was no longer contributing to the light Hubble measured. Only light from the star was being observed. If an observer subtracted the spectra taken after the eclipse of the planet from the spectra taken just before the eclipse, the remainder was the spectrum of the daylight hemisphere of the exoplanet. This was called the "eclipse method" and the spectra it provided were called, naturally enough, "dayside spectra."

Mark Swain and his colleagues obtained near-infrared dayside spectra with NICMOS of HD 209458b and HD 189733b, the two exoplanets discussed previously (Swain et al. 2009a, 2009b). In the spectrum of HD 209458b they identified water, methane, and carbon dioxide. In HD 189733b water, carbon monoxide, and carbon dioxide were needed to explain the observed near-infrared spectrum.

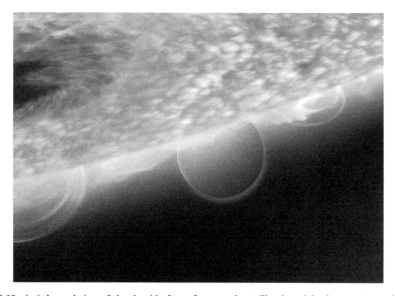

Figure 19.18. Artist's rendering of the dayside face of an exoplanet illuminated by its parent star just before it passes behind or is eclipsed by its parent star. (Credit: NASA, ESA, M. Swain, M. Kornmesser, STScI 2008-41.)

NICMOS was taken off line after Servicing Mission 4 in 2009 in large part because the IR Channel on WFC3 superseded it in its capabilities. In 2016 Hannah Wakeford from the STScI and her colleagues used the WFC3 IR grisms to observe three transits of the Sun-like star WASP-39 by an exoplanet that was about the mass of Saturn, orbiting the star every four days (Wakeford et al. 2018). It was a "hot Saturn." The resulting spectrum of the near-IR light from the star transmitted through the planet's atmosphere, combined with prior visible-wavelength spectral data from Hubble's STIS instrument and broad-band filter images obtained with the Infrared Array Camera (IRAC) on Spitzer, yielded the more or less complete spectrum from 400 to 5000 nm, shown in Figure 19.19.

A couple of features stand out in this graph. First, the strength of the absorption lines of sodium (589 nm) and potassium (766 and 770 nm) indicate that they must be formed deep in the atmosphere. Clouds did not obscure them. It was a clear day on WASP-39b. Second, the strong absorption by water vapor at 1400 nm, coupled with weaker water absorption features at 950 and 1200 nm, showed that water is very abundant in this atmosphere. There is about three times more water in the atmosphere of this hot Saturn than in our own Saturn. From this the astronomers inferred that WASP-39b must have formed much farther away from its parent star, out beyond the "snow line" where any water present would be frozen. Way out there in the protoplanetary disk, comets and planetesimals that carry abundant frozen water must have bombarded it. Later, as the WASP-39 system evolved, the gas giant

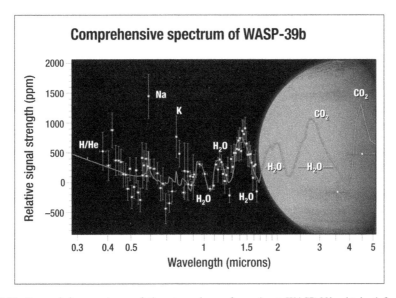

Figure 19.19. Transmission spectrum of the atmosphere of exoplanet WASP-39b obtained from transit observations with STIS and WFC3 on Hubble and with the IRAC instrument on Spitzer. The observations are plotted as yellow points. Vertical red lines represent the uncertainty in the individual data points. The wavelength span of each data point is shown with horizontal red lines. (Credit: NASA, ESA, H. Wakeford, G. Bacon, STScI 2018-09.)

planet must have migrated inward until it reached its current location, 20 times closer to its star than the Earth is from the Sun.

The final example I want to mention is that of the seven planets orbiting the small, cool red-dwarf star TRAPPIST-1 about 40 lt-yr from Earth. These planets range from Mars-size to Earth-size. All of them are closer to their parent star than Mercury is from the Sun. But their star is so small and so cool that the planets are not heated to high temperatures by its radiation. Three of them are in the star's habitable zone, the "Goldilocks zone"—not too hot and not too cold—where water can exist in liquid form.

Using the WFC3 near-infrared grisms, Julien de Wit and his team attempted to obtain transmission spectra of the atmospheres of four of these planets (de Wit et al. 2018). Their data revealed no prominent spectral signatures of the kind that had been found previously in the gas-giant "hot Jupiters or Saturns." The atmospheres were not the puffed up, hydrogen-dominated entities that had been previously observed with Hubble. These four planets may have atmospheres containing little or no hydrogen that, in fact, are too compact in vertical height for Hubble to detect. It is possible that they might be habitable. They will be prime targets for study with JWST.

As of 2019 August 30, 4109 extra-solar planets have been discovered.[1] Of these, 857 were detected with precision spectrographs at mountaintop observatories measuring the periodic variations of the Doppler shifts in their parent stars' spectra due to the planets' orbital motion.

The Kepler Space Telescope discovered the large majority of these exoplanets. Kepler monitored the brightness of hundreds of thousands of stars in a large fixed field of view on the sky for 9.6 yr from 2009 March to 2018 November. Its job was to detect the faint dimming of stars caused by transits of planets in orbit around them. The telescope observed 8054 transit-like events. After eliminating "false positives," caused for example by the detection of binary stars, star spots, etc, the Kepler team came up with 4034 planet candidates, of which 3789 were considered of high reliability. They confirmed a total of 2344 extra-solar planets.[2] It turns out that small planets, 1–3 times the size of Earth, are much more common than large planets. On average every Sun-like star has at least one planet. One in every five of these is an Earth-size planet in the habitable zone. So, in our galaxy there likely exist about four billion Earth-size planets under conditions that might support life.

Today, astronomers continue, in the words of the introduction to Star Trek, "to explore strange new worlds" with the Hubble Space Telescope and other modern observatories. This kind of research in the 21st century was unimaginable when Hubble was being built in the latter part of the 20th century—when experienced astronomers who were also "Trekies" dreamed that one day their new telescope might actually see a planet in orbit around a bright, nearby star.

---

[1] Extrasolar Planets Encyclopedia at Exoplanet.eu.
[2] Statistics taken from a 2019 public lecture by Natalie Batalha, University of California at Santa Cruz.

References

Brown, T., et al. 2001, ApJ, 552, 699
Charbonneau, D., et al. 2000, ApJ, 529, L45
Charbonneau, D., et al. 2002, ApJ, 568, 377
de Wit, J., et al. 2018, NatAs, 2, 214
Gáspár, A., & Rieke, G. 2020, PNAS, 117, 9712
Kalas, P., et al. 2008, Sci, 322, 1345
Marois, C. 2008, Sci, 322, 1348
Mayor, M., & Queloz, D. 1995, Natur, 378, 355
McCaughrean, M., & O'Dell, C. R. 1996, AJ, 111, 1977 and references therein
O'Dell, C. R., & Wen, Z. 1994, ApJ, 436, 194
Swain, M., et al. 2008, Natur, 452, 329
Swain, M., et al. 2009a, ApJ, 704, 1616
Swain, M., et al. 2009b, ApL, 690, L114
Wakeford, H. R., et al. 2018, AJ, 155, 1

Chapter 20

The Fathomless Universe: Dark Matter and Dark Energy

In 1933 a Swiss astronomer working at Caltech named Fritz Zwicky set out to measure the total mass of a rich, giant cluster of galaxies, the Coma Cluster, about 320 million light years from Earth. He compiled measurements made with a spectrograph of the radial velocities of the brightest galaxies in the cluster. He then subtracted these individual values from the average radial velocity of the galaxies. Some galaxies were moving faster than average, some were moving slower. This "dispersion" of velocities above and below average gave an indication of how fast the galaxies were moving randomly within the cluster. In fact they were, on average, moving very rapidly. All together the galaxies possessed a lot of kinetic energy. Zwicky then posed the question, how much mass must the cluster have to hold itself together gravitationally, given the high-velocity motions of its individual member galaxies? He estimated the masses of all the galaxies in the cluster by measuring their brightness or luminosity, assuming the brighter a galaxy was, the more stars it must contain and, therefore, the more massive it must be. What his calculations showed was astonishing. There was not nearly enough luminous mass among all the galaxies in the cluster to exert enough gravitational force to hold it together. The Coma Cluster should have flown apart eons ago. But here it was, a family of a thousand galaxies, serenely floating together in space. Zwicky postulated a way out of this conundrum. There must be a lot more mass present in the cluster than we can see, invisibly exerting enough gravity to keep the Coma Cluster from flying apart. He called the missing mass "dunkle Materie," German for "dark matter (Zwicky 1937)." In short order other astronomers obtained similar results for other clusters of galaxies.

In Zwicky's day and in the decades that followed other astronomers didn't know what to make of his discovery of a mass anomaly in a cluster of galaxies; it was just a cosmic curiosity. Fast forward to 1970. American astronomer Vera Rubin and her collaborator Kent Ford published their observations of the rotational velocity of M31, the grand spiral galaxy in Andromeda that is a close neighbor to our own Milky Way.

To make the observations Rubin and Ford used a highly sensitive spectrograph invented by Ford. It incorporated an electronic image intensifier to convert incoming photons of light into a stream of electrons that was then accelerated through a high voltage electric field and ultimately focused onto a phosphor screen. Where the electrons hit the phosphor it glowed, creating a bright image that could be photographed with a camera. This device anticipated some of the modern electronic image detectors of today, some of which are in instruments on board Hubble.

Without the image intensifier, the light from glowing nebulae in M31 would have been too faint to register on a bare photographic plate. With the intensifier, Rubin and Ford were able to measure the Doppler shifts, and thus the radial velocities, of 67 glowing clouds of ionized hydrogen gas scattered in and around the galaxy (Rubin & Ford 1970). From these measurements they were able to deduce the "rotation curve" for M31—a plot of rotational or orbital velocity as a function of distance from the center of the galaxy (Figure 20.1).

As had been the case with Zwicky and the Coma Cluster, what Rubin and Ford observed was astonishing. Within the central, luminous body of the spiral galaxy, the gas clouds traced circular orbits around the galaxy's center with velocities that made sense in terms of Newtonian gravity and Kepler's laws of motion. Their velocities were dictated mainly by the total mass of stars, dust, and gas enclosed within their orbits, and by their distance from the center of the galaxy. This is similar

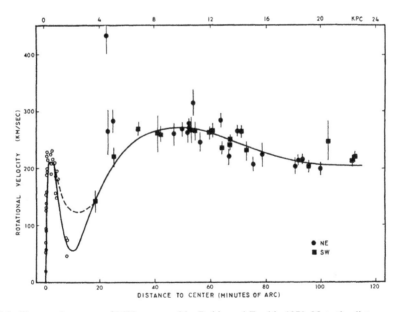

Figure 20.1. The rotation curve of M31 measured by Rubin and Ford in 1970. Note the distance scale at the top, extending out to 24,000 parsecs, or about 78,000 lt-yr, from the center of the galaxy. The solid and dashed curves are mathematical fits to the data. Rather than dropping to lower velocities as distance from the center of the galaxy increased as expected, the curve flattened out at larger radii (toward the right in the graph). It takes the gravity of a large dark matter halo to maintain this rotation and to keep the galaxy from flying apart. (Credit: Reproduced from Rubin & Ford 1970. © 1970. The American Astronomical Society. All rights reserved.)

to our solar system in which the mass of the Sun and the distance of a planet from the Sun dictate the planet's orbital velocity, following Kepler's second law of motion. But what was surprising was that the velocities Rubin and Ford measured far out from the galaxy's center did not drop off as the distance increased. Unlike the planets in our solar system, the rotational velocity curve of M31 remained more or less flat, the farther out from the main body of the galaxy their measurements were made. Stars, gas, and dust at the outer periphery of the galaxy were moving just as fast in their orbits as the material closer to its center. The mass of the central body of M31 was not sufficient to maintain such large rotation speeds. The galaxy should fly apart. Its continued existence implied that there was a large amount of unseen mass whose gravity was holding M31 together and producing this flat rotation curve.

By 1980 Rubin and Ford had measured rotation curves for 21 spiral galaxies. In all cases the rotation curves at distances far out from the center of the galaxy remained flat or rose with increasing distance. A massive dark matter halo had to be present in all of them (Rubin et al. 1980).

The high-resolution cameras on the Hubble Space Telescope enable an entirely different method for measuring the masses of galaxy clusters and revealing the gravitational influence of dark matter. But it requires some help from Albert Einstein—the bending of light across curved spacetime and the phenomenon of gravitational lensing.

General relativity posits that a mass bends or curves the spacetime around it. An object, or even of a photon of light, moves in a gravitational field along a trajectory defined by the curvature of spacetime in the vicinity of that mass. In Einstein's theory there is no distinction between gravity and the curvature of spacetime. They are the same thing.

Figure 20.2 is an attempt to depict this effect, with the red grid representing spacetime. Light coming from a distant source at the top of the grid follows a curved trajectory as it travels through spacetime in the vicinity of a massive object—the

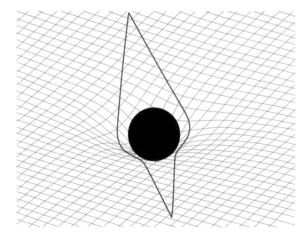

Figure 20.2. An artist's depiction of spacetime being curved near a massive object. Multiple light rays from an object beyond the mass will follow different paths and be deflected at different angles. In this case an observer at the bottom of the grid will not see the light coming from its true point of origin at the top, but rather from two different directions across the grid. This is the basis of gravitational lensing.

sphere. The light is deflected in a different direction from its original path. An observer standing at the bottom will see the light appearing to come from a different place on the grid than its true point of origin. Multiple rays of light (two in this case), following different paths in spacetime, are deflected in different directions, creating multiple images of the same distant source as seen by the observer in the lower corner.

The apparent position of a star in the sky will be altered slightly if its light happens to pass close to a massive body like the Sun on its way to Earth. This small shift can readily be measured during a total solar eclipse. Einstein had predicted the magnitude of the shift to be about 1.75 s of arc. To test Einstein's prediction, based on the general theory of relativity, the British astrophysicist Sir Arthur Eddington led an expedition to the island of Principe off the coast of western Africa to observe and photograph the total eclipse of 1919 May 29. Eddington took photographs of a field of stars close to the Sun during totality. For about 7 min the Moon's disk completely obscured the Sun, causing the sky to turn dark and the stars to be visible. He measured the stars' positions on his photographic plates and compared them to the positions of the same stars seen in the night sky at another time of the year. Einstein's predicted shift was confirmed. The mass of the Sun curved spacetime in its vicinity, and the photons of starlight were deflected from a straight line as they passed nearby.

Now, consider the scenario where the mass exerting the gravitational force, curving spacetime, is not a single star but rather a massive cluster of galaxies spread out over a large volume. The cluster's mass is made up not only by its luminous galaxies, but also by a large amount of dark matter such as Zwicky discovered in the Coma Cluster. In fact most if its mass is dark matter. In front of the cluster, at some distance, is the Hubble Space Telescope orbiting the Earth, pointed toward the cluster. On the other side of the cluster, are other sources of light—say a random assortment of galaxies distributed in space beyond the cluster in the direction opposite the Earth. What would the images acquired by the cameras on board Hubble look like?

Figure 20.3 is an image taken with the advanced camera for surveys' wide field channel of the massive cluster of galaxies, Abell 1689. The cluster is 2.2 billion light years from Earth. There are many arcs of light in this view, which are the focused but distorted images of numerous background galaxies along the line of sight, some of them at great distances far beyond Abell 1689. The distributed mass of the dark matter, plus a small contribution of mass from the luminous galaxies in this cluster, curves spacetime in a complex pattern creating a not very smooth or well-figured gravitational lens—the strongest such lens known. Figure 20.4 provides magnified views of two regions of the cluster, showing details of the lensed arcs. (The images produced by a gravitational lens are sometimes compared to what one sees when looking at a bright light through the clear bottom of an empty soda bottle.)

Figure 20.5 shows an example of how the mass of a cluster of galaxies is estimated using gravitational lensing. This method is entirely independent of other methods, such as those discussed previously having to do with galaxy rotation or motions within a cluster of galaxies. In the figure are two inset boxes that contain zoom-in

Figure 20.3. The massive galaxy cluster, Abell 1689, some 2.2 billion light years from Earth, as observed with Hubble's advanced camera for surveys. Note the large number of arc-shaped images. Those are the distorted images of distant galaxies far beyond this cluster as seen through the gravitational lens created by the mostly dark matter within the cluster. (Credit: NASA, ESA, N. Benitez, et al., the ACS Team, STScI 2003-01.)

Figure 20.4. Details of the gravitationally lensed images in Abell 1689. Each arc is the distorted image of one or more distant galaxies far beyond the cluster, created by the complex gravitational field of the mostly dark matter in the cluster. (Credit: NASA, ESA, N. Benitez, et al., the ACS Team, STScI 2003-01.)

views of two images of the same very distant background galaxy seen at different positions in the cluster. How this kind of geometry comes about in the presence of curved spacetime is explained in Figure 20.2 and the related discussion.

We know the two images are of the same distant galaxy for several reasons. The relative brightness of the two images is consistent. They have the same colors as seen through the four filters used to take this composite image. And, importantly, the redshift of each, measured with a spectrograph, is the same—$z = 4.86$. They are at the same distance, calculated with Hubble's law and with the current consensus cosmological model. Or, rather, they were at that distance when they emitted the

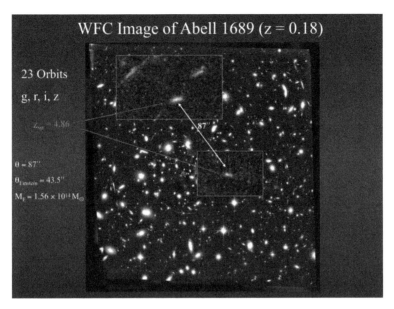

Figure 20.5. Estimating the mass of a cluster of galaxies using gravitational lensing. This is a portion of the same image shown in Figure 20.3, but the light from the brightest galaxies has been suppressed, allowing faint details in the image to be seen. The two inset boxes contain magnified images of the same galaxy situated far beyond Abell 1689. The curvature of spacetime has deflected light from the distant galaxy in two different directions, creating two separate images, brought into focus by the gravitational lens created by the mass of Abell 1689, as illustrated in Figure 20.2. (Credit: N. Benitez, K. Zekser, T. Broadhurst, D. Coe, and the ACS Team.)

light we now see—about 12.3 billion light years away. They are images of one and the same object.

The critical quantity in this calculation is the angular separation between the two images—87 s of arc. Half this amount, 43.5 s of arc, is called the Einstein angle. From it astronomers derive a value for the mass enclosed within an imaginary ring that passes through the two images. The radius of the ring is the length of the line subtended by the Einstein angle at the distance of the cluster. The derived mass is called the Einstein mass and, in this estimate, $M_E = 1.56 \times 10^{14}$ times the mass of the Sun.

Armed with new ACS observations of several rich clusters of galaxies and sophisticated computer codes that analyzed the gravitationally lensed images of numerous background galaxies, astronomers on the ACS Science Team began to construct fairly detailed maps of the distribution of dark and luminous matter within the clusters (Broadhurst et al. 2005; Coe et al. 2010; Zekser et al. 2006). This was made possible by the remarkably high resolution, wide field of view, and sensitivity to faint sources of light of the newly installed camera. The larger the number of lensed image pairs of the kind illustrated in Figure 20.5, the more detailed and accurate were the maps created by the computer algorithms. In the end they found that in Abell 1689 the mass of dark matter exceeded that of ordinary baryonic matter—stars, gas, and dust made from protons and neutrons—by a factor of about

Figure 20.6. Computer generated model of the distribution of dark matter (shown in gray) in the galaxy cluster Abell 1689 superposed on the ACS image shown in Figure 20.3. The model exactly reproduces the positions of 135 lensed images of 42 distant galaxies beyond the cluster. (Credit: NASA, ESA, D. Coe, et al., STScI 2010-37.)

6 to 1. For every gram of ordinary baryonic matter there were 6 g of dark matter, gravitationally felt but not seen, within the cluster.

Dan Coe from JPL and his collaborators used a computer code named "LensPerfect" to produce a map of the distribution of dark matter in Abell 1689 based on the ACS images coupled with ground-based observations (Figure 20.6). Their model exactly reproduced the positions of 135 gravitationally lensed images of 42 background galaxies at various distances beyond the cluster. The dark matter was not uniformly distributed, but it was smoother than the clumpy distribution of the individual galaxies in the cluster. It came as something of a surprise to the astronomers that the dark matter was so strongly concentrated toward the center of the cluster.

The curvature of spacetime induced by the dark and luminous matter spatially distributed within a rich cluster of galaxies creates a natural telephoto lens that can be used to augment the Hubble telescope's optics. Looking through such a telephoto lens, Hubble has detected objects far across the universe, whose light was emitted shortly after the Big Bang. The images were magnified and amplified in apparent brightness by the natural lens. The telescope would not be able to see them if the gravitational lens were not there.

In 2013 the Director of the STScI, Matt Mountain, conceived a program that would be a natural extension of the long series of Hubble Deep Fields, going back to 1995 (Chapter 11). The idea was to use Hubble's superb cameras, the ACS Wide Field Channel and the WFC3 IR Channel, to extend the telescope's imaging of the deep universe as far as nature and physics would allow it to go. This was the Hubble Frontier Fields (HFF) Initiative. Very deep exposures of six rich clusters of galaxies, each possessing strong gravitational lenses, would be taken through a selection of color filters. Additionally, simultaneously with the images of each cluster, a parallel field of view not associated with the cluster would be observed with the alternate camera. When WFC3 was observing a cluster, ACS would be observing a parallel field, and vice versa. There would be twelve deep fields in all.

As was the case for the earlier deep fields, the initiative would be a community-based project, overseen by an advisory board of recognized experts. The entire community would have immediate access to the calibrated and archived data. Matt Mountain provided 840 orbits of Hubble Director's Discretionary observing time spread over three years. The two other orbiting NASA Great Observatories, Chandra and Spitzer, would join in on the fun with deep observations of the Frontier Fields in X-rays and farther into the infrared, respectively. Interpretation of the gravitationally lensed images was strongly dependent on the availability of sophisticated theoretical models of the distribution of dark matter within each of the six clusters. Grant funding was allocated for the development of these computer-based models, based on competitive proposals submitted by experts in the community.

The HFF Initiative resulted in the discovery of thousands of previously unseen galaxies—the faintest and youngest ever detected. Figure 20.7 shows one excellent example. This is the HFF image of a rich cluster of galaxies with the awkward name MACS J0416-2403. MACS stands for "Massive Cluster Survey," a program carried out earlier with the Chandra X-ray Telescope. This cluster is called M0416 for short. The inset contains the image of a small, low luminosity galaxy likely still in the process of formation.

Its redshift was measured using the photometric technique described in Chapter 11. Long exposures were taken through a specific sequence of color filters with ACS and WFC3. Each filter was transparent to light in a limited band of wavelengths, ranging from blue all the way into the near infrared. The span of wavelengths covered was extended further into the infrared with exposures made through a single filter at the European Southern Observatory in Chile, and with two filters in a camera on board the Spitzer Space Telescope. In all, the filters covered wavelengths from 435 to 4500 nm.

Absorption of ultraviolet light by hydrogen atoms, the most abundant element in the universe, causes every galaxy to "go dark" below about 120 nm at what is called the Lyman break. Because of the expansion of the universe, the light from distant galaxies gets stretched toward redder wavelengths. In the case of the red object seen in the inset of Figure 20.7, its Lyman break got redshifted all the way to about 1333 nm. It could not be seen at all in any of the shorter wavelength filters. It appeared faintly in the WFC3 IR Channel filter centered on 1400 nm. It was clearly present in the WFC3 1600 nm filter. It was also seen in the two Spitzer infrared wavelength bands.

With the helping hand from Spitzer, the astronomers were certain that this object was real and was at a redshift of about 10.1. Plugging this redshift value into the

Figure 20.7. Hubble Frontier Field image of the rich cluster of galaxies, MACS J0416.1-2403. The inset is a zoom-in on a small, intrinsically faint galaxy still in the process of formation. At a redshift $z \sim 10$, it is perhaps the most distant and youngest galaxy every observed, having emitted the light we see 300–400 million years after the Big Bang. (Credit: NASA, ESA, L. Infante, et al., STScI 2015-45.)

cosmological model now widely accepted by astronomers tells us that this tiny toddler of a galaxy emitted the light we now see when the universe was only about 330 million years old, a bit over two percent of its present age. It's about the size of the Large Magellanic Cloud (LMC), a small satellite galaxy orbiting the Milky Way. However, it is producing new stars ten times faster than the LMC. Gravitational lensing by the dark matter in M0416 has magnified the image of this very distant object by a factor of 20, allowing Hubble and Spitzer to detect it.

The team of astronomers that discovered this object, led by Leopoldo Infante from the Pontifical Catholic University of Chile, nicknamed it Tayna. That is a word from the Aymara language spoken by indigenous people in the Andes and Altiplano (high plateau) regions of South America. Translated into English it means, "first born." Tayna was the most distant of the 22 young galaxies discovered by Infante's team in M0416 (Infante et al. 2015).

The evidence for a dark matter component of the universe goes back many decades and has taken on several independent forms—the rotational velocities of spiral galaxies, the random velocities of galaxies in clusters, and the strong gravitational lensing seen in clusters of galaxies. But what is dark matter anyway? And since we can't see it directly, does it really exist? Could there be an alternative version of the law of gravity that could reproduce the observed phenomena? Astronomers and theoretical physicists have pursued other possibilities for a long time. None of them

Figure 20.8. Two examples of colliding clusters of galaxies. In the left frame is the Bullet Cluster (IE 0657-558). In the right frame is the cluster MACS J0025.4-1222. In both cases the collision has segregated the dark matter and galaxies of the original clusters (blue areas) from their intergalactic gas. The latter was stripped out and heated (emitting X-rays) during the collision (pink areas). The dark matter and galaxies passed through seemingly unaffected by the collision. Most of the mass in each system is concentrated in the two blue areas and is dominated by the dark matter. (Credit: NASA, ESA, D. Clowe, A. Gonzalez, M. Markevitch, M. Bradac, S. Allen, Chandra X-Ray Center, STScI 2006-39 and 2008-32.)

have panned out. For example, could there be large populations of dense, low luminosity objects like brown dwarfs (not quite massive enough to be a star), black dwarfs (white dwarfs that have cooled down and become very dim), undetected black holes, etc. These are objects made from ordinary baryonic matter, from neutrons and protons. The answer is that there is no plausible way that such objects could exist in sufficient numbers to produce the enormous total mass needed to hold galaxies and galaxy clusters together or to produce the remarkable gravitational lensing of background galaxies observed in rich galaxy clusters. Alternative theories of gravity also don't hold water; they've never been able to explain simultaneously the full panoply of observations. The leading hypothesis is that dark matter is made of what are humorously called WIMPs—non-baryonic weakly interacting massive particles—left over from the Big Bang. Their name is simply a description of the basic properties they must possess. Major efforts continue at the Large Hadron Collider in Europe and at other particle accelerators to detect them—so far with no joy.

The most compelling evidence to date that dark matter is real and has certain properties—it doesn't emit or absorb electromagnetic radiation (hence the designation "dark"); it interacts with itself only very weakly, if at all; and it exerts gravitational force on ordinary matter—can be found in the two images in Figure 20.8. These show two separate examples where two clusters of galaxies have collided. The pink areas show observations taken with the Chandra X-ray Telescope. Here the intergalactic medium in one cluster—mostly hydrogen gas that fills the space between the galaxies—has rammed into the intergalactic medium of the other. The large bodies of colliding gas have compressed, and heated up to very high temperatures with a resulting emission of X-rays. The hydrogen has had its electrons ripped away from its protons, creating hot plasma. The two clusters had been approaching each other from opposite directions, each with their own particular velocity and momentum. Ram pressure built up as the two bodies of gas collided, and their original velocity and momentum was greatly reduced. In the Bullet Cluster (the left frame in the figure), the plasma from one cluster actually blew through the plasma from the other, forming a bullet-shaped shock wave seen to the right in the image. In the example in the right-hand frame, the two bodies of hot plasma simply merged, with no shock wave having formed (Bradac et al. 2008; Clowe et al. 2004).

Meanwhile, the dominant mass from within each cluster—the dark matter accompanied by the cluster galaxies—continued to move unimpeded, as though a collision had never taken place. As a result each cluster moved beyond the cloud of hot plasma; each cluster was stripped of what had originally been its intergalactic medium. The blue areas to the left and right in both frames show the location of the dark matter and the cluster galaxies after the collision. During the collision, there apparently was no interaction between the plasma and the dark matter. Moreover, the two bodies of dark matter originally associated with the two galaxy clusters passed through each other with no evidence that they interacted with or impeded each other's motion.

How do astronomers know that there is an abundance of dark matter within the two blue areas in each image? They observe some evidence of "strong lensing"—arcs and such. They combine that with measurements of the small distortions in the shapes of distant galaxies seen beyond each blue area. The latter is called "weak lensing." When a heavy concentration of dark matter distorts spacetime within a rich galaxy cluster, it strongly lenses the light from background galaxies, resulting in arc-shaped and other bizarre-looking images. Weak lensing results when the mass of dark matter is not so concentrated, is more spread out, for example at distances far from the center of a cluster. The distortions in the apparent shapes of distant galaxies caused by weak lensing are small, but statistically significant.

In these two examples, what were originally two separate entities, two clusters of galaxies, have been segregated into three parts during a collision. There is a region containing hot, X-ray emitting plasma made of baryonic matter, but containing few (or no) galaxies. The body of hot plasma separates two regions composed of non-baryonic dark matter, and ordinary galaxies, containing little (or no) left-over intergalactic gas. The overall center of mass of the system is located between the two blue regions in each case, but is not centered on the pink areas. Maps of the distribution of mass show that in both cases, the dominant masses are centered on the blue regions. In the analysis of MACS J0025.4-1222 astronomers concluded that the galaxies contribute about 1% of the total mass and the hot plasma contributes about 9%. The remaining 90% of the mass is in the form of dark matter. Those numbers are consistent with the content of normal clusters of galaxies that have not been subjected to a cosmic train wreck.

While we do not know what dark matter is, the consensus among the experts is that it constitutes about 85% of the total mass in the universe. Ordinary baryonic matter, that make up the visible universe—the Sun, the stars, the galaxies, you and me—contributes the remaining 15%. The dark matter is called "cold" because if it were warm or hot, it would be less likely to occur in dense concentrations of the kind we see in galaxy clusters. The jargon of cosmology commonly includes references to "CDM." That's shorthand for "Cold Dark Matter."

But that is not the whole story. It gets stranger.

In an interview for this book, astrophysicist and Nobel laureate Adam Riess, made the following statement:

> *"At the end of the day, I believe that the arc of science is long, but it bends toward the truth."*

That quote is adapted from Dr Martin Luther King, Jr., who once said, *"The arc of the moral universe is long, but it bends toward justice."* Whether pursuing justice or scientific truth, one's path may not be straight and the search may not be swift. But in the end, hopefully we will get to the right answer. I think that captures the essence of the final science story I'm going to tell—the puzzle of the Hubble constant and the accelerating universe.

In Chapter 10, I discussed the pursuit by Wendy Freedman and her Key Project Team of the current rate of expansion and the current age of the universe. That was one of the primary objectives for the Hubble mission from its inception. But when the observations were in and the analysis of the data had been completed, what we had was a puzzle. The universe was currently expanding more rapidly than many had expected. The Key Project Team measured the expansion rate, the Hubble constant, H_0, as about 72 km s^{-1} Mpc^{-1}. For every million parsecs (3.26 million light years) farther out in space one looked, the velocity of expansion increased by 72 km s^{-1}. At that rate it took a relatively short time for the universe to expand to its present size. The universe must be young—somewhere in the range 9–12 billion years.

But that presented a conundrum. The ages of the oldest stars in our own galaxy had been calculated, using sophisticated theoretical models of stellar evolution, to be around 15 billion years or older. How could the universe be younger than the oldest stars within it? Some new physics was needed to solve the problem.

Fortunately, the required new physics was at hand. It was discovered first in ground-based observations, which were later greatly improved in scope and accuracy with data from Hubble. The expansion of the universe was speeding up, not slowing down. Taking that acceleration into account increased the expansion age of the universe derived by the Key Project Team to 13 billion years with an uncertainty of about 10%. Refinements in the details of calculating the ages of the oldest stars reduced those to about 12 billion years. So, the story now seemed consistent. But the long arc of science hadn't yet reached its final truth. There was much more to be learned, and Hubble was the vehicle that was carrying us along on this amazing voyage of discovery. Why was the expansion of the universe accelerating? Had it always been accelerating? Based on what we now know, what will be its ultimate fate? Is there any more new physics involved in this tale that we haven't yet learned?

In conversation, Adam Riess seems like just a regular guy from New Jersey—that is until you start discussing his work on how the universe works. Then you begin to see the brilliance, enthusiasm, and fixation on getting the details right that drive his research. He refers to Hubble as a "high precision instrument," and uses it almost exclusively in probing the properties of spacetime. "It's perfect for what I do," he says.

Adam got interested in physics late in high school when he participated in the New Jersey Governor's School of the Sciences in 1987. He went on to MIT as an undergraduate, majoring in physics. In graduate school at Harvard, he was unsure in the beginning of where his main interests lay, but was attracted to astrophysics. He read Frank Shu's famous text, "The Physical Universe: An Introduction to Astronomy,"[1] and that got him excited about cosmology. In particular the idea of the expansion of the universe intrigued him. At the time people were arguing about the Hubble constant at a level of a factor of two—was it 50 or 100 km s^{-1} Mpc^{-1}? Was the universe very old, or very young? "How can that be?" he asked himself. "This is amazing. I'd like to contribute to this in some way."

So, Adam started working with his thesis advisor at Harvard, supernova expert Bob Kirshner, on a new tool for accurately measuring distances far out across the universe—the Type Ia ("one-a") supernovae. In its simplest terms a supernova is a star that undergoes a catastrophic explosion, ejecting most of its mass into space, and emitting a prodigious amount of light.

The precursor to a Type Ia supernova is believed to be a white dwarf, the highly dense remnant core of a low-to-medium mass star that has exhausted all its nuclear fuel and has shed its outer envelope as a planetary nebula. The white dwarf emits light only by virtue of its remaining thermal energy. It glows because it is still hot, though in the process of cooling down. Most white dwarfs are made of carbon and oxygen, left over as the products of nuclear reactions that powered their parent stars. Most white dwarfs have masses in the range 0.5–0.7 times the mass of our Sun. However, there is a theoretical limit—the Chandrasekhar limit of 1.4 solar masses—which a carbon–oxygen white dwarf cannot exceed without becoming unstable and blowing itself apart. A Type Ia supernova results when a white dwarf is in a binary pair and the companion star is dumping material onto it. If enough material accretes onto the white dwarf to push its mass over the Chandrasekhar limit, it will explode spectacularly.

A Type Ia supernova has a characteristic light curve. Its brightness varies with time in a more or less consistent way, rising to a maximum in about 20 days, falling rapidly over a period of 3–4 weeks, and then dimming more gradually for weeks thereafter. The energy producing the light comes from the radioactive decay of unstable heavy isotopes, primarily Ni^{56} and Co^{56}, produced during the thermonuclear explosion. The classical light curves from a Type Ia, SN 1994ae, are shown in Figure 20.9. In this illustration, the light curve in blue (B), visible (V, green–yellow), red (R), and deep red (I) are shown. The curves vary in shape depending on the wavelength band in which they were observed because in different colors we are

---

[1] University Science Books, Sausalito, CA, 1982.

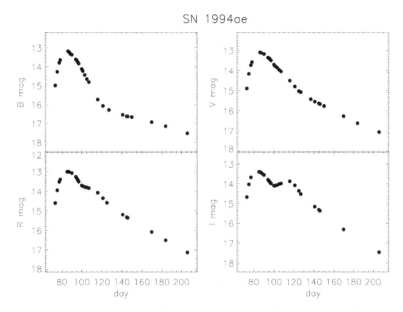

Figure 20.9. Measured light curves in four different wavelength bands for a well-observed Type Ia supernova. Usually only the blue band, upper left, is used for galaxy distance determinations. The unit "mag" or magnitude is a logarithmic scale of brightness, where smaller numbers denote brighter light intensity. (Credit: A. Riess.)

observing different regions in the ejected mass from the supernova. Usually only the B-band light curve is used for measurements of the distances to far-away galaxies.

Not all Type Ia B-band light curves are identical in shape and peak brightness. Those that reach a brighter maximum fade more gradually, than those with a less luminous peak. But astronomers have found simple ways to relate peak brightness to the width of the light curve, or to its overall shape, or to the rate of dimming over the first 15 days following the maximum. Those relationships allow all the Type Ia supernovae to work as accurate standard candles in measuring the distances to galaxies far across the cosmos. And measuring distances in this way, as accurately as possible, became the focus of Adam Riess' professional life.

As Adam was finishing up his PhD thesis at Harvard, the High-Z Supernova Team was getting started. It consisted of a combination of Bob Kirshner's group at Harvard, a group at the large mountaintop observatories in Chile, and a few other astronomers. Meanwhile, across the continent at Lawrence Berkeley National Laboratories in California, a second, completely independent group was setting out on a similar quest, to probe the past history and properties of the universe using Type Ia supernovae as standard distance indicators. These were two different cultures. The Harvard team was made up of observational astronomers who knew all about supernovae and how to use them to measure distances accurately. The team at Lawrence Berkeley was a particle physics group, ultimately led by Saul Perlmutter, who wanted to measure supernovae to understand the basic physics of the universe. Adam was one of the early, young members of the High-Z team who

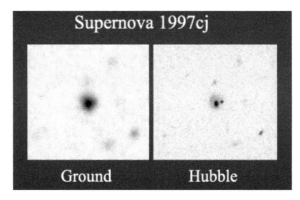

Figure 20.10. Detection of Type 1a supernova 1997cj as seen from a ground-based observatory and from Hubble. In the ground-based observation, the supernova image is blended with that of the host galaxy. Its brightness can only be measured by subtracting from this image an image of the host galaxy taken before or long after the supernova exploded. The noise in such measurements is high. In the Hubble image, the supernova is clearly resolved and separated from the image of the host galaxy. The light curve can readily be measured without significant contamination by light from the galaxy. (Credit: A. Riess.)

was going to help with actually finding supernovae and measuring them. Heading the team was Brian Schmidt, who had overlapped with Adam as a grad-student at Harvard.

It was unusual, as a matter of policy, for the Director of the STScI—Bob Williams in this case—to accept two separate proposals from two independent teams to perform the same kind of observations with Hubble with the same scientific objectives. Observing time on Hubble was very valuable, and normally such duplication would be considered wasteful. However, as had been the case with the Comet Shoemaker–Levy 9 campaign in 1994 and the Hubble Deep Field observations in 1995, Bob Williams proved that he had great insight and the courage to manage Hubble's science program "outside the box." He recognized that the expansion history of the universe was an especially important scientific problem. It was important to support two independent groups, so that the results could be crosschecked.

The High-Z team started looking for supernovae in 1995. By 1996 they were already using Hubble to measure the light curves of the supernovae. The advantages of using Hubble over trying to do the measurements from the ground were obvious and compelling (Figure 20.10).

Hubble observing time was always fully subscribed weeks in advance. But no one could predict weeks in advance when a supernova would go off. So both Brian Schmidt's High-Z Supernova Search Team and Saul Perlmutter's Supernova Cosmology Project Team used a novel approach, in cooperation with the STScI, in scheduling their observations. Each team would nail down a specific time a week or so in advance when their observations would commence. In the observing plan for that week, Hubble would be commanded to point to some placeholder position somewhere within a one square degree patch of the sky at that specific time. Meanwhile, the teams would use ground-based telescopes to search for one or more supernovae within their chosen one square degree. There would be so many galaxies in a field that large that the probability of finding at least one supernova during a few days of observing was very high.

Having identified the potential targets, the observers would send their celestial coordinates on to team members at other observatories in Chile, Hawaii, Arizona or elsewhere. Those astronomers would use spectrographs on various large telescopes to record the spectra of the candidate supernovae, both to verify that they had the spectral signature of a Type Ia and also to measure their redshifts (velocities of recession). Having verified that the targets were of the correct type, the teams would send the pertinent celestial coordinates, within the pre-arranged one square degree of sky, and exposure time information on to the STScI. The observing schedule would then be updated with these specific details and the Hubble cameras would acquire the data at the appointed time. The only rub was the weather at the ground-based observatories. Bad weather at any one of them could interrupt this flow and delay acquiring the needed information.

Through 1996 and 1997 the High-Z team obtained light curve for 16 high-redshift Type Ia supernovae, which they combined with prior observations of 34 nearby, low-redshift ones. Adam Riess led the analysis of this first large sample and was the lead author on the seminal paper from the High-Z team (Riess et al. 1998). The result was shocking to Adam and to the rest of the team. All of the high-z supernovae were fainter than expected, about 21% fainter on average.

Common wisdom held that the universe should be decelerating. Its expansion should be slowing down under the mutual tug of the gravity exerted by all the matter within it, dominated by the cold dark matter. The expansion rate long ago would have been faster than the current rate in the nearby universe. So, the universe would not have grown as large as if the higher expansion rate from long ago had continued unabated to the present time. The high-z supernovae would be closer to us and would appear brighter than they would have if the expansion rate had retained the same constant value over all that time. But the actual observations showed that the supernovae were dimmer, and hence farther away from us, than would have been the case with a constant rate of expansion. The universe had grown a bit larger than expected. The expansion of the universe must have speeded up, not slowed down—acceleration rather than the expected deceleration.

Both the High-Z team and the Supernova Cosmology team reached the same conclusion, entirely independently of one another (Perlmutter et al. 1999; Perlmutter 2003). Both teams determined that the expansion of the universe is accelerating at the present time.

Figure 20.11 is a Hubble diagram (velocity versus distance) that includes data from both studies. The data points in the box at lower left are for nearby, low-z galaxies and the straight line that fits those data tells us the current expansion rate. The slope of that line is the current value of the Hubble Constant, about 73 km s^{-1} Mpc^{-1}. In the figure that straight line is extended to higher redshifts. If the expansion rate were constant, if it had been the same, say, 5–6 billion years ago as it is today, all of the observations would have fallen on, or scattered around, that extended straight line. However, the high-z observations show a strong tendency to fall below the line.

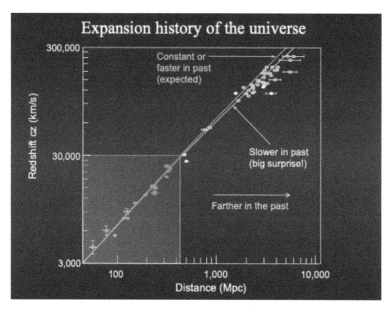

Figure 20.11. The measured expansion history of the universe over the past six billion years or so. Observational data from two separate teams working independently of each other are combined in this plot. Each symbol is the average of the measurements of multiple supernovae. (Credit: A. Riess.)

The velocities of recession of those supernovae that exploded many billions of years ago were slower than the astronomers had expected. Since then the expansion of the universe has speeded up.

What repulsive force could cause the expansion of the universe to accelerate? The answer is, no one knows. It's one of the great mysteries of modern physics. But astrophysicists have given it a name, "Dark Energy." The leading candidate is Einstein's cosmological constant. That was a kind of repulsive gravity Einstein introduced into general relativity when he thought the universe might collapse in on itself. He felt he needed to do something to keep the universe static. Of course, after he learned from Edwin Hubble that the universe was in fact expanding, Einstein removed the cosmological constant from his equations, calling it his "biggest blunder." In modern physics the cosmological constant is usually equated with a quantum energy density of the vacuum. That raises a problem, however. The calculated value of the energy density of the vacuum exceeds what is necessary to cause the observed acceleration of the universe by a factor of 10^{120} (1 followed by 120 zeros). In another formulation the excess is only a factor of 10^{55}. The enormous discrepancy is commonly called the "cosmological constant problem." Indeed!

Based on what is known so far, Dark Energy must make up about 73% of the total mass-energy budget of the universe. Cold Dark Matter constitutes about 23%. Ordinary baryonic matter, the atoms that stars and planets and people are made of, fills in the remaining 4%. I've heard some people say that just shows how insignificant we are. We're only 4% of the cosmos. I personally believe we are the most significant 4%. We are sentient creatures swimming in an ocean of dark energy and dark matter.

In the late 1990s and early 2000s creative people in the community began to postulate alternative explanations for the observations of the two supernova teams. Perhaps there was some kind of fog or dust of a mysterious kind in the distant universe that caused the high-z supernovae to look fainter than they actually were. It would have to be a gray fog (or dust), by which is meant material that scatters or absorbs all wavelengths of light equally. It could not be like ordinary interstellar dust that selectively scatters away blue light, leaving the light from a star looking redder. Or, perhaps the Type Ia supernovae were intrinsically fainter in the early universe than they are now. Perhaps they had evolved in their properties over time.

The best way to address the skeptics would be to acquire a body of observational data about very high redshift supernovae. The original group of Type Ia supernovae analyzed in 1998 by Adam Riess and his team only extended to a maximum redshift $z = 0.62$, corresponding to a look back time of about 5.6 billion years. To probe what the cosmic expansion was doing significantly farther back in time, he needed light curves for supernovae at redshifts greater than $z = 1$, at look back times of 8 or 9 billion years or more. Finding supernovae that far away and that faint required Hubble. They were beyond the reach of even large ground-based telescopes.

Adam serendipitously stumbled onto his first $z > 1$ Type Ia supernova in 2001. Two astronomers, Ron Gilliland at the STScI and Mark Phillips at the Carnegie Institute, had re-observed the original Hubble Deep Field (HDF) in 1997, actually looking for supernovae. By comparing their new HDF images with the originals from 1995, they discovered two supernovae, one of which was in an elliptical galaxy whose photometric redshift was 1.7. It was given the official name SN 1997ff. Adam got excited about this. It must be a Type Ia, he reasoned, because that is the only type of supernova that is known to occur in elliptical galaxies. The problem was that Phillips and Gilliland didn't pursue their discovery any further by measuring the light curve. So, it wasn't possible to calculate the supernova's distance.

Frustrated, Adam started going through the data archives at the institute, looking for other observations of the Hubble Deep Field around that same time. He was in luck! Shortly after NICMOS was installed during SM2 in 1997, Principal Investigator Rodger Thompson used it to take a series of infrared images of the HDF. His objective was to study galaxy evolution by pushing the HDF to higher redshifts than had been achieved in the original visible-light version. The NICMOS observations were spread out over a couple of months. But the NICMOS field of view covered only 15% of the original HDF. Luckily, the elliptical galaxy containing the Type Ia supernova was there in one corner of the NICMOS field.

Adam was able to go back to the archive, extract the light curve from the NICMOS observations, measure the supernova's distance, and put the first point on the Hubble diagram at a redshift of $z = 1.7$. The data point fell above the straight-line extension of the present day Hubble law (above rather than below the straight line plotted in Figure 20.11). It appeared brighter than expected, suggesting that about 9 billion years ago, the expansion of the universe was slowing down, decelerating. That would make the universe smaller at that time than it would have been without the deceleration. Galaxies would be closer together and appear brighter. That single data point raised the possibility that gravitational attraction was dominant over the repulsive force of dark energy in the early universe.

The new supernova gave the lie to the idea of gray dust or fog making distant supernovae appear too faint. This object at an even greater distance than the supernovae in the two original studies was too bright. Moreover, its light curve looked perfectly normal, compared to nearby supernovae. Based on this one observation, at least, supernovae in the early universe were not intrinsically fainter than those nearby. There was no sign of evolution in its properties.

The story made the front page of the New York Times. "Farthest Supernova Shows Universe Decelerating, Shows Dark Energy Is Probably Real" the headline proclaimed.

Now it became really important to collect other samples of Type Ia supernovae at large redshifts, to fill in the Hubble diagram for the early universe. It was around that time that the astronauts had installed Holland Ford's Advanced Camera for Surveys on Hubble during SM2. At last the observatory had a sensitive camera with a wide enough field of view to make searches for distant supernovae practical. Also, NICMOS had been brought back into operation with the new cryocooler. That was important because the higher redshift supernovae have their blue light, the color band in which the light curves are measured, shifted by cosmic expansion into the near infrared.

Adam Riess led a group called the Higher-Z team to go looking for the most distant Type Ia supernovae. Beginning in 2002 his team won a lot of Hubble observing time to collaborate with another team, the GOODS Team, to systematically search for supernovae.

The GOODS (Great Observatories Origins Deep Survey) program was a follow on to the original Hubble Deep Field in the northern hemisphere and the similar Chandra Deep Field in the southern hemisphere. The idea was to create a survey of the sky in fields of view surrounding the original HDF and CDF that would go almost as deep, but also cover a much wider area—160 arcmin^2 as opposed to the original 5 arcmin^2 in each hemisphere. The most powerful space and ground-based observatories would participate in surveying these same areas of the sky in their own unique wavelength bands. For Hubble the GOODS images were taken with the ACS in four color filters. The GOODS survey required the telescope to be pointed at about 30 separate fields in each hemisphere to create a mosaic covering a large area on the sky.

Adam and his team helped plan the GOODS observations so that Hubble would return to the same field about once every 45 days. That is about the time, as measured on Earth, that a Type Ia supernova in the distant universe takes to rise to its maximum brightness after the initial explosion. Nearby supernovae take about 20 days to reach peak brightness. Why the difference? It comes from the time dilation effect in Einstein's Special Relativity Theory. Observing the light curve of a supernova is like watching a clock. The expansion of the universe is carrying that clock away from us at a particular velocity, following Hubble's law. The clock appears to an observer on Earth to run slower, the farther away it is and thus the

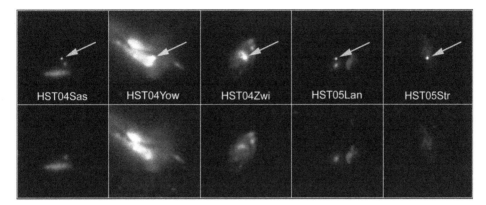

Figure 20.12. Five high redshift Type Ia supernovae in the context of their host galaxies discovered by the Higher-Z team in images taken with Hubble's Advanced Camera for Surveys during the GOODS survey. Across the top, arrows point to the supernovae. Across the bottom, are images of the host galaxies taken before the supernovae exploded. (Credit: NASA, ESA, A. Riess, STScI 2006-52.)

faster it's moving. For a supernova at a redshift z, the light curve will appear to be spread out in time by a factor of $(1 + z)$, thanks to time dilation. So, if the time for a nearby Type 1a supernova to rise to its peak brightness is 20 days, a similar supernova in the distant universe, say at a redshift $z = 1$, will appear to us to take twice as long—40 days.

Adam needed to time the return visits to each Hubble pointing in the GOODS survey to maximize the chances of catching a new supernova while it was still rising to its peak brightness. By 2006 the Higher-Z team had discovered 24 of the most distant Type Ia supernovae known and had measured their light curves and distances. Five of these, together with their host galaxies are shown in Figure 20.12.

The GOODS survey sample of Type Ia supernovae encompassed about 10 billion years of cosmic history. What the data showed was that long ago, when the universe was smaller and the concentration of matter—mostly cold dark matter—was high, gravity dominated the repulsive force of dark energy. The tug of gravity caused the cosmic expansion to decelerate. But about 5 billion years ago (around a redshift $z = 0.5$) the trend reversed. There was a relatively rapid switch from deceleration to acceleration. As the universe expanded, grew larger, the density of matter in it (in grams per cubic meter) grew smaller. The matter was more spread out and its cumulative mutual gravitational attraction diminished. Dark energy caught up with and overtook dark matter as the dominant force controlling the expansion.

The observations also showed that the spectra of Type Ia supernovae in the young universe were essentially identical to what we see today. There was no evolution in their properties. They were indeed reliable standard candles (Riess et al. 2007).

In 2011 Adam Riess, Brian Schmidt, and Saul Perlmutter were awarded the Nobel Prize in Physics for the discovery of the acceleration of the expansion of the universe. For many that might seem the culmination of a successful career—but not for Adam. There was more to do. The problems were meaty. They lent themselves to observations, either to get clues or to being discovered in the first place, as was the case with dark energy. The problems required deep plunges into theoretical physics for explanations. Adam did fundamental physics using the tools of astronomy. And he believed passionately that the best tool for what he did was the highly calibrated, high precision Hubble Space Telescope.

Adam went back to work developing new observing techniques that would squeeze the last drop of precision out of the instruments—particularly ACS and WFC3. Instead of holding the telescope pointing steadily on a field of stars, he had the spacecraft drift at a precisely known rate so that the point images of stars came out as long, straight lines. Instead of each image being focused on a few pixels, it was drawn out over thousands of pixels. This allowed the position of a star, relative to other stars in the field of view, to be measured to a remarkable precision of 1/1000 of a single camera pixel. Using this technique, he measured the trigonometric parallaxes of Cepheid variable stars in the Milky Way. That is, he triangulated their distances and thus determined their absolute brightness with greater precision and accuracy than had ever been achieved before.

The Cepheids were the primary standard candles used by Wendy Freedman and her Hubble Constant Key Project Team (Chapter 10). Wendy's objective was to measure the Hubble Constant with an uncertainty of 10%. Adam's goal was ultimately to measure it to 1% accuracy. He found nearby galaxies that contained a large number of Cepheids and also at least one Type Ia supernova whose light curve had previously been measured (Figure 20.13). The Cepheids, with their intrinsic brightness more accurately calibrated with Hubble, gave him more accurate distances to the galaxies. In turn that allowed the supernovae peak brightness to be measured with unprecedented accuracy. The supernovae could be detected in galaxies at far greater distances than could the Cepheids, and now the distances to those far away galaxies could be measured with greater accuracy than ever before.

Adam did this same thing—improved the precision and accuracy of standard candles and used those to improve the precision and accuracy of the Hubble constant—in a number of different ways. He refined the brightness measurements of Cepheid variables in the Large Magellanic Cloud, a small companion galaxy orbiting the Milky Way. He used trigonometric parallaxes of Cepheids in the Milky Way, measured with high precision by the European Space Agency's Gaia satellite, to determine their distances and recalibrate their absolute brightness. He recalibrated the light curves of Cepheids in the galaxy NGC4258. The distance to that galaxy had been determined with high accuracy using radio observations of masers in clouds of water molecules orbiting around a massive black hole at its center.

With each refinement in the measured distances of nearby Cepheid variable stars, Adam was able to improve the accuracy of the period–luminosity relation for those Cepheids—the tight relation between the length of the cycle of their periodic brightness variations and their intrinsic brightness. That gave him more accurate distances to the more distant Cepheids. With those, he was able to re-derive the distances to the Type Ia supernovae in his ever-growing database, and finally to derive a more accurate value for the Hubble constant.

Figure 20.13. Example of a galaxy in which a large number of Cepheid variable stars, whose intrinsic brightness had been recalibrated with high precision with Hubble, were used to more accurately measure the galaxy's distance. A Type Ia supernova, SN 2011by, had previously gone off and its light curve recorded. The newly calibrated Cepheid distance to the galaxy allowed the peak brightness of the supernova to be re-determined with much improved accuracy. Yellow circles indicate the positions of Cepheid variables. The position of the supernova is marked with a white plus sign. (Credit: NASA, ESA, A. Riess, STScI 2018-12.)

By 2019 Adam and his collaborators had reduced the uncertainty in the Hubble constant, as determined from the distances and recessional velocities of Type Ia supernovae, to about 1.9%. His best value was $H_0 = 74.03 \pm 1.42$ km s^{-1} Mpc^{-1} (Riess et al. 2019).

In discussing the nature and properties of dark energy, astrophysicists have expressed it in terms of an "equation of state," a simple relationship between the pressure and density of whatever the stuff is. It's expressed in terms of a quantity, W, that represents the ratio of pressure to density. As it is observed, dark energy exerts negative pressure, so W has a negative sign. The value of W that corresponds to Einstein's cosmological constant, the quantum energy of the vacuum is $W = -1$. To date, multiple estimates of the value of W seem to be converging to -1.0 to within 5%. It didn't have to be that way. For the universe to accelerate it was only necessary that W be less than $-1/3$. And there remains the issue of whether W varies with time, whether it has been constant over the entire history of the universe. But, for now, it seems Einstein's arbitrary cosmological constant was correct—a brilliant intuition.

Adam's original purpose in steadily working to reduce the uncertainty in the Hubble constant was to try to measure W to high precision, and perhaps to determine if W varied with time. He could do that by combining his measurements of the Hubble constant with certain measurements of the Cosmic Microwave Background (CMB), the relic radiation left over from the Big Bang, made by other space experiments such as NASA's Wilkinson Microwave Anisotropy Probe (WMAP) and the European Space Agency's Planck Observatory. But a puzzling new finding, of potentially cosmic importance, superseded that quest.

The Planck Observatory, launched in 2009, was designed to measure the finely detailed structure of the CMB with greater precision and resolution than the earlier NASA WMAP satellite. The Planck data contained information about the properties of the universe, as it was a few hundred thousand years after the Big Bang. One application of the observations was to derive what the Hubble constant should be today by extrapolating the Planck data from the earliest epochs of cosmic time to the present day, using the best cosmological model we currently have for the universe. That is the ΛCDM (Lambda CDM) model with the geometry of spacetime being flat. In this model the universe is dominated by the cosmological constant or dark energy, symbolized by the Greek letter Λ, and cold dark matter (CDM). The conclusion from the Planck observations, extrapolated with the ΛCDM model, is that the Hubble constant should be $H_0 = 67.4 \pm 0.5$ km s^{-1} Mpc^{-1}.

So, the properties of the early universe, shortly after the Big Bang, are telling us that the Hubble constant we should be measuring today, in the modern universe, is supposed to be 67.4 km s^{-1} Mpc^{-1}. But the extraordinary efforts of Adam Riess and his colleagues to measure the value with high precision and accuracy, using the light curves of Type Ia supernovae, have yielded a current value of 74.03 km s^{-1} Mpc^{-1}. The difference between the two numbers is statistically significant—it can't be reconciled by the size of the error bars attached to each measurement. Exhaustive efforts to find errors in both numbers led Adam to draw the conclusion that the probability that this discrepancy was a fluke, the result of some observational or computational error, was about 1 in 100,000. The "tension" between the two values seemed to be very real.

In July of 2019 a conference was held at the Kavli Institute for Theoretical Physics on the campus of the University of California, Santa Barbara, to provide a forum where physicists and astronomers could discuss their recent work on "Tensions between the Early and the Late Universe (Conover 2019)." At the conference six recent new results for the present day Hubble constant were presented, including Adam Riess' newest work. The new values were each measured independently using a variety of techniques. The results were remarkably similar, with an overall "consensus" average of 73.4 (+0.9, −0.8) km s^{-1} Mpc^{-1}. The reduced magnitude of the uncertainty relative, for example, to Adam's $\pm$ 1.42 km s^{-1} Mpc^{-1} in his individual study, results from the averaging of values that are in close agreement with each other. With this small error estimate, the gap between the Planck value, extrapolated from the early universe to the present time, and the late universe value actually measured at the present time has become a yawning gulf. In the terminology of statistical error analysis, the difference is 6.0σ (six sigma). That's huge, highly significant.

There's something missing in our understanding of how the universe operates. There's apparently some new physics at work here that we don't yet understand. And that's what makes science so much fun.

References

Bradac, M., et al. 2008, ApJ, 687, 959 and references therein
Broadhurst, T., et al. 2005, ApJ, 621, 53
Clowe, D., Gonzalez, A., & Markevitch, M. 2004, ApJ, 604, 596
Coe, D., et al. 2010, ApJ, 723, 1678
Conover, E. 2019, The Expanding Question, Science News, September 14
Infante, L., et al. 2015, ApJ, 815,
Perlmutter, S., et al. 1999, ApJ, 517, 565
Perlmutter, S. 2003, PhT, 56, 53
Riess, A., et al. 1998, AJ, 116, 1009
Riess, A., et al. 2007, ApJ, 659, 98
Riess, A., et al. 2019, ApJ, 876, 85
Rubin, V., & Ford, K. 1970, ApJ, 159, 379
Rubin, V., Ford, K., & Thonnard, N. 1980, ApJ, 238, 471
Zekser, K., et al. 2006, ApJ, 640, 639
Zwicky, F. 1937, ApJ, 86, 217

Life With Hubble
An insider's view of the world's most famous telescope
David S Leckrone

Chapter 21

Epilogue

On 2020 April 24 we celebrated the 30th anniversary of the launch of the Hubble Space Telescope. For the first 3 years of its operations, the observatory produced good and important science, but it was not what we expected. It suffered from the malady of spherical aberration. Over the following 27 years of operation, astronomers from around the world have exploited the remarkable capabilities of the telescope and its scientific instruments to revolutionize our understanding of the universe. This is, in no small measure, due to the fact that Hubble was refurbished and upgraded with new technology in five servicing missions, conducted by astronauts and large support teams on the ground. Hubble never got old or out of date.

It has been over 11 years since the last servicing mission. At 30 years of age Hubble remains reasonably healthy and vigorous. As might be expected, the only significant failures during the last 11 years have been among the six gyros (Chapter 14). After SM4 in 2009, Hubble had six new gyros. We had planned to install five of those with enhanced flex leads and one with the old fashioned standard flex leads. The old flex leads were the Achilles heels of the gyros. They tended to corrode and break. But there were problems in the installation of the new gyros during SM4 (Chapter 18). One of the new gyro packages proved impossible to install. It just didn't fit. A flight spare package was substituted for the balky unit. We ended up with three gyros with the new and improved flex leads and three with the old standard ones. Since then the three with standard leads have all failed as expected given the nominal lifetime of those flex leads. The three gyros with enhanced leads are still operating.

Each of the three functioning gyros has experienced other anomalies unrelated to its flex leads. These have long been familiar to the Hubble engineers. For example, every once in a while a tiny contaminant particle, one that isn't supposed to be there, finds its way into the gyro rotor housing, interfering with the gyro's ability to spin and causing the motor current that drives the gyro to increase. That has affected two of the three working gyros, but they still continue to operate.

The gyros also suffer from instability in the calibration of their orientation in space called "bias drift." The Hubble operations engineers have created and are now implementing new techniques for managing the gyros, in conjunction with the other parts of the spacecraft's pointing control system, to tame this problem.

Years ago the operations team created new software that would allow Hubble to conduct its observing program with only one operational gyro. The current plan is for the spacecraft to switch immediately to this one-gyro mode upon the next failure of a gyro. Two gyros will continue to operate, but only one of them will be in use to help point and stabilize the spacecraft. The engineers don't want to turn off the second gyro out of concern that it might not come back on when needed. How much longer the remaining three gyros will last is unknown.

Other than the gyros, all other spacecraft systems are continuing to operate well. They all still have redundancy so that, if a failure occurs, there is a backup option.

Each of the active science instruments is healthy and operating well. The lineup remains ACS, COS, STIS, WFC3, and the FGS used for astrometry. NICMOS is still on board Hubble, but was retired from active use after SM4. One of the COS light detectors continues gradually to lose sensitivity, a result of heavy use. But that was expected and has not as yet interfered with its ability to perform great science. There is a tiny electronics glitch in a memory chip within the advanced camera for surveys that can be controlled by keeping the chip from getting too cold. That's easily done.

There is no known reason why Hubble can't continue to produce pioneering, cutting edge science for many years to come. It's reasonable to expect that there will be years of overlap between Hubble and JWST, once the latter gets launched, deployed, stationed at its orbit a million miles from the Earth, and brought up to full science operations.

Over Hubble's lifetime about 7600 different science investigators have used the observatory for their research. They have come from over 48 countries, from Armenia to Venezuela. Of course most of the investigators, over 4000 of them, have come from the United States. Other heavy usage (over 100 users per country) has originated from the United Kingdom, Germany, Italy, France, the Netherlands, Spain, and Switzerland, all member states of the European Space Agency, NASA's partner in the development and operation of Hubble over the years. Astronomers from Australia, Canada, Chile, and Japan have also been heavily involved with Hubble observing.

Those astronomers have been exceedingly productive. As of the end of 2019, they had published over 22,000 peer-reviewed scientific papers in professional journals, based on observations obtained with the twelve scientific instruments that have been on board the observatory at one time or another over the course of its lifetime in orbit.

Since it was installed in the spacecraft in 2009, WFC3 has been the instrument most heavily in demand by astronomers. As of 2019, the infrared channel on WFC3 alone accounted for 31% of all of Hubble's on-target observations, while the WFC3 ultraviolet-visible channel added another 20%. By itself the grism built into the infrared channel was used for 9% of all on-target observations, mostly in the study of the atmospheres of exoplanets. Often when WFC3, COS, or STIS is the primary instrument observing a specific target, the ACS will be snapping an image in parallel of whatever field in the sky it happens to be pointing to. Those parallel, serendipitous observations aren't counted in these statistics; hence the emphasis on "on-target" statistics. Overall, ACS was used for 21% of on-target observations, COS for 14% and STIS for 14%.

I take some gratification from the heavy demand for the WFC3 infrared channel. We had to fight hard to get that capability added to the instrument and to keep it from being taken away. The popularity of the infrared channel for boundary-stretching observational astronomy speaks well of the good scientific judgment of the community in demanding that it be included in the design of WFC3.

The Hubble Space Telescope is the paradigm for a long-lived "mountain top" observatory in space. The best illustrations of this are the numerous areas of research that originated, evolved, and matured through Hubble observations over many years.

Bob O'Dell's discovery in the early 1990s of the proplyds (proto-planetary disks) in the Orion Nebula complex started our walk down the path of research on the planet-forming environs around a large number of stars. Hubble observations of exoplanets transiting their parent stars provided measurements of the size, mass, and density of those planets and set the stage for the future work of the Kepler Space Observatory. Finally, Hubble pioneered in detecting the atmospheres of exoplanets that transit or are eclipsed by their parent stars and in measuring their chemical composition and physical properties. This thread of research has spanned three decades so far (Chapter 19).

Bob Williams' courageous sponsorship of the first Hubble Deep Fields in 1995 and 1998, which reached across the universe to a redshift $z \sim 4$, led inexorably through the ACS ultra-deep field ($z \sim 7$) in 2003–2004, and the combined ACS plus WFC3 HUDF 2009–2010 ($z \sim 8$–9), to Garth Illingworth's XDF, which, with help from the Spitzer Space Telescope, achieved $z \sim 10$–11 (Chapter 11). The latter surveys revealed the universe, as it was only a few hundred million years after the Big Bang, when galaxies were much smaller and more disorganized than they are in the nearby universe as we see it today. From these deep fields astronomers determined how the rate of new star formation changed over the eons, reaching its peak at about $z = 2$, 10 billion years ago.

Then there was Wendy Freedman's Key Project team whose objective was to measure the Hubble constant, H_0, the current rate of expansion of the universe, to an accuracy of 10%. They started their project in 1994 and continued for about seven years thereafter. Their conclusions about the expansion age of the universe only made sense in the context of the newly discovered acceleration of that expansion (Chapter 10). Hubble observations played a major role in the latter discovery. For their discovery of the acceleration of the expansion of the universe Adam Riess, Brian Schmidt, and Saul Perlmutter received the Nobel Prize in Physics in 2011. But the story didn't end there.

Adam Riess continued to push hard on the accuracy of the measured Hubble constant, seeking to reduce its uncertainty to about 1%. His goal was to use that highly accurate value to gain insight into the force at work that causes the acceleration—so-called Dark Energy. That work may now, in 2019, be revealing new physics. The value of H_0 Adam has determined using Hubble observations, and the values determined by other scientists using different techniques, are in good agreement, averaging 73 km s^{-1} Mpc^{-1}. But that value differs by a statistically highly significant amount from the value obtained from observations of the microwave background formed in the early universe, extrapolated with the best cosmological model available to the present time by astronomers using the European Planck Observatory, 67 km s^{-1} Mpc^{-1}.

I'm citing these examples from Hubble research to illustrate that scientific progress sometimes requires decades to develop and evolve to a natural culmination. The ability to both stimulate new areas of research and continue to propel their evolution over a period of decades is a unique virtue of a very long-lived, technologically up-to-date observatory.

I'd like to return to some words that I originally touched on in Chapter 20 from our Nobel Laureate, Adam Riess. These have bearing both on Hubble's longevity and on the research culture Hubble engenders:

"Hubble has all the latest gear in it, but also, it's now so well calibrated. When I told you about this project [that requires] measuring the positions of stars to 1/1000 of a pixel, you can't do that unless you have mapped the geometric distortion of the camera to incredible precision. Otherwise that's all you're going to be discovering—the geometric distortion of the camera. So, there are projects that we do with Hubble now that you just couldn't do years ago, even though the equipment was there. It just wasn't well enough calibrated. One of the things that is most unique about Hubble, that people don't often appreciate, is that its high precision calibration is what enables a lot of science. People otherwise just wouldn't try to do really hard things or really precise things, if that weren't the case. For the kind of work that I do, which is fundamental physics with astronomy tools, Hubble is just perfect."

"What's incredible about Hubble is, not only is it arguably the greatest scientific instrument human kind has ever had, but it has also been available for use in the most democratic way imaginable. In my case, when I got started, I was like every young person starting out applying to use Hubble—a nobody. Writing proposals that were selected and doing the science was the way that I was able to grow my career and do exciting things. And that is more on the basis of merit-of-project than on who you are. It used to be, back in California in the 1920s or 1930s or 1940s or 1950s, for a long time, that if you were not inside the right institutions, you didn't have access to the big glass [big telescopes]. That was not very democratic."

"Hubble is a completely different model. I've been a tremendous beneficiary of that. I'm very appreciative of that. I would like people to understand how available Hubble was. You had to write a really good proposal, and do something with it [if you were selected], but still...."

"I showed up in astronomy just when Hubble was getting fixed. I went to graduate school in 1992. One of the first symposia I attended was Jeff Hoffman [one of the SM1 astronauts] talking about the repair. And at one of the first American Astronomical Society meetings I went to [in 1994 January], I saw the first images from WFPC2. I saw the first color-magnitude diagrams of globular clusters. People were gasping and applauding. Fortunately, I didn't have to deal with the, 'oh my God, it's not working right' problems."

Adam's words sum up two important and not widely understood aspects of what has made the Hubble Space Telescope great—it's long term stability and calibrated precision, and the openness of the observatory to all astronomers around the world. All those astronomers had to do was to submit carefully framed observing proposals whose quality would be judged to fall at or near the top of a ranked list of such proposals. They also had to be a bit lucky. The over-subscription rate for Hubble observing time was typically about five or six to one and in some years higher than that. For every proposal that got accepted, four or five or more had to be rejected. Often the latter were very good proposals. There has never been enough observing time on Hubble to go around.

That's the story of the Hubble Space Telescope to date, as I have observed and experienced it, along with hundreds of colleagues and friends. I hope it is a story that will still be familiar to people a century from now. It represents one of the great achievements of humanity. And its voyage of discovery goes on.